What Every Engineer, Scientist, and Technologist Should Know About Finance

This book bridges the gap between engineering expertise and financial acumen, equipping technical professionals with essential knowledge of corporate finance. This comprehensive guide transforms complex financial concepts into accessible insights, enabling engineers, scientists, and technology specialists to confidently navigate financial documents and understand the financial drivers behind corporate decisions. The book:

- Provides comprehensive coverage of professional financial context.
- Presents self-contained explanations of all financial terminology with no prior accounting knowledge required.
- Offers practical guidance for interpreting budgets and common financial documentation.
- Discusses cost-based project metrics, control mechanisms, and risk management strategies.
- Aligns with PMI's Project Management Professional exam topics in Project Cost Management, Business Environment, and Risk Management.
- Includes forward-looking chapters examining emerging financial technologies including cryptocurrencies, FinTech innovations, open banking, and AI applications.

Written specifically for technically-minded professionals seeking to enhance their financial literacy, this essential resource serves as both a practical handbook and a strategic guide. Whether you're an engineer looking to better understand how financial considerations impact your projects or a technical team leader needing to communicate effectively with finance departments, this book provides the knowledge framework to make informed financial decisions that complement and enhance your technical expertise throughout your career.

Phillip A. Laplante is Professor Emeritus of Software and Systems Engineering at Penn State with more than 40 years of experience as a professor, practitioner, academic administrator, and consultant. He has worked in avionics, CAD, and software testing systems, and he has published 40+ books and more than 300 scholarly papers. He is a licensed professional engineer in the Commonwealth of Pennsylvania and a Fellow of IEEE, SPIE, and the Washington Academy of Sciences. From 2010 to 2017, he led the effort to develop a national licensing exam for software engineers. He received his B.S., M.Eng., and Ph.D. from Stevens Institute of Technology and an MBA from the University of Colorado.

What Every Engineer Should Know

Series Editor: Phillip A. Laplante, Pennsylvania State University

What Every Engineer Should Know About Cyber Security and Digital Forensics
Joanna F. DeFranco and Bob Maley

Ethical Engineering: A Practical Guide with Case Studies
Eugene Schlossberger

What Every Engineer Should Know About Data-Driven Analytics
Satish Mahadevan Srinivasan and Phillip A. Laplante

What Every Engineer Should Know About Reliability and Risk Analysis
Mohammad Modarres and Katrina Groth

What Every Engineer Should Know About Smart Cities
Valdemar Vicente Graciano and Mohamad Kassab

What Every Engineer Should Know About Digital Accessibility
Sarah Horton and David Sloan

What Every Engineer Should Know About Python
Raymond J. Madachy

Diversity in STEM: Analyzing Inequities and Improving Opportunities in Education and the Workplace
Benjamin Blocker

Ethics for Engineers: Toward Ethical Behavior within Engineering Organizations
Seppo J. Ovaska and Andrew T. Brei

Technical Writing: A Practical Guide for Engineers, Scientists, and Nontechnical Professionals, Third Edition
Phillip A. Laplante and Chris Laplante

What Every Engineer Should Know About Artificial Intelligence and Big Data
Satish Mahadevan Srinivasan and Raghvinder S. Sangwan

What Every Engineer, Scientist, and Technologist Should Know About Finance
Phillip A. Laplante

For more information about this series, please visit: www.routledge.com/What-Every-Engineer-Should-Know/book-series/CRCWEESK

Series Statement

What every engineer should know includes an overwhelming catalogue of information. Regardless of discipline, engineering intersects most scientific fields and modern technologies. Practicing engineers however must also navigate managerial, socio-economic, and even political concerns. The engineer discovers soon after graduation that any curriculum omits important and thorny issues of daily practice—for example, problems concerning new technologies, scientific advances, business practices, legal implications, and team dynamics.

With the *What Every Engineer Should Know* series of concise, easy-to-understand volumes, every engineer can access primers on important subjects across a broad range of knowledge areas including intellectual property, contracts, software, business communication, management science, and risk analysis, as well as more specific topics such as embedded systems design. These books are very accessible to every engineer, scientist, and technology professional and are necessary to remain competitive in this dynamic, global economy.

What Every Engineer, Scientist, and Technologist Should Know About Finance

Phillip A. Laplante

CRC Press
Taylor & Francis Group
Boca Raton London New York

CRC Press is an imprint of the
Taylor & Francis Group, an **informa** business

Designed cover image: © 2026 Shutterstock

First edition published 2027
by CRC Press
2385 NW Executive Center Drive, Suite 320, Boca Raton FL 33431

and by CRC Press
4 Park Square, Milton Park, Abingdon, Oxon, OX14 4RN

CRC Press is an imprint of Taylor & Francis Group, LLC

ISBN: 978-1-041-22767-0 (hbk)
ISBN: 978-1-041-22766-3 (pbk)
ISBN: 978-1-003-73385-0 (ebk)

DOI: 10.1201/9781003733850

Typeset in Times
by Apex CoVantage, LLC

Disclaimer

This book is intended to provide general overview information only and not investment, taxation or business advice. I am not a lawyer, accounting professional, tax advisor, nor financial advisor. You must consult these professionals before undertaking any personal business or investment decision.

Contents

Preface

This book is an introduction to finance for engineers, scientists, and technologists. It introduces basic terminology and concepts, project-based finance and estimation, financial management of a collection of projects, and much more. It is intended to make you a more financially aware team member and to prepare you for increasing financial leadership in the workplace. The text will also help you better understand how financial considerations drive corporate decision-making at the senior executive and board level.

While the book does not cover personal finances, many of the concepts from business carry over, such as budgeting, investing, and risk management. So, this book may help you in managing your own finances in a way that is compatible with a dynamic engineering career.

The early chapters introduce the basics and familiarize you with the language of finance. The middle section introduces metrics and day-to-day management of financial matters. The final chapters introduce large-scale and strategic finance in engineering, science, and technology companies. The rigor and difficulty increase with each chapter, so I recommend that you address the chapters in order. However, if you have some financial experience, you can read the book in any order or simply use it as a reference.

By the time you have finished reading this book, it is my hope that you will have a high level of financial literacy, with special emphasis on engineering, technology, and science companies and strategies. You should be able to "speak the language of finance."

As a bonus, you can carry this literacy into your personal finances. Many of the terms, concepts, and strategies apply in both domains. So, you will likely find that you are also more savvy with your own finances.

CHAPTER OVERVIEW

The book starts with basic concepts that increase in scope and complexity with each succeeding chapter. The following provides brief descriptions of each chapter.

Chapter 1

Chapter 1 establishes the fundamental necessity of financial literacy for engineers, scientists, and technologists, bridging the gap between technical expertise and strategic organizational leadership. Core concepts such as compensation evaluation, the financial skill ladder, and the use of SEC 10-K filings to assess corporate health and strategic risks are also introduced. It then covers analytical tools like Fermi estimation and key metrics such as gross margin and the Rule of 40, that EST professionals can use to optimize their career trajectories and effectively communicate the economic impact of their technical initiatives.

Chapter 2

This chapter establishes the foundational accounting and financial principles necessary for EST professionals to navigate the mechanics of organizational money management. Here we explore how core tools—including the accounting equation, double-entry bookkeeping, and the three primary financial statements—provide a transparent "dashboard" for evaluating technical operations. By mastering metrics such as net present value (NPV), internal rate of return (IRR), and essential liquidity ratios, you will learn how to justify capital-intensive projects and assess the financial health of partners and competitors.

Chapter 3

An engineering, science, or technology company can be viewed as a kind of black box "business system." Chapter 3 demystifies this system by examining how legal structures, accounting standards like GAAP and IFRS, and macroeconomic factors such as tariffs influence technical execution. This chapter explores the use of the SEC 10-K as a "system architecture diagram" to identify strategic leverage points and potential failure modes within a corporate entity. By mastering essential metrics like burn rate, runway, and the Rule of 40, you will be able to navigate complex regulatory landscapes and build resilient, profitable organizations.

Chapter 4

This chapter explores the strategic business models and commercial frameworks required to transform technical innovation into sustainable revenue streams. Here, we will analyze competitive strategies such as cost leadership, differentiation, and focus alongside diverse pricing models like fixed-price, cost-plus, and performance-based agreements. By examining disruptive innovation, overhead recovery, and collaborative structures such as teaming and public–private partnerships, you will discover how to align engineering excellence with long-term economic viability.

Chapter 5

This chapter examines risk and opportunity management, framing uncertainty not as an adversary but as a fundamental force to be harnessed through strategic foresight and resilient design. It also introduces quantitative diagnostic tools—including the Altman *Z*-score and Piotroski *F*-score—to evaluate organizational health alongside strategic frameworks like the Balanced Scorecard and diversification to optimize project portfolios. By mastering these concepts, you will be able to mitigate the unintended consequences of incentives and effectively align technical risk with long-term organizational stability.

Chapter 6

Here, the financial tools and frameworks necessary to evaluate, budget, and manage technical projects are discussed. These help ensure that engineering objectives remain synchronized with economic realities. The chapter then details an approach to estimation—utilizing Work Breakdown Structures (WBS), analogous and parametric models, and risk-based contingencies—while leveraging Earned Value Management (EVM) to monitor performance against a cost baseline. By mastering analytical tools such as Total Cost of Ownership (TCO) and break-even analysis, you will be able to navigate complex make-versus-buy and lease-versus-buy decisions to optimize resource allocation across the project lifecycle.

Chapter 7

A structured framework for engineering-driven firms to synchronize strategic goals with financial execution through disciplined budgeting, forecasting, and capital allocation is given. This chapter also explores how evidence-based models, including Stage-Gate, real options, and technology readiness-based assessments, enabling organizations to manage technical uncertainty and sequence investments effectively. By examining diverse financing pathways from venture capital to nondilutive grants like SBIR/STTR, you will discover how to secure the resources necessary to bridge the gap between research and commercialization while preserving organizational liquidity.

Chapter 8

This chapter establishes a disciplined framework for building financial models and utilizing data analytics to drive strategic investment decisions. The coverage provides a modular architecture for spreadsheets—integrating income statements, balance sheets, and cash flows—while leveraging advanced techniques such as Monte Carlo simulation, sensitivity analysis, and copula-based probability modeling to quantify technical and market uncertainty. By emphasizing Exploratory Data Analysis (EDA), valuation frameworks like DCF, and rigorous governance standards, the chapter ensures that financial models serve as auditable, transparent assets rather than static spreadsheets.

Chapter 9

Finally, we examine emerging technologies such as blockchain, artificial intelligence, and quantum computing, providing engineers, scientists, and technical professionals with a strategic framework to navigate the next generation of financial management. The text explores how these tools drive operational efficiency and innovation while simultaneously addressing critical challenges in cybersecurity, regulatory compliance, and ethical governance. By integrating concepts like artificial intelligence, FinTechs and open banking, the chapter will equip you to leverage automated, data-driven systems to streamline financial and technical decision-making across complex environments.

WHY THIS BOOK?

Over my many years as a project manager, board member, staff member, and a professor, many working engineers expressed to me that they did not understand why "management" would make certain project decisions. It was clear to me that many engineers struggle to understand financial drivers behind management decisions—an issue this book directly addresses.

It has also been my experience that most engineers, scientists, and technology professionals choose to avoid accounting and financial courses because they are not as interesting as engineering, mathematics, and science. But accounting and finance courses are very interesting and informative. I often encouraged my colleagues, students, and friends to take these courses by noting that "if you like money, then you'll like accounting." So, I ask you to consider taking at least one basic accounting or introduction to finance course, either as a continuing education course at your local college, a for-credit course, or even as some kind of online self-study course.

Even if you never prepare a budget or manage a budget, it is my hope that this book helps you understand better why certain management decisions are made by your company—whether at a project, department, division or corporate level. Many times, these decisions seem "obviously wrong" or "stupid," but after understanding certain underlying financial realities, these decisions may seem more reasonable to you.

HOW TO USE THIS BOOK

This book is intended for an undergraduate or professional graduate course in Finance for Engineers or a similar title. However, this book could also be used as a secondary text in Engineering Management programs, particularly in a Technical Project Management course, which I have taught many times. In these cases, the chapters and topics are organized in a logical progression, moving from very basic concepts to the more complex topics in the latter chapters.

If you are using this book as a reference guide, I don't expect you to read it linearly. Instead, you can simply select the topics and terms of interest to study and review as you encounter them at work. However, if you have no background in accounting or finance, I suggest that you read the book in chapter order as the early chapters define the terms and introduce the principles that will appear in later chapters. The first six chapters will be of particular interest if you are a project team member or manager, as they focus on individual project finances. Chapters 7, 8, and 9 are the most complex as they involve corporate-level finance and startups.

While reading the chapters, you will find italicized terms and acronyms—these are defined in-line and/or in the glossary of Appendix A. This comprehensive glossary includes terms, definitions, and acronyms for many business, financial, and related engineering and technical terms. Keep this book on your work bookshelf for reference at any time you encounter a financial term or situation that you don't understand, or if you are asked to prepare some form of cost estimation or analysis for a project.

This book contains more than 200 examples and case studies—many are short and in line with the text. Longer examples are numbered and set out with a borderline for easier reference and location. The most extensive examples and case studies are set out in their own subsections.

With nearly 150 exercises, this text is classroom-ready. Exercises range from spreadsheet builds and calculation problems to qualitative analyses, case studies, and professional communication simulations, such as drafting pitches to a CFO or designing AI governance policies. Even if you are using this book for professional reference, I highly recommend you try some of these exercises to reinforce and check your learning.

Finally, this text can be used to help you prepare for the Project Management Institute's Project Management Professional (PMP) exam. Chapters 4 through 7 map well into the exam's financial topical areas. However, you would need to supplement this text with a standard project management guide to master the non-financial aspects of schedule management. For the latter, please consider my other book, Colin J. Neill, Phillip A. Laplante, and Joanna DeFranco, *Antipatterns: Identification, Refactoring, and Management, Second Edition*, Taylor & Francis, 2012.

About the Author

Dr. Phillip A. Laplante is Emeritus Professor of Software and Systems Engineering at Penn State and is internationally renowned for his contributions to the field of software and systems engineering. Throughout his career, he has made significant impacts through his research, teaching, and service in areas such as artificial intelligence, critical infrastructure systems, image processing, and real-time systems.

A prolific writer, Dr. Laplante has published more than 40 books and 350 scholarly works, including definitive texts in Requirements Engineering, Real-Time Systems and Technical Writing. These and many of his other books are used in universities around the world. His recent text on Chaos, Fractals, and Dynamical Systems was written for use in homeschooling.

Dr. Laplante has more than 40 years of experience as a senior technology leader in development, research, teaching, and administrative roles (including as board member, president, vice president, and director) in industry, academia, and government and has managed organizations of more than 100 people and multi-million-dollar budgets. Phil has also served on many non-profit and corporate boards.

Laplante earned B.S., M.E., and Ph.D. degrees in engineering and computer science from Stevens Tech and an MBA with a concentration in finance from the University of Colorado at Colorado Springs. He is a licensed professional engineer in the Commonwealth of Pennsylvania. He also holds black belts in Krav Maga, Aikido, and Brazilian Jiu Jitsu, one of the handful of people in the world to earn this rare combination.

Acknowledgments

Many friends, family, and colleagues have supported me through the development of this book. Most importantly, I'd like to thank my longtime friends and editors at Taylor & Francis, Allison Shatkin and Nora Konopka, for their enthusiastic support on this project and the many previous ones.

The following friends lent their expert opinions in helping to develop the general outline for this book.

- Hugh Connelly
- Dr. Jeff Daniels
- Dr. Joanna DeFranco
- Ron Herzog

Many other former friends, colleagues, students, and even strangers have provided ideas and inspiration for the diverse topics in this book. No work of any meaning is a singular accomplishment—it is simply a recasting of knowledge created by many others. Thank you to all of these influences, both known and subconscious.

Throughout the writing of this book, I have been accompanied, entertained, and occasionally annoyed by my beloved dogs—Henry, Miles, and Molly. I want to thank them for their companionship, unconditional love, and tolerance of my crankiness when they needed to be let out while I was preoccupied during some deep thinking.

Finally, I want to thank my wife, Nancy, for her endless patience and encouragement during this, and every other project over these many years.

1 Why Engineers, Scientists, and Technologists Need Financial Literacy

Years ago, I read a satirical, allegorical essay, describing a grand commerce ship, once captained by visionary engineer-leaders who understood how the vessel functioned in detail and plotted bold courses. Over time, however, control shifts to accountants, who prioritize short-term profit (counting "beans") over innovation or long-term strategy. These new leaders, "the bean counters," are ignorant of the ship's engineering. They delegate technical matters to underlings, while enforcing strict cost controls and stifling creativity. The engineers are eventually exiled to the engine room below deck, where they are kept ignorant of the ship's destination and the reasons for the many changes to the equipment that are being demanded. The engineers ask to meet the captain to make suggestions for improvements to the vessel. But an assistant to the first mate is delegated to meet with the engineers. The assistant scoffs at the engineers—"You know nothing of beans!" and they are sent back to the engine room without further regard for their suggestions [Lucky 1993].

This allegorical tale struck me as funny yet truthful. When I first read it as a young engineer, I was not well-versed in finance (beyond the simple balancing of a checkbook and managing my student loans and mortgage). I had avoided accounting and all business courses as an undergraduate. These seemed undignified to me as an aspiring engineer. Later, I realized my mistake.

Throughout my career, I have observed that many engineers and technologists would rather ignore financial matters and focus on engineering and problem-solving. This is an unfortunate perspective and one that I hope to help you overcome through this book.

1.1 THE INTERSECTION OF ENGINEERING AND FINANCE

Engineers, scientists, and technologists need financial literacy for the same reason that anyone else does—to make smarter decisions with money in both their personal and professional lives. Engineering, science, and technology (EST) all have mathematical foundations, so these professionals hope to find a similar level of mathematical rigor in finance. There is such rigor, but in general, not as much as you might like. Let me illustrate my point with a personal story.

Many years ago, as an academic administrator, I realized that even though I had been managing budgets on the order of several million dollars, I lacked financial

DOI: 10.1201/9781003733850-1

sophistication. I began taking some graduate finance courses, enjoyed them, and they were filling the gaps in my education. Eventually, I decided to matriculate into a Master of Business Administration (MBA) degree program at the University of Colorado at Colorado Springs (one of the first remote, accredited programs in the United States). Even though I had a PhD in Computer Science and aced all the courses, I was still required to apply for the program and meet the admissions requirements, like anyone else. One of these requirements was having completed a college calculus course. This was no problem—I had taken many such courses and had taught graduate-level mathematics. I later learned that many of my fellow MBA students had some difficulty meeting this requirement and that some feared taking any courses that might be calculus-based. I was looking forward to such courses, but was later disappointed that not a single course that was required in the program used calculus. Nonetheless, I was very satisfied with what I had learned in the program, and I earned my MBA in 1999.

The mathematics used in everyday finance is straightforward:

- linear equations,
- probability and statistics,
- finite series,
- difference equations, and
- basic calculus.

But mostly, you will use spreadsheets for financial calculations.

Yet, you can find more sophisticated mathematics in finance, for instance, in portfolio and market analysis and even in individual investments. Let's see just one example.

A *stock option* is a contract that gives the holder the right to buy or sell a specific stock at a predetermined price within a certain period. Consider equation (1.1):

$$\frac{\partial V}{\partial t}+\frac{1}{2}\sigma^2 S^2\frac{\partial^2 V}{\partial S^2}+rS\frac{\partial V}{\partial S}-rV=0 \tag{1.1}$$

This partial differential equation is known as the Black–Scholes model. It predicts stock option price V over time, where r is the risk-free interest rate, and σ represents the volatility of the stock [Black and Scholes 1973]. A stock option is a kind of *derivative*, that is, a contract whose value depends on an underlying asset, index, or benchmark. There are derivatives for stock, bond, commodities, currencies, interest rates and more. Many sophisticated individual investors and most institutional investment houses trade in derivatives.

Hedge funds tend to invest heavily in derivatives. The Black–Scholes equation can be used to model other kinds of derivatives, and it is widely used in computer-driven (program) trading and arbitrage. Clearly, this equation represents sophisticated mathematics, but it is unlikely that you are ever going to use it directly.

Arbitrage is the practice of simultaneously (or near-simultaneously) buying and selling related assets in different markets to exploit temporary price discrepancies. For instance, suppose the price of a stock on the New York Stock Exchange is infinitesimally lower than that of the same stock on some other stock exchange. We know that this discrepancy won't last long—it will be identified and corrected in seconds, perhaps milliseconds.[1] But, if we can push through many trades (via online trading) before the window of the price discrepancy closes, we can make enough profit to cover the transaction costs and perhaps make a substantial profit.

The large-scale, rapid transactions needed for this kind of activity can only be achieved by massive computing power. Using large-scale computing power to conduct arbitrage, or any high-volume trading, is called *program trading*. Program trading was pioneered by computational biophysicist David E. Shaw, who was a professor of Computer Science at Columbia. He built his namesake, powerhouse company by hiring PhDs in physics, math, and astrophysics, so-called "rocket scientists" instead of MBAs. That company, D. E. Shaw, is still a leading force in financial markets [Celarier 2018].

The emergence of cryptocurrencies such as Bitcoin, Ethereum, and Dogecoin has transformed many aspects of finance. Of course, cryptocurrencies are grounded in mathematical computation as "proof-of-work," which makes them very appealing to engineers, scientists, and technologists. Cryptocurrencies and related technologies such as blockchain are at the intersection of engineering and finance and deserve some attention in this book. I will discuss these in more detail in Chapter 9.

So, although financial modeling and analysis can be quite rigorous, the mathematics I will use in this book will be mostly simple algebraic formulas, probability and statistics, and perhaps some set theory. I won't be looking to formalize any financial concepts with higher-level mathematics unless it is necessary or interesting.

1.2 THE VALUE OF FINANCIAL DECISION-MAKING IN TECHNICAL CAREERS

As an engineer, scientist or technologist you are probably at the higher end of the earnings scale. Over the course of your career, you will likely earn millions of dollars[2] and have many opportunities that could greatly increase your wealth. Throughout your adult life you will make many financial decisions, such as those related to dealing with student loans, purchasing a home, dealing with unexpected expenses, and saving for retirement.

Other financial decisions that you are likely to encounter include:

- taking a home or car loan,
- refinancing debt,
- when and how much to save,
- acquiring insurance, and
- investing versus paying off debt.

With all these activities, financial literacy can help prevent you from making a disastrous decision. Without such knowledge, you might overspend, under-invest, or lose significant income to taxes. You might also make mistakes of omission when offered financial opportunities, such as being offered a partnership or equity, investing in a new company, or starting your own.

Many engineers and technologists also dabble with day-trading, real estate, or side businesses. These endeavors require a different kind of financial literacy than personal finances, and the economies of scale make this type of financial activity different from managing, say, an engineering department budget. Yet there are certain basic principles that all these types of financial activities have in common, such as the time value of money.

During your career, there may be times when you are unemployed or furloughed. Technology cycles can lead to waves of "feast or famine" employment in the workplace. Part of your financial literacy is to plan for these contingencies and to deal with them if they occur. Basic strategies such as maintaining a 6- to 12-month emergency fund, diversifying your investments, and maintaining certain types of disability and life insurance can save you and your loved ones from financial stress.

When you retire, there are new financial decisions to be made, such as effectively managing expenses without a regular salary, withdrawal of retirement funds, managing your increasing healthcare costs, tax management, and wealth distribution upon your passing.

In the next sections, I will discuss some of the more significant financial events during your working career.

1.2.1 The Finance of Compensation

It helps to know your marketplace value, whether you are evaluating a job offer or trying to make the case for a promotion or raise. But how does a company determine the economic value of an engineer, scientist, or any other technologist? Let's look at an example.

Suppose an engineer, Bob, is a consultant and his employer can bill him out to a client at \$300 per hour. Assuming Bob can be billed for 40 hours a week at 50 weeks a year, then Bob generates \$40 × \$50 × \$300 =\$600,000 in revenue for his employer. Let's say Bob's hourly "cost" to his employer for wages, taxes, and overhead is \$200. Then Bob's annual cost is \$40 × \$50 × \$200 = \$400,000. Thus, Bob generates a net profit (revenue minus cost) to his employer of \$100,000 per year.

But, not all engineers, scientists, and technologists are consultants, and therefore it is not that easy to attribute direct revenue to them. While surveys can help determine expectations of engineering pay, salaries are generally based on the market for those with that job title and these salaries are highly influenced by the cost of living in the geographic location of the job.

When you receive a job offer, you may have to evaluate competing offers with different base pay, and possibly signing or periodic bonuses, stock options, retirement fund matches, and different combinations of sick time, vacation time, and personal days off. If the job offer includes some form of stock or equity (which is often the case for startups or in pure technology firms), you need to be able to compare this offer to another. You also need to know how to factor in risk, liquidity, and taxes. Without understanding vesting schedules, dilution, or tax implications, you could choose the wrong job and reduce your total compensation.

If you are going to be required to travel, you need to consider the travel time, hidden costs, and inconvenience (unless you view travel as a positive and then account for it differently). And if you compare two different job options, you should formulate them in a way to make a real comparison, for example, by converting each offer to an hourly wage.

EXAMPLE 1.1 EVALUATING A JOB OFFER: AN EXAMPLE FROM MY CAREER

When I was the president of a small, two-year technical college, my salary was the equivalent of about $250K in today's dollars. I was working an average of 80 hours per week (including many weekends). The stress was terrible, and I had no time to exercise or relax. The job really took a toll on my marriage and relationship with my children, so I began to look to extricate myself from the situation.

Unexpectedly, I was offered a position on the faculty at Penn State. The pay was far lower—the equivalent of about $150,000 per year in today's dollars. But, I wouldn't have to work 80 hours per week. The employer's match to my retirement was significantly more, and I could earn more money by teaching a few courses over the summer. My children would also receive a deep tuition discount if they chose to matriculate at Penn State.

When I converted my president's salary to an hourly wage, it worked out to be about $62 dollars per hour (in 1990 dollars). The Penn State job worked out to be about $75 per hour (when you accounted for my children's tuition discount). So, I took the Penn State position—and worked there for almost 25 years.

As a very significant bonus, I was able to spend more time with my wife and children—which improved those relationships, and I could spend time exercising, doing hobbies, and relaxing, and soon, my health improved.

So, when you are comparing two job opportunities, to get to an apples-to-apples comparison, I recommend converting all forms of compensation to an hourly rate, then consider physical and mental wellness from reduced stress, and improvement in personal relationships.

1.2.2 Your Career Path to Project Manager and Higher Roles

As you advance in your career and choose to climb the management ladder (as opposed to a purely technical progression), your financial responsibilities will increase. Table 1.1 shows the increase in financial responsibility as an engineer/scientist/technologist moves up the organization from an individual contributor to a C-level position such as Chief Technology Officer (CTO).

As an individual contributor, you have little direct financial responsibility, but you may have a small software tool or equipment budget or a small personal development

TABLE 1.1
Financial Responsibilities as You Climb the Management Ladder

Level	Financial Responsibility	Typical Financial Decisions
Individual Contributor	\$0–\$50k	Choose tools, justify travel to conferences, courses, and \$5k–\$20k experiments.
Manager	\$50k–\$500k	Headcount planning, project budgeting, team travel, vendor negotiations.
Senior Manager/Director	\$500k–\$5M	Multi-team programs, CapEx requests, contingency planning, and annual budget planning.
Vice President	\$5M–\$50M	R&D % allocation, mergers and acquisitions due diligence, hiring freezes, unit economics ownership, and company-level operational expenditures.
C-Level, for example, Chief Technology Officer	\$50M–\$500M+	Capital raise influence, new construction, P&L responsibility, and investor narratives.

allowance, which you can justify for use for travel to conferences, technical meetings or for training. You may also have a small experimental project budget. But you can learn to think financially, for instance, by justifying every \$5k experiment you conduct or design decision you make, even if it is just for your own satisfaction.

At the next level, for example, as a department or project manager, you could run a \$50k–\$500k mini-budget. You might be involved in vendor negotiations and in headcount planning for your unit, or in approving those travel requests for your staff. A larger responsibility might be if you are asked to prove your team returns three to five times its fully loaded cost.

By the next level, say as a director or senior manager, or whatever the title is in your organization, you could be allocating millions across programs. You will be managing the budgets for multiple teams and programs and approving capital expenditure (CapEx) requests. You will conduct contingency planning and likely have to fight for your budget during the annual planning process.

As a vice president or equivalent, you become a true business executive, perhaps owning research and development (R&D) as a significant percentage of corporate revenue. You will co-own company-level operational expenditures (OpEx) and sometimes must make large-scale personnel decisions such as hiring freezes. You will run mergers and acquisitions (M&A) technology due diligence, moving company-wide unit economics or gross margin by measurable points.

Finally, as a Chief Technology Officer, Chief Information Officer, or similar C-level equivalent you'll oversee a wide swath of responsibility and unit profit and loss (P&L) statement as a standalone business perhaps, with nine-figure budgets. You will directly influence capital raises and stock price through capital efficiency and be judged by the board with the same financial rigor as the CFO.

TABLE 1.2
Financial Skills Required (by Level)

Level	Required Financial Knowledge
Individual Contributor	None
Manager	TCO, ROI, break-even analysis, contingency analysis
Senior Manager/Director	CapEx versus OpEx, earned value management
Vice President	P&L, unit economics, M&A
C-Level, for example, Chief Technology Officer	Capital allocation, real options, government financial reports

In short, your organizational chart level positively correlates to the size of the financial decisions the company trusts you to make correctly. With each higher level of responsibility comes greater mastery of associated skills. The financial skill ladder shown in Table 1.2 illustrates the knowledge that separates engineers who stay mid-level from those who reach VP and CTO.

As a first-line manager, you must master the basics such as total cost of ownership (TCO) for an asset, return on an investment (ROI) and break-even point analysis for some activity. At the next level, you will need to understand the finances of capital and operational expenditures and manage multi-million-dollar program budgets. At the vice president level, you will need to understand large budgets, business unit economics, and be able to support M&A activities.

Finally, at the C-level (e.g., CEO, CTO, CIO, and CSO), you operate at the corporate-finance level, dealing with capital allocation, writing government-required financial reports, and using advanced financial tools. You will need a deep understanding of advanced concepts like economic value added and real options because the board and investors judge you on the same capital-efficiency metrics they use for the CEO and CFO.

1.2.3 Investing in Technology Companies

When considering a business arrangement with some company, whether as a customer, client, vendor, or partner, it is prudent to understand their financial situation. Are they financially healthy enough to purchase your company's products or services (or will they default on payment)? Do they carry too much debt or crippling financial obligations that can interfere with their efficiency? Are they stable enough to deliver what you purchase from them on schedule?

It is usually a part of any large business contract to require certain financial disclosures from all parties involved. Having the ability to read and understand these disclosures is an important part of your role as you move up the engineering management ladder.

An SEC 10-K is the annual financial and operational report that publicly traded US companies are required to file with the Securities and Exchange Commission

TABLE 1.3
International Approximate Equivalents to US SEC 10-K Filing

Country	Equivalent Filing
Australia	Annual Financial Report (under Corporations Act)
Canada	Annual Information Form (AIF) + Management's Discussion & Analysis (MD&A)
China	Annual Report (A-Share for mainland-listed)
European Union	Annual Financial Report (per Transparency Directive)
Hong Kong	Annual Report (under Companies Ordinance)
India	Annual Report (per Companies Act 2013)
Japan	Yuho (Annual Securities Report)
United Kingdom	Annual Report & Accounts (under Companies Act 2006)

(SEC). Think of this filing as the "owner's manual" for a company. It's a 100–300-page document that reveals the inner workings of the business. It is not a glossy brochure-type marketing pitch—it contains very detailed information about the company's finance.

Other countries require filings of reports that are like the SEC 10-K (see Table 1.3). These will be discussed in more detail in Chapter 3.

The ability to read SEC 10-K disclosures, their international equivalents, and similar financial documentation provides engineers, scientists, and technologists with a career advantage. These documents give a strategic systems view of a company. These filings reveal risks, costs, revenue targets, and leverage points (such as R&D spending) that directly shape engineering, scientific, and technical decisions at project, program, and strategic levels. For engineers, scientists and technologists, being able to read the 10-K can be helpful when seeking to understand a prospective customer, seeking a supplier or M&A partner or for due diligence on a prospective employer. I will cover more on reading financial statements in Chapter 3.

I mentioned in the introduction that this book is not intended for advice on your personal finances. But understanding a company's financials is just as important as understanding the products and services it sells and the overall market if you are considering investing in that company at any level, from start-up to maturity to a stock exchange-listed company.

1.2.4 Starting My First Company

In the early 1980s, when I was just a few years out of college, I started a company (an S-Corp—see Chapter 3) to provide service and support to small businesses as they began to use personal computers and business software in their operations. I also sold bundled PCs and software. I was in between regular jobs, finishing my doctoral dissertation, and living off my savings. I named the company "OmniSys Technologies" because I did not want to limit its scope. But limiting the scope is what I should have done.

At first, I was doing well—there was no job too big or too small for me to turn down. Soon, I had to start hiring help with sales and service. I was growing too fast and having difficulties with cash management. I knew nothing about profit or sales margins, commissions, tax allocations, leasing terms, and more—I was a tech guy who knew how to build PCs and install software.

Eventually, even though my sales were growing, I was losing too much money to stay in business, and I closed OmniSys. As a consolation prize, however, I took what I had learned about building and repairing PCs and authored a book about it—my first book [Laplante 1992].

I also learned a valuable lesson about starting a business. Unless you have a partner with some financial background, you are your own CFO. If you have no financial education, you are likely going to fail or at least suffer financially.

But even if you manage to survive and learn about finances along the way, those early decisions you make about financing the business, especially concerning equity sharing, will have huge implications downstream. For example, even if the company is successful, you may not make as much money as you could have if you had better planned your financial structure. So, if you are going to start a business, get a financial partner or educate yourself about finances first.

Some of the financial knowledge you will need to start and grow your own company includes:

- cash management,
- raising capital,
- stocks and dilution,
- asset allocation,
- operating business accounts,
- lines of credit,
- loans,
- compensation,
- commission structures,
- bonuses,
- business valuation, and
- local, state, and federal tax obligations.

Sometime during your career, you probably will do some consulting and/or freelance work. The financial knowledge here involves setting rates, managing taxes, and knowing how to correctly deduct home office expenses. All of these require some basic accounting knowledge. Even if you plan to hire an accountant to help you with these things, and you should, it is immensely helpful if you are also familiar with the basics of finance and some accounting concepts.

1.3 UNDERSTANDING FINANCIAL GOALS WITH AN ENGINEER'S MINDSET

One way an engineer/scientist/technologist can look at business finances is as a complex system that must be optimized. For example, with project and resource management, whether leading a team or contributing, you'll need to justify headcount, tools, or spending on outside contractors. Do you conduct a cost–benefit analysis for everything you do? For instance, if you are a software engineer, should you refactor legacy code (over 6 months, at a cost of \$300k) or keep patching (ongoing cost is \$50k/year)? You need to be able to model this financially, not just technically.

Whatever you do in your technical job, when it comes to finances, you should apply the same rigor to the analysis of revenue, costs, margins, and capital. Financial literacy lets you speak the language of return on investment, net present value and payback periods, making your informal requests or formal proposals more likely to get approved. You can think of financial goals as the key performance indicators (KPI) of the whole system.

Let's consider five financial metrics (and an associated icon) that every company uses to measure their progress against goals using some engineering analogies (Figure 1.1).

Revenue growth is the corporate version of "increase power output," for some machine, electrical system, or thermodynamic system. Gross margin might be thought of as an incremental increase in power after accounting for increases in energy input, friction, drag, and other factors.

Operating leverage is a measure of how a company's fixed and variable costs affect its profits when sales change, which is like increasing power output without increasing the input needed to produce the power (by, say, reducing friction).

Free cash flow is another important system metric. It is the cash in hand left after spending on operational expenditures (OpEx) and capital expenditures (CapEx). It might be thought of as the useful energy produced by some system (useless energy such as latent heat and vibration).

Finally, the *rule of 40* is the meta score for a financial system:

$$\textit{growth percentage} + \textit{free-cashflow margin} \geq 40 \tag{1.2}$$

This is a composite system metric that is widely used.

Here is another interpretation of the connection between engineering and finance found in *The Goal,* a 1984 book by Eliyahu Goldratt, which has been mandatory reading in business programs for decades. The book tells the story of a plant

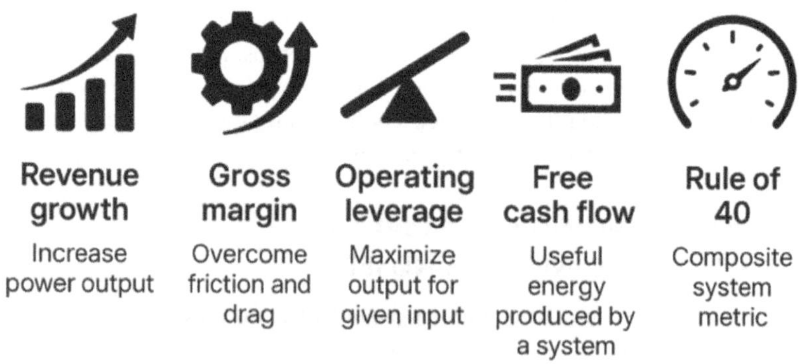

FIGURE 1.1 Five financial metrics, iconic representation, and engineering analogy.

manager whose factory is losing money and faces closure. Guided by his physicist-turned-consultant friend, the manager discovers that traditional efficiency metrics (e.g., keeping every machine busy) are wrong.

The real goal of any business is to make money, measured only by three numbers: throughput (the rate at which the system generates money through sales), inventory (this is money retained inside the system), and operating expenses (money spent turning inventory into product). Using the Theory of Constraints (TOC), which Goldratt explains in detail, the manager identifies the plant's single biggest bottleneck, subordinates everything else to it, and systematically improves it—dramatically increasing throughput while slashing inventory and lead times.

Within three months, the plant goes from near shutdown to the company's top performer, proving that optimizing the whole system around its constraint beats optimizing every part locally. This principle applies equally to factories, software engineering, hospitals, and entire companies [Goldratt and Cox 2012].

Once you see these parallels between engineering and finance, every engineering decision, hiring plan, and services invoice influences one of these five metrics. You can also use a financial perspective to impress your VP or CFO. For example, if you could phrase a proposed initiative as "this will add three points to gross margin" or "this extends runway enough to push us over the Rule of 40 this quarter." Going forward with every line of code you write, every design decision you make, every feature you add you will be able to answer the question "How does this make or save money for my company?"

1.4 ENGINEERING ESTIMATION AND FINANCE

Renowned physicist Enrico Fermi made significant contributions to many disciplines, but notably, theories related to atomic energy. Fermi was both a theoretical and experimental physicist—he was capable of world-class mathematical rigor, but he was also a proponent of informal estimation.

One incident highlights this seeming dichotomy. Fermi was an observer at the testing of the first atomic bomb. After the blast, he scattered some confetti in the air and observed the distance it was blown by the blast wave. From this distance, he estimated the bomb to be equivalent to 10K tons of TNT. After months of calculations, the actual result was computed to be 18.6K tons. While he was off by 86%, his back-of-the-napkin estimate was still close enough to give early feedback on the effects of the bomb and to provide a "sanity" check on the eventual result.

Fermi was a strong proponent of the value of rough estimates, and he used to show his classes their value by solving seemingly unsolvable problems using this approach. In a famous example, he asked the students to estimate the number of piano tuners in Chicago, given only the city population. When the students decried that the problem was impossible, he would show them that a good estimate could be obtained rather easily using the following logic:

1. From the US Census, we know that Chicago has about three million people.
2. If an average family has four members, then the number of families in Chicago must be about 750,000.

3. Suppose 20% of families own a piano, then there are about 150,000 pianos in Chicago.
4. Suppose the average piano tuner can service four pianos a day and works 5 days a week, 50 weeks a year, then each piano tuner serviced 1,000 pianos per year.
5. Since there are 150,000 pianos in Chicago, there must be about 150 piano tuners in Chicago [NASA 2025].

This method is an approximation only since it requires many assumptions. But Fermi's approach does establish a reasonable estimate and serves as a check on a more rigorous calculation.

You can use the same, "Fermi Estimation" approach in any of your work activities to estimate its financial impact. You can calculate such metrics as gross margin percentage, payback period, total engineering fully loaded cost, or return on investment multiple created by an engineering activity.

EXAMPLE 1.2 ESTIMATING PAYBACK PERIOD FOR A MAJOR INFRASTRUCTURE PROJECT.

Suppose you are part of project team building a new 50-Megawatt electric substation at a total cost of approximately $60M. It is believed that this project will save your company $25M per year in reduced congestion charges. This means it will take approximately 2.5 years to recover the initial investment.

Here is another example.

EXAMPLE 1.3 DETERMINING REQUIRED VALUE CREATION FOR A SENIOR ENGINEERING MANAGER

Many CFOs look for at least five times return on the loaded cost for senior staff and managers. Suppose an engineering manager has a total loaded cost of $350k-per year. Then the needed value returned is:

5×$350k = $1.75M/year = ~ $145k/month = ~$36K/week = ~$900 per hour

This means that the manager needs to find or create ~$900k of value every working hour.

You can use this approach to justify your time spent on side projects—for example—if I am successful, one week of my time will result in $100K annual savings to the company.

These examples show that the ability to make million-dollar decisions with ten-minute napkin math is a very valuable financial skill for any engineer, scientist, or technologist.

Whether at work or in your daily life, you should practice financial estimating—do it all the time—at the store, estimate the total of the items in your cart before ringing out at the cashier. At the restaurant, estimate the bill before it is presented. As you hone your financial estimation skills, these act as a "sanity check" against the more formal calculations—just as Fermi guessed the power of the atomic blast within a reasonable margin before confirming it via extensive calculation.

1.5 EXERCISES

1.1 Build a spreadsheet called "My Career Budget" and include the following columns:
   a. Year
   b. My Fully Loaded Cost to Company
   c. $ Impact I Drove/Influenced
   d. ROI = $impact/Cost

   The rows should be the previous two years, the current year, and the next three years (forecasted).

   Your goal is to always have a personal ROI > 3 after your first year of employment.

1.2 Take any feature, component, or asset you personally touched in the last 6 months and describe the money flow in the following format:

   "Where did the dollar come from (revenue source) → how did my work change throughput, inventory, or operating expense → where did the dollar go (profit, reinvestment, shareholder)?"

1.3 Retrieve your company's latest 10-K. If your company does not file a 1-K, use a peer or similar company. In the 10-K form, find and highlight the information that provides:
   a. Primary revenue driver this year
   b. Largest cost driver
   c. Explicit multi-year financial target (growth %, margin %, Rule of 40, etc.)
   d. A risk factor that could kill a major project or significantly affect gross profit.

1.4 Calculate how many hours you need to save the company (or how much extra revenue you need to generate) to personally break even (cover your salary) this year.

1.5 Imagine you are handed 100% ownership of the company and all current cash for your company. Write down the first three engineering-related moves you would make in the next 90 days and exactly how many basis points you believe each would add to gross margin or growth rate (with justification).

1.6 Using the financial ladder in Table 1.2, identify your current level and the gap (in tools and dollar scope) to the next rung.

   Write one concrete action for this quarter that closes that gap (e.g., "Present first NPV model to finance," "Read next 10-K cover-to-cover," and "Calculate break-even on my team").

1.7 Write a single paragraph email to your CFO (real or imaginary) titled "How my organization will move one of the five corporate financial goals this year." Discuss at least one number from the 10-K (or equivalent), one constraint you own, and one metric.
1.8 Imagine the Chief Financial Officer (CFO) of your company corners you in the hallway and says, "I'm cutting next year's CapEx in half. Convince me in one minute why your biggest 2026 initiative should survive." Write out your one-minute pitch.
1.9 Using the "Fermi" estimation approach, calculate how much economic value you personally need to create in the next 12 months to justify your existence at least five times over.
1.10 Using the "Fermi" estimation approach, estimate your company's annual revenue.
1.11 Suppose you are your company's Chief Technology Officer. Your CEO and Board of Directors have given you $100M to spend over three years on new research and development. Write a script for a 60-second pitch of how you would use this money including:
    a. Rough cost breakdown
    b. Likely return on the investment
    c. Year-3 annual benefit

NOTES

1 It is often said that "stock markets predict the future, they don't care about the past." This is a common perspective in financial discourse, often attributed to the *efficient market hypothesis*, which suggests that stock prices reflect all available information and expectations about future performance.
2 Or whatever currency applies to you. Throughout this book, "dollar" is just a placeholder for whatever currency you wish.

REFERENCES

Black, F., & Scholes, M. (1973). The pricing of options and corporate liabilities. *Journal of Political Economy, 81*(3), 637–654. https://doi.org/10.1086/260062

Celarier, M. (2018, January). How a Misfit group of computer Geeks and English majors transformed wall street. *New York Magazine*. https://nymag.com/intelligencer/2018/01/d-e-shaw-the-first-great-quant-hedge-fund.html

Goldratt, E. M., & Cox, J. (2012). *The goal: A process of ongoing improvement* (40th Anniversary ed.). North River Press.

Laplante, P. A. (1992). *Easy PC maintenance and repair*. Windcrest/McGraw-Hill.

Lucky, R. W. (1993). The bean counters. In *Lucky strikes . . . Again: (Feats and foibles of engineers)*, IEEE (pp. 151–153). https://doi.org/10.1109/9780470545164.ch33

NASA. (2025). *Fermi's piano tuner problem*. https://www.grc.nasa.gov/WWW/K-12/Numbers/Math/Mathematical_Thinking/fermis_piano_tuner.htm, accessed 11/29/2025.

2 Basic Accounting and Financial Concepts

I never took accounting as an undergraduate, and when I matriculated for my MBA program, even after I had aced several finance courses, I was required to take an undergraduate Accounting 101 course. This requirement came from the program accrediting body and was non-negotiable. So, I reluctantly took an undergraduate, basic accounting course remotely. It was an interesting situation because this was in the late 1990s and internet-based distance learning courses were rare. Each week's assignments were mailed to the professor, and feedback received via a telephone message.

I did well on the course and more importantly, I learned a lot. I also realized why I needed to take this course—even though I was familiar with high level concepts of finance and managing budgets, I was not familiar enough with basic accounting terminology. This gap knowledge was essential to understanding some of the nuances of higher-level finance.

I am not going to try to embed a basic accounting course in this book. But I do want to define and review some basic accounting terminology to fill in those gaps. I encourage you to take a basic accounting course if you have not already done so. If you enjoy earning and managing your own money, then you will very much appreciate this basic accounting knowledge.

The second half of the chapter builds these accounting concepts up to the level of finance. My goal here is to convey a basic understanding of financial concepts for all kinds of work environments, whether public, private, government, small, large and so on. This treatment is not comprehensive but is intended to enhance your overall understanding of how money works in the world of engineering, science, and technology, and it provides a basis for understanding later concepts in this book.

2.1 BASIC ACCOUNTING TERMS AND CONCEPTS

In 1494, Franciscan friar and mathematician Luca Pacioli published *Summa de Arithmetica*. The book spread the Venetian merchants' method of "double entry" accounting across Europe and earned him the title "Father of Accounting." Ever since, double-entry accounting has been the foundation of global finance.

Accrual accounting records revenues and expenses when they are earned or incurred, not when cash changes hands. This concept helps engineers, scientists, and technologists align project costs and revenues with project timelines, especially for long-term projects where payments may lag work performed. Contrast this approach

DOI: 10.1201/9781003733850-2

with *cash accounting*, which is generally used for personal finances and sometimes in small businesses.

2.1.1 Debits and Credits

Double-entry accounting records all business transactions in a *ledger*, which is the central book or digital record where all financial transactions are organized into accounts, showing debits, credits, and balances. These accounts are organized into five categories:

- assets,
- equity,
- revenue,
- expenses, and
- liabilities.

An *asset* is anything of value. Assets include inventory, facilities, raw materials, equipment, and more. *Equity* represents the stored value of the company after all liabilities are paid. *Revenue* is earnings from selling products or services. An *expense* is any cash outflow to run the business. Sometimes that money is spent such that it is converted into an asset, for example, purchasing a new facility or piece of equipment. Other times, that expense does not create an asset, for instance, paying salaries, taxes, or purchasing office supplies. Finally, a *liability* is money owed to other entities including other companies, governments, and individuals. Let's see how these accounts are maintained on a day-to-day basis.

A *debit* is an entry recorded on the left side of a ledger that usually increases assets or expenses and decreases liabilities or equity. A *credit* is an entry recorded on the right side of a ledger that usually increases liabilities, equity, or revenue, and decreases assets or expenses.

Assets and expenses increase with debits (recorded in the left column of an entry) because they represent value flowing into the business. Liabilities, equity, and revenues increase with credits (recorded in the right column of an entry) because they represent value owed or earned (Figure 2.1).

To illustrate, the effects of debits and credits on some asset accounts are given in Table 2.1.

Account	Increased By	Decreased By
Assets	debit	credit
Expenses	debit	credit
Liabilities	credit	debit
Equity	credit	debit
Revenue	credit	debit

FIGURE 2.1 The effect of debits and credits by type of account.

TABLE 2.1
Example Effects of Debits and Credits on an Asset Account

Transaction	Account Affected	Debit or Credit	Effect on Asset
Buy equipment for cash	Equipment	Debit	↑ Increase
Record Inventory shrinkage	Inventory	Credit	↓ Decrease
Invoice a customer for sale on credit	Accounts Receivable	Debit	↑ Increase
Receive cash from a customer	Cash	Debit	↑ Increase
Pay off a bank loan	Cash	Credit	↓ Decrease

In *double-entry accounting* every financial transaction is recorded in at least two accounts: one as a debit and the other as a credit, ensuring that total debits always equal total credits.

EXAMPLE 2.1 BUSINESS TRAVEL

Suppose an engineering consulting firm pays $800 for an airline ticket for an employee to visit a project site. Because the ticket is paid immediately using the company's corporate debit card, cash leaves the company at the time of purchase. This transaction affects two accounts:

	Debit	Credit
Travel Expense	$800	
Cash		$800

This little example shows the counterbalancing effect: one account goes up (Travel Expense), while another asset goes down (Cash). The total balance remains consistent, which is the essence of double-entry accounting.

EXAMPLE 2.2

Suppose the firm purchases a $2,400 high-performance laptop for use by a project engineer. The laptop is paid for using the company's corporate debit card, so cash decreases at the time of purchase. This transaction affects two accounts:

	Debit	Credit
Asset		
Electronics	$2,400	
Cash		$2,400

Because the laptop is a tangible asset with a multi-year useful life, it must be capitalized and depreciated rather than recorded as an immediate expense.

Depreciation is a systematic way to account for the reduced usefulness of a tangible fixed asset over its useful lifetime. Accumulated depreciation behaves similarly to a liability but is considered a *contra-asset.*[1] It is recorded as a debit or reduction in the value of the asset, balanced by correspondingly increasing (or crediting) the reduction value in a liability account. You will need to understand depreciation to calculate the total cost of ownership for capital equipment and make informed buy-versus-lease decisions (more on this in Chapter 6).

Now suppose you use your corporate credit to pay for a $45 a meal at a restaurant. In these cases, your purchases would be recorded as a $45 debit to "Meals Expense" and a $45 credit to the Credit Card Payable" account. While you do not record these transactions in the company's ledger, you do make these kinds of entries when you submit an expense report to your employer, which is just another kind of ledger.

This concludes the introduction to debits and credits, as it becomes more complicated with special kinds of accounts like contra accounts, prepaid taxes or insurance. But that is why I recommend you take a basic accounting course.

2.1.2 Chart of Accounts, Journal, and Ledger

Moving up the conceptual hierarchy, there are three important transactional records that every company must have. The three records are the chart of accounts, the journal, and the ledger. In the pre-computing era, these were all maintained manually; now they are powerfully enabled by accounting software at the desktop to enterprise scale. These three records are closely related but distinct parts of the accounting system (Figure 2.2).

The *chart of Accounts* is a structured list of all financial accounts used by an organization to record transactions, categorized into assets, liabilities, equity, revenues, and expenses. Each account has a unique number and name (e.g., 1001 Cash, 2001 Accounts Payable, and 4001 Sales Revenue). For EST, it is critical in project budgeting and cost tracking, as it organizes costs like materials, labor, and equipment into specific accounts for accurate financial reporting. For example, in a construction project, the chart of accounts might include accounts for "Concrete Costs" (expense), "Equipment Depreciation" (asset), and "Project Revenue" (revenue), enabling precise cost allocation and performance monitoring, which are reported to stakeholders.

The *journal* is the place where transactions are recorded chronologically. It captures the details of each transaction (date, accounts affected, debit/credit amounts). For instance,

> December 7, 2025—Debit Electronics $800, Credit Cash $800
> December 8, 2025—Credit Sales Revenue $2300, Debit Cash $2300.

The journal serves as the first step before posting to the ledger.

The *ledger*, or *master* or *general ledger*, as we have seen, is the place where debits and credits are posted in an account. The ledger is also known as the *second book of entry* because transactions are first recorded in the journal and then posted to the ledger for classification and summarization. For example, the "Cash account" in the ledger will show all journal increases and decreases in cash over time. In most organizations, the ledger is usually decomposed into *subsidiary ledgers* that provide greater detail. The ledger is also used as the basis for preparing financial statements.

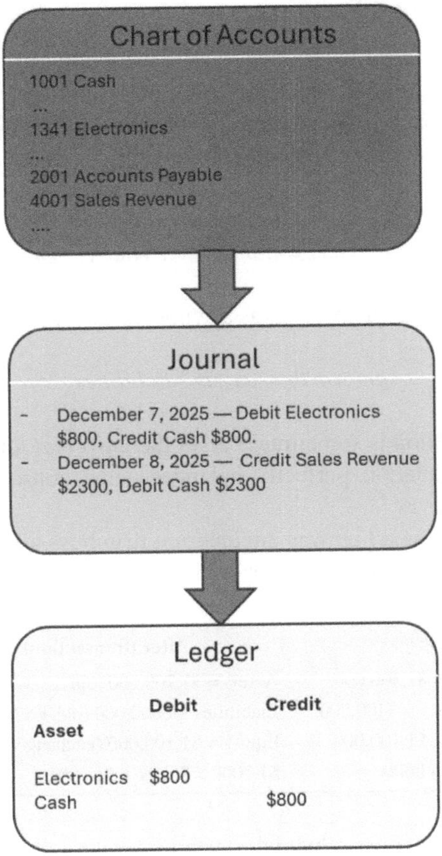

FIGURE 2.2 Relationship between the chart of accounts, journal, and general ledger (or simply "Ledger").

For EST, think of the ledger as an aggregated dashboard. You might not touch it directly, but your project's costs (e.g., equipment purchases and travel) eventually flow here, influencing metrics like gross margin.

On a day-to-day basis, the ledger is used for high-level oversight.

- Accountants post summarized entries (often at the end of the day or week).
- Managers review it for trends like rising expenses.
- It is the source material for generating financial statements.

Understanding these three important accounting artifacts—chart of accounts, journal, and ledger—enables engineers, scientists, and technologists to integrate financial considerations into project planning and execution, bridging the gap between technical and financial decision-making. We will discuss project planning and financial decision-making in Chapter 6.

2.1.3 The Accounting Equation

As we move up one more level in the accounting conceptual ladder, and as the last step before discussing financial principles, consider the most important fundamental equation in accounting and finance, aptly named, the *accounting equation*. This equation illustrates the relationship between assets, liabilities and equity (2.1):

$$assets = liabilities + equity \tag{2.1}$$

Or rearranged and using "net worth" as a synonym for "equity," we have:

$$net\ worth = assets—liabilities \tag{2.2}$$

The accounting equation is sometimes called the *universal law*. Every single transaction keeps this equation perfectly balanced—like conservation of energy, or Kirchhoff's Current Law.[2]

To illustrate, suppose a highway engineering firm buys a $200,000 survey drone on credit, then the accounting equation holds as:

Before	After (transaction)
Assets = $1,500,000	Assets = $1,700,000 (old assets + drone)
Liabilities = $400,000	Liabilities = $600,000 (old + $200k loan)
Equity = $1,100,000	Equity = $1,100,000 (unchanged)
Equation holds	$1,700k = $600k + $1,100k

As another example, suppose a biotech startup invoices a pharma partner $5M for a milestone, then the accounting equation holds as:

Before	After (transaction)
Assets = $8,000,000	Assets = $13,000,000 (old + $5M receivable)
Liabilities = $2,000,000	Liabilities = $2,000,000 (unchanged)
Equity = $6,000,000	Equity = $11,000,000 (old + $5M profit/retained earnings)
Equation holds	$13M = $2M + $11M

Finally, suppose a software company pays $120,000 cash for new servers. Then the accounting equation applies as follows:

Before	After (transaction)
Assets = $3,000,000	Assets = $3,000,000 (cash ↓ $120k, servers ↑ $120k)
Liabilities = $500,000	Liabilities = $500,000
Equity = $2,500,000	Equity = $2,500,000

Cash flows out of one asset account and into another asset account (servers). Therefore, the total assets, liabilities and equity are unchanged, and the equation is still balanced.

The accounting equation is the most important concept you must understand to read standard financial reports confidently. The next section describes those reports.

2.1.4 The Three Major Financial Reports

Moving to the next level of financial reporting, we have the three major reports that provide a picture of a company's financial situation: the income statement, the balance sheet, and the statement of cash flows. These reports are created from information in the chart of accounts, journal, and ledger. Think of them as three different dashboards for the same machine. Let's view them in more detail.

2.1.4.1 Income Statement

The *income statement*, sometimes referred to as "profit and loss" statement or P&L, shows if the business entity made or lost money during the accounting period, typically monthly, quarterly and annually. The income statement is like a power meter or speedometer, and it shows.

The income statement includes the following elements:

- gross revenue,
- cost of goods sold,
- gross profit,
- gross margin,
- operational expenses,
- amortization, and
- taxes.

Gross revenue or simply *revenue* is the income from the company's operations including the sale of goods and services produced. Creating revenue, along with keeping costs under control, is the key to high profits. Good managers and their staff understand this reality. Companies exist not just to deliver the best products or services—the company must increase revenue. Even a non-profit organization, which exists to make some positive societal impact, cannot continue to exist if it is in a perpetual state of economic loss.

The *cost of goods sold* (COGS) are direct costs (e.g., computing services, materials, and power consumed). Your efficiency in managing these directly affects gross margin.

Gross profit or simply *profit* is revenue, less the cost of goods sold (equation 2.3).

$$gross\ profit = revenue—costs \tag{2.3}$$

The higher the gross profit, the more the company can spend on R&D, salaries, bonuses, or as a return to investors as dividends.

The *gross margin* is simply the ratio of revenue to profit (Equation 2.4):

$$gross\ margin = gross\ profit/revenue \tag{2.4}$$

Operational expenses (OpEx) are any expenditures needed to run the business. These include salaries, equipment rental, computing services, professional services, property rents and leases. Good managers watch these expenses closely, as do the "bean counters."

Amortization is an accounting way of expensing the cost of intangible assets (like patents, copyrights, trademarks, software, or licenses) over the period they provide economic benefit. Contrast this with depreciation, which has already been defined and applies to tangible assets such as buildings and equipment.

Taxes on any aspect of the business need to be managed carefully. Certain taxes such as business income taxes or value-added taxes (VAT) may not be computed until the end of the year reports. Tax management can drive many business decisions in ways that may not be clear to those who are not accountants or senior managers.

These income statement elements may be further expanded in the income statement; for example, R&D expenses may be broken out as a separate category of expenses. Sometimes, when more detail is needed, subsidiary income statements by category are generated. The format and inclusion of elements in the income statement vary widely from industry to industry and even within different companies in the same industry. But the key elements just noted are always there.

The following page shows a sample income statement for A-One Engineering, Inc., a small, fictitious engineering consulting firm that specializes in highways, roads, bridges, and related infrastructure.

Income Statement (unaudited)
For the Quarter Ended March 31, 2026

Line Item	Amount	Notes
Revenue		
Engineering design & consulting services	$430,000	2 highway projects
Construction supervision fees	$75,000	1 continuing state project
Total Revenue	$505,000	
Cost of Services (Direct Costs)		
Salaries—project engineers & technicians	$214,000	6 staff × avg. $107k annualized
Subcontractors (geotech, surveying)	$99,000	
Materials & reimbursable expenses	$27,000	
Equipment rental & field vehicles	$18,000	
Total Cost of Services	*$358,000*	
Gross Profit	*$147,000*	29.1% gross margin
Operating Expenses		
Salaries—management & admin	$80,000	
Office rent & utilities	$12,000	
Software licenses	$14,000	
Insurance (professional liability + general)	$20,000	
Marketing & business development	$1,000	
Depreciation (equipment & vehicles)	$14,000	
Other overhead	$8,000	

Total Operating Expenses	$149,000	
Operating Income (EBIT)	($2,000)	Near break-even in growth quarter
Interest expense (equipment loan)	$3,000	
Income Before Taxes	($5,000)	
Income tax expense (benefit) @ 21%	($1,050)	
Net Income (Loss)	*($3,950)*	Expected in heavy hiring/investment quarter

EBIT is an acronym for earnings before interest and taxes. You can see on the income statement that taxes and interest are subtracted from EBIT afterward. EBIT is an important dashboard metric for managers, accountants, and senior executives. A related, earnings before interest, taxes, depreciation, and amortization (EBITDA) is equally as important.

Here are some key takeaways from A-ONE Engineering's Income Statement

- Gross margin of 29% is typical for highway engineering firms (20–35% range) [Withrow 2025].
- You might deduce that A-ONE Engineering is investing aggressively (as evidenced by high operating expenses and depreciation). This implies short-term loss but positioning for larger state contracts in the last three quarters of the year.
- Every $100k of additional billings at 29% gross margin drops ~$15k–$20k direct to the bottom line after fixed overhead is covered.

These are the kinds of observations that Board Members and senior managers would make.

Certain information from the income statement would flow into the equity section of the balance sheet and drive cash flow planning for the next bidding cycle. The end of year income statement also is input to the company's annual report to shareholders and government entities, such as taxing authorities.

2.1.4.2 Balance Sheet

The *balance sheet* is a financial statement that shows the assets, liabilities, and owners' equity of a company. Thus, it shows the details on which the accounting equation is based. It also helps answer the questions "what do we own and owe at a point in time?" or "are we solvent and what assets do we have to work with?"

The format of a balance sheet is relatively simple and consistent across entities. An example for A-One Engineering, Inc. is provided below.

A-ONE Engineering Inc.—Balance Sheet (unaudited)
As of March 31, 2026

ASSETS	
Current Assets	
Cash and cash equivalents	$410,000
Accounts receivable (net of allowance)	$321,000
Unbilled revenue (work in progress)	$138,000
Prepaid expenses & insurance	$25,000
Total Current Assets	*$894,000*

Property, Plant & Equipment	
Surveying & CAD equipment	$670,000
Vehicles & field trucks	$225,000
Office furniture & leasehold improvements	$38,000
Less: Accumulated depreciation	($58,000)
Net PP&E	*$875,000*
Other Assets (deposits, software licenses)	$12,000
Total Assets	*$1,781,000*
LIABILITIES & EQUITY	
Current Liabilities	
Accounts payable	$249,000
Accrued payroll & benefits	$93,000
Current portion of equipment loans	$3,000
Deferred revenue (retainers)	$54,000
Total Current Liabilities	*$399,000*
Long-term equipment loans (net of current)	$318,000
Total Liabilities	*$717,000*
Equity	
Common stock	$550,000
Retained earnings (beginning + current loss)	$518,000
Net loss for the quarter	($3,950)
Total Equity	*$1,064,050*
Total Liabilities & Equity	*$1,781,000*

Notice that the balance sheet is cumulative to the reporting date, while the income statement was for the calendar year quarter at the reporting date. The Statement of Cash Flows will also be for the quarter. You will also note that the accounting equation holds nicely. If it did not, it means that a calculation error was made.

Here are some observations from this Balance Sheet:

- Working Capital = $894k – $399k = $495k. This is healthy liquidity for bidding on new state contracts.
- Debt-to-Equity ratio = 0.67. This is conservative leverage, and typical for engineering firms.
- Equipment-heavy balance sheet (≈ 49% of assets). This is normal for highway engineering; depreciation will be a steady non-cash expense [Withrow 2025].
- Unbilled revenue $138k. This means that the company has earned this revenue as a percentage-of-completion contract; it will convert to cash as milestones are approved.

This balance sheet, together with the earlier income statement and general ledger sample, gives a complete snapshot of a typical mid-sized highway engineering firm in early-stage growth mode.

2.1.4.3 The Statement of Cash Flows

The *statement of cash flows* (or *cash flow statement*) is a financial report that shows management where money came from and where it went for the accounting period.

The statement answers questions like "where did the actual cash go this period" or "are we burning or generating cash?" Remember, in any business, "cash is king" because it is liquid and can be spent more freely than securities or other assets.

Here is a sample statement of cash flows for A-ONE Engineering Inc. Remember that the format of any of the three major financial statements varies quite significantly from company to company, but the basic elements that it contains are ubiquitous.

STATEMENT OF CASH FLOWS
For the Quarter Ended March 31, 2026

CASH FLOWS FROM OPERATING ACTIVITIES		**Amount**
Net loss		($3,950)
Adjustments to reconcile net loss to net cash:		
+ Depreciation & amortization		$14,000
+ Increase in accounts receivable		($62,000)
+ Increase in unbilled revenue (work in progress)		($48,000)
+ Increase in prepaid expenses		($4,000)
+ Increase in accounts payable		$118,000
+ Increase in accrued payroll & benefits		$34,000
+ Increase in deferred revenue (retainers)		$55,000
Net cash provided by operating activities		*$103,050*
CASH FLOWS FROM INVESTING ACTIVITIES		
Purchase of surveying & CAD equipment		($150,000)
Purchase of field trucks		($35,000)
Net cash used in investing activities		*($185,000)*
CASH FLOWS FROM FINANCING ACTIVITIES		
Proceeds from new equipment loan		$90,000
Principal repayments on existing loans		($18,000)
Net cash provided by financing activities		*$72,000*
Net decrease in cash during the quarter		*($9,950)*
Cash & cash equivalents—January 1, 2026	$418,950	
Cash & cash equivalents—March 31, 2026	$409,000	

Here are some observations that we can make from this statement of cash flows, even without knowing anything else about the situation:

- The company is investing $185k in equipment and field vehicles, this implies that the company has landed some big jobs starting in the second quarter. This is a positive sign for the company.
- Borrowing $72k to fund the purchase of new equipment is reasonable as it preserves cash reserves.
- Free cash flow = operating cash flow—capital expenses = $14k—a positive free cash flow is excellent, especially during a growth phase.

This cash flow statement reconciles with the earlier income statement and balance sheet.

2.1.4.4 Using the Three Major Financial Documents

The three main financial documents are typically prepared by the Office of the Chief Financial Officer, or, as Bob Lucky called them, the "bean counters" (see Chapter 1). These reports are prepared monthly for internal management review, quarterly for board of directors (or trustee) review, and annually for tax purposes, preparation of 10-K reports (in the case of US public companies) and for review by governing boards.

The relationship between the three primary financial reports is rather complex—information such as cash spent, earnings retained, assets bought and sold, and so forth is updated on the relevant statements. Then any shared information is transcribed from one report to another. Here are a few examples of how everyday business transactions affect one or more of these documents.

EXAMPLE 2.3 LINKING NET INCOME TO CASH FLOW AND BALANCE SHEET UPDATES

As net income for an activity is reported, it is recorded to the cash flow statement, and that number is adjusted for non-cash items (such as depreciation) and adjusted for working-capital changes.

As another example, when the company spends cash, say, to buy an asset, that asset appears on the Balance Sheet, but the spent cash is recorded in the cash flow statement. If that asset is a durable item, then it is expensed via depreciation on the Income Statement over time.

EXAMPLE 2.4 HOW PROFIT FLOWS INTO RETAINED EARNINGS

When profit is recorded for a reporting period on the Income Statement, this is recorded as retained earnings (a component of owners' equity) on the Balance Sheet.

EXAMPLE 2.5 EFFECTS OF BORROWING ON ALL THREE FINANCIAL STATEMENTS

When the company takes on some debt, for example, a loan on its operational credit line, cash increases on the balance sheet (as does the loan amount as a liability). But as interest is incurred it is recorded in the Income Statement (as an expense), and as payments on the loan are made, these are recorded in the Statement of Cash Flows.

As an engineer, scientist, technologist, etc., on staff, you won't be aware of any of these transactions, as you are not watching or maintaining these financial reports. But these reports contain important information that reveals why certain management decisions are or are not taken.

For instance, if a manager wishes to justify an increase in staff headcount in the budget, they might look at the gross margin trend in a series of Income Statements, then check the free cash flow in the Statement of Cash Flows. Obviously, positive trends could justify the increase in headcount.

EXAMPLE 2.6 DIAGNOSING CASH TIGHTNESS DESPITE REPORTED PROFITS

Suppose you hear a rumor at your company that cash is tight despite well-publicized high profits. To understand this contradiction, you would look at the Statement of Cash Flows, focusing on the cash flow from operations—is it too high relative to other costs such as debt service? Are there changes in capital expenditures (e.g., plant maintenance) that are burning cash?

EXAMPLE 2.7 UNDERSTANDING INVESTMENT DECISIONS DESPITE LOW REPORTED PROFITS

Your company might be investing in a new $100M plant even though you hear that profits are slim. But the answer could be found on the Balance Sheet, where your company is loaded with cash (e.g., accumulated retained earnings unspent from previous years) and shows a positive cash flow on the Statement of Cash Flows.

These are just a few examples of analysis to help you appreciate that there is much more to these kinds of decisions than is generally known to those who are not privy to the detailed financial data.

2.1.5 Cash Management

There is a critical linkage between the principles of the time value of money and risk, reward, and return on investments. A company's cash is managed very carefully. It is true that cash sitting in a bank account will only earn enough interest to keep up with inflation. But operating capital (cash) is the lifeblood of a business, as it can be spent freely and quickly ("is *liquid*") and is stored in the bank at very low risk. Making that cash earn more than it can in a bank account and exceeding inflation requires inquiring calculated risks. Risk, Return, and Trade-offs help describe individual and organizational monetary behavior and are discussed in Chapter 5. But there are two important ways that businesses manage their cash to squeeze every dollar out of it—lines of credit and sweep accounts.

A *line of credit* is a flexible, short-term loan arrangement that lets a business borrow money as needed up to a preset limit. Borrowing money (drawing from) a line of credit is done, for example, when accounts payable are high and cash is low. Businesses borrow carefully against the line, however, as loan interest reduces profits. Banks will watch the amount of borrowing that a business does against its line of credit and alert if too much is being borrowed, and sometimes, extend additional credit if conditions warrant—just as a credit card company might do with an individual.

A related banking service called a *sweep account*, automatically transfers ("sweeps") excess funds from a checking or operating account into another account (e.g., a money market fund, investment account, or loan balance) to optimize interest earned or reduce interest owed. Sweeps typically occur daily and help ensure idle cash is put to work while maintaining adequate liquidity for transactions. Large EST firms have sophisticated sweep arrangements as the large sums involved can generate $10s of thousands per day, but even small firms will use sweep accounts to keep their money working all the time.

2.2 BASIC PRINCIPLES OF FINANCE

Now that we have reviewed some basic accounting and banking concepts, let's discuss some terms, concepts and principles of finance and financial planning. First, we start with financial principles for individual investments and financial activities. The principles that will be introduced here apply to the finances of both individuals and business entities, so you can use these ideas to make personal investments or workplace financial decisions.

After these, we will look at the most important financial metrics for organizations, which incorporate the basic metrics and require new ones for the aggregate behaviors of business activities and the engineers, scientists, and technologists who conduct these activities.

Underlying all of these are two very important notions—the time value of money and risk-reward principles. Everything else, all economic (some non-economic) behavior, can be largely explained by these two factors. For engineers, scientists, and technologists, these principles can frequently explain why companies make decisions that seem irrational—at least from an engineering or scientific perspective.

2.2.1 The Time Value of Money

The real value of money depends on when that money is paid. For example, when there is a period of inflation, a dollar paid today is worth more today than if the same dollar is paid in the future—if you were to go back in time 60 years, you would pay around 20 cents per gallon for gas. *Inflation* is the general increase in prices across an economy over time. Inflation is normal in an economy as it is a function of supply and demand, economic growth, rising production costs, expansion of the money supply and other factors. But severe inflation can be very destructive to an economy; for instance, Venezuela suffered 136% inflation in 2025. During the 1920s, Germany experienced extreme *hyperinflation*, when the rate of inflation increased by at least 50% per month.

Deflation, the opposite of inflation, is an unusual situation where there is a general decrease in prices across an economy, which increases the purchasing power of money. Deflation can be caused by falling demand, tight credit, overcapacity, and banking crises such as multiple failures. Therefore, it is dangerous because it causes consumers to delay purchases, businesses to cut wages, and the cost of investment and debt to increase in real terms. Deflation has only been experienced a few times in history, such as the US Great Depression (1929–1933), and the Post-World War I Recession (1919–1921) in the United States and Europe. Japan's "Lost Decades" (1990s–2010s) is considered the modern case study for deflation risk.

When there is no inflation nor deflation, the dollar today holds its value over time, but this situation only exists at the inflection point between inflation and deflation. Most *central banks*, the banking entity of a country that manages banking policy, try to maintain an inflation rate of 2–3% through manipulation of the discount rate, quantitative easing and tightening.

The *discount rate* is the interest rate charged by a central bank when lending to commercial banks. This rate affects all other kinds of loans, such as mortgages, equipment loans, working cash loans (or lines of credit) and so forth.

Quantitative easing is when a central bank creates new money and uses it to buy back its debt in the form of bonds to lower long-term interest rates, increase liquidity, stimulate lending and investment and prevent deflation.

Quantitative tightening is the opposite of quantitative easing. In *quantitative tightening*, the central bank reduces the money supply by selling bonds back into the market or letting bonds mature without reinvesting to raise long-term interest rates, slow down borrowing, reduce inflation and cool an overheated economy.

For the remainder of this book, we will assume that there is a normal inflationary situation in the economy.

2.2.2 The Finances of Investing

In the next sections, we will examine how to evaluate a business investment, for example, purchasing equipment and tools, investing in R&D initiatives, or building a new plant. We will look at the key methods for evaluating investments with a focus on practical applications, real-world examples and show engineers, scientists, and technologists can use them to justify projects like infrastructure upgrades or AI initiatives.

These concepts can help you compare risky investments against safer alternatives by quantifying returns, recovery time, and profitability. This ensures that decisions align with financial goals.

2.2.2.1 Net Present Value (NPV)

The *net present value (NPV)* formula is used to calculate the current value of a future payout of an investment, asset, or cash flow. NPV is based on the principle, previously discussed, that a dollar today is worth more than the same dollar in the future due to inflation. A positive NPV signifies a profitable project.

It's worthwhile to derive the NPV formula. First, consider an investment, asset, or cash flow with present value *P*, an inflation rate *r*, held for *Y* years. Then the future value (FV) of the investment is given by Equation (2.5):

$$FV = P(1+r)^Y \tag{2.5}$$

To find the present value (P) of a future asset (FV), rearrange Equation (2.5), yielding Equation (2.6):

$$NPV = P_Y = \frac{FV}{(1+r)^Y} \tag{2.6}$$

To illustrate, here is a little problem that I used in my project management classes to illustrate NPV. Suppose you win a $1,000,000 lottery prize. You can choose between receiving $1,000,000 in 20 equal annual payments ($50,000 per year) or $800,000 immediately. Assuming that the inflation rate is 3% per year, and ignoring taxes, which option is better?

To answer, calculate the NPV of the 20 annual payments:

The first payment ($50,000) is received immediately, therefore

$$P_0 = \frac{\$50,000}{(1+0.03)^0} = \$50,000$$

The second payment's NPV is:

$$P_1 = \frac{\$50,000}{(1+0.03)^1} = \$48,544$$

The third payment's NPV is:

$$P_2 = \frac{\$50,000}{(1+0.03)^2} = \$47,130$$

. . .

Continuing in this manner, we see that the last payment's NPV in year 19 is:

$$P_{19} = \frac{\$50,000}{(1+0.03)^{19}} = \$28,514$$

Summing the present values for all 20 payments yields approximately $766,189. Since $800,000 exceeds $766,189, the lump sum seems to be the better choice.

But there may be reasons to opt for the annual payments. For instance, your income tax rate may be high that year, and you expect it to decline in succeeding years. You may have trouble disciplining yourself with such a large sum and prefer to put yourself on a budget. On the other hand, there are instances where lottery awarding entities have gone bankrupt or stopped paying distributions for other reasons—leaving the prizewinners short-changed. So, there is also the element of risk and lost opportunity in making the decision, in addition to the simple cash equivalent comparison.

The previous example was for a personal investment, but the approach to evaluating the real value of some asset or activity has applications for engineers, scientists, and technologists. Consider the following.

EXAMPLE 2.8 COMPARING INVESTMENT OPTIONS USING NET PRESENT VALUE (NPV)

Suppose a project manager must decide between purchasing a $250,000 testing tool that saves $500,000 immediately or hiring testers for $250,000, with expected savings of $750,000 in two years (inflation rate = 10%). The NPVs are:

Testing Tool:

$$PV_{tool} = \frac{\$500,000}{(1+0.10)^0} = \$500,000$$

Hiring Testers:

$$PV_{hire} = \frac{\$750,000}{(1+0.10)^2} = \$619,835$$

The hiring option has a higher NPV, suggesting it is preferable, though risks (e.g., project cancellation and departure of testers) must be considered.

Since the NPV is positive, the initiative is financially.

Here is one more example.

EXAMPLE 2.9 CALCULATING THE NET PRESENT VALUE FOR A PROCESS REVIEW INITIATIVE

A process improvement initiative costs $50,000 and saves $100,000 in rework costs in two years (inflation = 10%). Using Equation 2.6, the NPV is:

$$NPV = \frac{\$100,000}{(1.10)^2} - \$50,000 = \$82,645 - \$50,000 = \$32,645$$

Therefore, this initiative is viable.

A *cash flow* refer to the actual movement of money into and out of a business (or individual finances) during a specific period. In other words, cash flows show where money came from and where it went, which is a direct measure of liquidity.

The lottery question or any series of cash flows can be represented as in Equation (2.7):

$$NPV = \sum_{t=1}^{n} \frac{CF_t}{(1+r)^t} - C_0 \tag{2.7}$$

where

CF_t is the cash flow at time (t)
r: Inflation or interest rate
t: Time period
C_0: Initial investment

This format is more convenient to implement in software.

There are some assumptions in the NPV calculations given. For example, inflation does not remain fixed, and NPV does not account for opportunities lost if the expenditure were made on something else. Therefore, you must consider all factors when making some form of investment decision, whether personal or for your organization, not just the raw numbers.

2.2.2.2 Internal Rate of Return (IRR)

The net present value (NPV) is an indirect measure that requires specifying the market opportunity cost (inflation or interest rate) of the capital involved. In contrast, the *internal rate of return (IRR)* is defined as the discount rate in the NPV equation that causes the calculated NPV to be zero. It is not the same as the return on investment (ROI). IRR is a popular metric for financial analysis because it does not require prior knowledge of the cost of capital.

To calculate the IRR, we need to find the rate r that causes the present value P to go to zero in the future value (FV) equation:

$$0 = \frac{FV}{(1+r)^Y} - c$$

where c is the cost of the alternative being considered. Solving for r, we rearrange the equation:

$$c = \frac{FV}{(1+r)^Y}$$

Multiplying both sides by $(1+r)^Y$

$$(1+r)^Y = \frac{FV}{c}$$

Taking the Y th root of both sides:

$$1+r = \left(\frac{FV}{c}\right)^{\frac{1}{Y}}$$

Finally, subtracting 1 from both sides:

$$r = \left(\frac{FV}{c}\right)^{\frac{1}{Y}} - 1 \tag{2.8}$$

where r is the internal rate of return.

EXAMPLE 2.10 CALCULATING THE INTERNAL RATE OF RETURN (IRR) FOR A CODE REVIEW INITIATIVE

Consider a code reading initiative expected to cost \$50,000, with returns of \$100,000 in reduced rework two years in the future. What is the IRR for this activity?

Using equation (2.8)

$$r = \left(\frac{100,000}{50,000}\right)^{\frac{1}{2}} - 1 = (2)^{\frac{1}{2}} - 1 = 1.414 - 1 = 0.414 = 41.4\%$$

To decide on this initiative, you would compare the IRR of an alternative with higher return or similar return but lower risk. If the IRR is low, a lower-risk option such as paying down debt might be preferable.

This derivation of IRR is appropriate for a single future payment.

For a cash flow—a series of future payments—the internal rate of return is the discount rate that makes NPV = 0. That is, the activity's expected compound annual rate of return. The formula is solved iteratively:

$$0 = CF_0 + CF_1/(1+IRR) + CF_2/(1+IRR)^2 + \ldots + CF_n/(1+IRR)^n,$$

where CF is cash flow and n is periods. If IRR exceeds the cost of capital, the project is viable.

The IRR can be computed iteratively using a spreadsheet or other software program and then compared to the cost of capital. Cost structures and the cost of capital are discussed in Chapter 3.

EXAMPLE 2.11 CALCULATING THE INTERNAL RATE OF RETURN AN ENERGY-EFFICIENCY INVESTMENT

Suppose a solar panel installation costs \$500,000 and generates \$120,000 annually for 5 years. At a 5% interest rate, NPV is calculated from equation (2.7) as:

$$NPV = \frac{\$120,000}{1.05} + \frac{\$120,000}{1.05^2} + \frac{\$120,000}{1.05^3} + \frac{\$120,000}{1.05^4} + \frac{\$120,000}{1.05^5} - 500,000 \approx \$19,537$$

Since the NPV > 0, the project is financially viable.

But what is the IRR of this project and is it preferable to some other investment or activity? Using a spreadsheet, it is easy to plug values into equation 2.8 until the IRR is determined to be about 6.4%.

Engineers, scientists, and technologists can appreciate IRR because it converts the return of some investment to a percentage output. But beware of multiple IRRs in non-conventional cash flows (e.g., alternating positive/negative).

2.2.2.3 Payback and Discounted Payback Period

The *payback period* is the time to recover an initial investment in an asset or an initiative. The formula is given in Equation (2.9). Here, the *annual cash flow* represents the one-time return on the investment or initiative:

$$\textit{payback} = \textit{initial investment/annual cash flow} \tag{2.9}$$

This formulation is simple, but useful for quick risk assessment. That is, a shorter payback period means lower exposure to uncertainty. Here are a couple of examples:

EXAMPLE 2.12 CALCULATING THE PAYBACK PERIOD FOR AN LED LIGHTING RETROFIT

Suppose an engineering firm's owners wish to retrofit all incandescent fixtures with LED lighting. The initial cost is $120,000 for the fixtures and installation. It is projected that the conversion will yield annual electricity savings of $35,000:

payback = $120,000/$35,000 = 3.43 years

Thus, the investment recovers in just over 3 years. If the owners' criterion is to recover the investment in less than four years, then the project would be acceptable.

EXAMPLE 2.13 CALCULATING THE PAYBACK PERIOD FOR A CLOUD MIGRATION INITIATIVE

Suppose a technology company wishes to convert some of its applications to a cloud-based solution at a cost of $450,000 for migration fees, training, temporary parallel execution during the transition. It is projected that the annual savings will be $180,000 from reduced server maintenance and power consumption. Then

payback = $450,000 ÷ $180,000 = 2.5 years

This investment recovers in 30 months, which is great for a tech firm where cloud efficiency is strategic.

The payback formulation of Equation (2.9) may be too simplistic, however, for long-term projects like R&D, where big payoffs come later. For example, suppose a $400k conveyor upgrade generating \$100k/year pays back in 4 years. But what if the timeframe is for it to recover in less than 3 years? You would have to reject this initiative despite the potential \$500k post-payback profits. So, for uneven payoffs, you need to accumulate the cash flows until the cumulative value equals or exceeds the initial investment. Here is an example of that situation.

EXAMPLE 2.14 DETERMINING THE PAYBACK PERIOD FOR A MASS SPECTROMETER INVESTMENT

Suppose a biotech firm wishes to purchase a new mass spectrometer that costs \$800,000. The annual savings/benefits of the new device are estimated to be as follows:

Year	Savings	Cumulative Savings	
1	\$150K	\$150K	
2	\$250K	\$400K	
3	\$300K	\$700K	*Break-even occurs in*
4	\$350k	\$1,050K	*about 3.286 years*

We can see from interpolating the cumulative savings that payback occurs in about 3.3 years. A quick rule of thumb is that any return in less than 2 years is a sure thing; a 2 to 4 years return is likely a good investment, and longer than that requires significant justification for other reasons. Thus, investing in this new spectrometer would be a good investment.

The simple payback period formula of equation (2.9) determines the time required to recover the initial investment from cumulative cash flows, but it ignores the time value of money and post-recovery cash flows. To account for these, we need another formula.

The *discounted payback period* uses discounted cash flows, accounting for the time value of money. A finite period indicates recovery of the investment and capital costs, aligning with a non-negative NPV. Discounted payback improves on the standard by incorporating the time value of money, discounting future cash flows C_F to their present value before cumulating.

Discounted cash flow is given in Equation (2.11):

$$PV = \sum_{t=1}^{n} \frac{CF_t}{(1+r)^t} \geq initial\,investment \tag{2.11}$$

where r is the interest or discount rate and t is time in years.

EXAMPLE 2.15 CALCULATING THE DISCOUNTED PAYBACK PERIOD FOR A CONVEYOR SYSTEM

Suppose a $400k conveyor system is financed at a 10% rate (via a commercial loan). The conveyor is expected to produce improvements in productivity. The resulting PV cashflows are $90.9k (year 1), $82.6k (year 2), $75.1k (year 3), $68.2k (year 4), $62.0K (year 5), and $56.4k (year 6), at which point, by Equation 2.11

PV = $435k ≥ $400k.

In fact, the cumulative PV exceeds $400K at about 5.5 years, which is the approximate discounted payback period.

Engineers, scientists, and technologists should use discounted payback for capital-intensive projects (e.g., chemical reactors) where a finite period confirms recovery of investment plus opportunity cost, aligning with non-negative NPV thresholds. However, discounted payback addresses basic payback's time value of money limitation, but retains the problem of ignoring post-recovery value.

2.2.2.4 Average Rate of Return

The *average rate of return* (ARR), sometimes called the *accounting rate of return*, estimates profitability using average accounting income over the investment's life, often as a percentage of initial or average investment. Common formulations are given in equation (2.12) a through c:

a. average profit / initial investment

b. average projected after-tax profit / average investment (2.12)

c. average projected after-tax profit / initial investment

where *average investment = (initial + salvage value)/2*

Other variations follow a similar format.

EXAMPLE 2.16 CALCULATING THE AVERAGE RATE OF RETURN (ARR) FOR AN INTERNAL SOFTWARE BUILD

Suppose your company is going to undertake a $1.2M internal software build. The company expects a $300k average annual profit over 5 years. Using Equation (2.12a) yields

ARR = $300k/$600k average investment = 50%.

Thus, this initiative is attractive if the target return is >20%, but you must adjust this result for non-cash depreciation, inflating profit.

Metric	Definition	Acceptance Criterion	Issues
Average Rate of Return (ARR)	$\frac{\text{Average Income}}{\text{Average } I}$	> some standard	Arbitrary standard; no time value; not cash flow;
Discounted Payback	Years to recover I plus capital costs	< ∞	Same as payback (except time value)
Internal Rate of Return (IRR)	discount rate such that NPV=0	$IRR > r$	Scale; different results with alternating cash flows; reinvestment rate
Net Present Value (NPV)	$PV - I$	$NPV > 0$	
Payback	Years to recover I	< cutoff	Arbitrary cutoff; no time value; no credit beyond cutoff; scale problem?
Profitability Index (PI)	$\frac{PV}{I}$	$PI > 1$	Scale

FIGURE 2.3 Summary of financial measures for a single investment or initiative, where r is the interest or inflation rate.

For EST situations, ARR is useful for non-discounted comparisons but requires expert input for adjustments like tax shields. ARR aligns with accrual accounting (including depreciation) but varies by definition and ignores TVM or cash timing, and tax implications. Therefore, you should cross-reference with IRR and NPV calculations for robust decisions and get accounting expert input.

Just to wrap up this section, Figure 2.3 provides a summary of the investment and initiative assessment metrics that we have been discussing.

Consider Figure 2.3 to be a "cheat sheet" for your use in evaluating project initiatives going forward.

2.2.3 Metrics for Assessing a Company's Financial Position

Now, let's explore some metrics that accountants, financial types, and management look at to assess a company's financial position. These metrics can be used to assess your own company's situation, or to assess potential clients, customers, suppliers, and partners. Investment firms and individual investors also use these metrics.

2.2.3.1 Some Key Metrics

Let's consider eight key metrics that assess a company's value and leverage. Several key valuation and leverage metrics offer a clear window into a company's financial structure, risk level, and market perception.

I am not going to be giving strict guidance on interpreting these metrics; however, this is a very sophisticated business. Instead, I just want to define these metrics and discuss how they are used in decision-making. In a few cases, I will suggest some healthy ranges, but only for the purposes of understanding scale, not for absolute decision-making.

The *debt-to-equity ratio* (D/E), calculated as total debt divided by shareholders' equity, essentially determines how much of the company is financed by loans versus owners' funds. You can think of it in terms of "how much of our equipment is bought on credit?" A simple formula is given in Equation (2.13):

$$\textit{debt-to-equity ratio} = \textit{total debt/shareholders' equity} \tag{2.13}$$

Healthy ranges for these metrics are highly dependent on the industry type, but are generally expected to be between 1.0 and 1.5 for established companies. For example,

D/E < 1.0 implies the company is using relatively little debt
D/E ≥ 2.0 is often considered high leverage unless the industry is capital-intensive

Industries with heavy equipment or infrastructure naturally run higher D/E ratios, while service-based or tech companies tend to run lower. Table 2.2 reflects how much debt is structurally required to operate in various sectors.

The next metric, the *total debt to EBITDA* ratio, shows how many years of cash profit it would take to pay off all debt:

$$\textit{total debt to EBITDA} = \textit{total debt/annual EBITDA} \tag{2.14}$$

Banks and bonding agencies will look at this metric. Typically, if the metric is under 2.0, the company is considered very safe, 2.0 to 3.5 is normal, and over 4.0 may be considered risky [Kenton 2025].

Next, *interest coverage*, given in Equation (2.15)

$$\textit{interest coverage} = \textit{EBIT/interest expense} \tag{2.15}$$

tests whether operations can easily handle interest payments. Interpreting this number is very complex, but obviously, the higher the number, the better, and factors much higher than unity (e.g., 5 or more) are much preferred.

The *price-to-earnings (P/E)* ratio, given in Equation (2.16), shows how many years of current earnings investors are willing to pay for the stock:

$$\textit{P/E} = \textit{market price per share/earnings per share} \tag{2.16}$$

TABLE 2.2
Examples of Typical D/E Ranges

Industry	Typical Healthy D/E Range
Tech/Software	0.1–0.5
Consumer Goods	0.5–1.0
Manufacturing	1.0–2.0
Utilities	1.5–3.0+
Telecom	1.5–2.5
Financials	Highly variable—often >2.0 is normal

Source: [Eqvista 2025].

In essence, this metric answers the question "how many years of today's profit to pay the current stock price?" This metric can range from the low tens to the hundreds in high-growth industries. But its interpretation is very tricky and requires expert advice.

Forward P/E, using forecasted next-12-month earnings per share in Equation (2.16), provides a more useful forward-looking view, typically 15% to 30% lower than trailing P/E in expanding sectors.

The *enterprise value to EBITDA ratio* indicates how many years of cash profit it would take to acquire the whole company. This ratio is simply:

$$\textit{EBITDA ratio} = \textit{enterprise value/EBITDA} \tag{2.17}$$

For public companies, the enterprise value is the company's *market capitalization*, that is, the total of the shares (at current price) of the company. For privately held companies, it is based on a *business valuation*, which is an assessment of the value of the company by experts based on numerous factors. It answers the question "How many years of cash profit to buy the entire company, debt and all?" This metric is useful for mergers and acquisitions comparisons.

Next, the *price-to-sales* metric measures investor willingness to pay per dollar of revenue. This metric is especially useful for pre-profit companies. It is determined as:

$$\textit{price-to-sales} = \textit{market capitalization divided by annual revenue} \tag{2.18}$$

This metric will be much higher for early-stage ventures such as pre-revenue biotech, somewhat lower for software companies and much lower for asset-heavy sectors such as infrastructure construction firms.

Next, *Price-to-book* (market price divided by book value per share) assesses whether the market values the company above its accounting net worth, particularly relevant for asset-heavy sectors.

Finally, the *PEG* ratio is the price-to-earnings (P/E) ratio divided by annual earnings per share (EPS) growth percentage. That is:

$$PEG = \frac{P / E}{\textit{annual EPS growth}} \tag{2.19}$$

The PEG ratio evaluates if a high P/E is justified by growth, with an ideal generally below 1.0 to 1.5.

These metrics can guide management decisions—from assessing leverage risks to spotting growth potential in certain areas.

Note that there are significant international considerations for financial metrics like debt-to-equity, P/E, EV/EBITDA, and others. These arise from differences in accounting standards, tax regimes, capital structures, market maturity, cultural financing preferences, and regulatory environments. While some metrics (e.g., EV/EBITDA) are designed to be more comparable across borders by neutralizing tax and capital structure effects, others (e.g., D/E or P/E) can vary substantially, requiring adjustments for fair cross-country analysis. Further discussion of these is beyond the scope of this text.

2.2.3.2 Extended Example of Metrics Calculation

ESTech Solutions, Inc. was founded by a team of engineers who set out to modernize infrastructure operations using advanced analytics, automation, and high-reliability engineering tools. Table 2.3 depicts some metrics for the company in year 10.

Let's assess ESTech's financial health using this data.

First, calculate Total Debt to EBITDA using Equation 2.14

$$\frac{40M}{12M} = 3.33$$

A ratio of 3.33 falls in the *normal to elevated* range (2.0–3.5 typical, >4 considered risky).

Next, using Equation 2.15, calculate the Interest Coverage Ratio

$$\frac{8M}{2M} = 4.0$$

An interest coverage of 4.0 is reasonably strong since values well above 1 are preferred [Finance].

Next, calculate Price-to-Earnings (P/E) Ratio using Equation 2.16

$$\frac{50}{2.5} = 20$$

A P/E of 20 is typical for a moderately growing tech/engineering firm.

Now, using Equation 2.17, calculate the EV/EBITDA ratio

$$\frac{150M}{12M} = 12.5$$

12.5× EBITDA implies that acquiring the firm would cost 12.5 years of its current cash profit. This is common in higher-growth EST sectors.

TABLE 2.3
10-Year Financial Metrics for ESTech Solutions, Inc.

Item	Value
Total Debt	$40 million
EBITDA	$12 million
EBIT	$8 million
Annual Interest Expense	$2 million
Market Price per Share	$50
Earnings per Share (EPS)	$2.50
Enterprise Value (EV)	$150 million
Annual Revenue	$60 million
Book Value per Share	$30
Expected Annual EPS Growth	12%

Next, using Equation 2.18, we compute the price-to-sales ratio. To use the formula, assume:

- Shares Outstanding = 10 million
- Market Price = \$50
- Market Cap = 10M × \$50 = \$500 million

Thus we have

$$\frac{500M}{60M} = 8.33$$

A ratio of 8.33 is high but not unusual for tech-oriented EST companies emphasizing growth over current profits.

Finally, we compute the PEG Ratio using Equation 2.19.

Here:

- P/E = 20 (previously computed)
- EPS Growth = 12%

So we have

$$\frac{20}{12} = 1.67$$

A PEG below 1–1.5 is typically desirable; a PEG of 1.67 suggests the stock may be slightly expensive relative to expected growth.

Table 2.4 summarizes the metrics we calculated.

Based on these metrics, ESTech Solutions, Inc., appears to be a financially stable, moderately leveraged, and growth-oriented EST firm with generally solid fundamentals, though with some indications of elevated valuation relative to earnings.

2.2.3.3 Current Ratio, Quick Ratio, and Related Liquidity Metrics

Liquidity metrics are essential for assessing a company's short-term financial health, particularly in the EST sectors where cash flow can be unpredictable due to

TABLE 2.4
Summary of Financial Metrics for ESTech, Inc.

Metric	Result	Interpretation
Debt/EBITDA	3.33	Normal–slightly high leverage
Interest Coverage	4.0	Good ability to service debt
P/E Ratio	20	Reasonable for a growth-oriented EST firm
EV/EBITDA	12.5	Higher valuation, growth premium typical
Price to Sales	8.33	High; reflects market optimism
PEG Ratio	1.67	Slightly overvalued relative to growth

project cycles, R&D spending, or inventory needs. Here are some of the key liquidity metrics. Here, ranges are given for a sense of scale, not for decision-making purposes. To use these for decision-making requires a much deeper understanding and context.

The *current ratio* measures overall short-term liquidity and is calculated as

$$\textit{current assets/current liabilities} \tag{2.20}$$

Where *current assets* include cash, receivables, inventory and *current liabilities* are payables and short-term debt. A ratio above 1 indicates assets cover liabilities, with an ideal range of 1.5 to 2.0 (a *current ratio* greater than 3 suggests inefficiency).

The *quick ratio (or acid test)* provides a stricter view focused on the most liquid assets. The quick ratio is given by:

$$\textit{quick ratio} = \frac{\textit{current assets} - \textit{inventory} - \textit{prepaid expenses}}{\textit{current liabilities}} \tag{2.21}$$

with a value above 1 considered healthy and an ideal of 1.0 to 1.5 (though tech firms often run 0.8 to 1.2 due to fast receivables).

The *cash ratio* is a strict liquidity test that measures only cash and cash equivalents divided by current liabilities. This metric signals ultimate solvency with > 0.5 indicating safety and an ideal of 0.2 to 0.5 (and rarely over 1 in growth firms). However, the ideal ranges vary significantly by sector.

Finally, the *operating cash flow ratio* evaluates cash generation against obligations. It's computed quarterly or annually and given by the following:

$$\textit{operating cash flow ratio} = \frac{\text{cash flow from operations}}{\text{current liabilities}} \tag{2.22}$$

Above 1.5 for this metric is considered ideal for sustainable business operations.

EXAMPLE 2.17 CALCULATING LIQUIDITY RATIOS FOR A-ONE ENGINEERING

To illustrate the calculation and use of these metrics, let's use the financial statements previously shown for A-One Engineering (Highway Engineering Firm). From these statements, we have the following data:

Current Assets = $894k,
Current Liabilities = $399k,
Inventory = $0 (this is a service-based business),
Cash = $410k.

Calculating some ratios from this data, we have:

Current Ratio = \$894k/\$399k = 2.24 → healthy; covers liabilities 2.24 times.
Quick Ratio = (\$894k—\$0)/\$399k = 2.24 → same as current ratio since they have no inventory.
Cash Ratio = \$410k/\$399k = 1.03 → strong immediate liquidity.

Interpretation: A-One Engineering can easily handle short-term bills but should monitor receivables (i.e., not allow them to increase significantly) to avoid dips in these metrics below 1.5.

These liquidity ratios matter for those working in EST firms. Low ratios signal risks such as delayed client payments (common in civil engineering projects) or R&D overruns (common in biotech), while high ratios highlight opportunities for investment (e.g., new lab equipment). Managers should track quarterly against benchmarks. Pair these metrics with cash flow trends for a full picture of the organization, using tools such as Excel or QuickBooks to automate.

As with the previous set of metrics, there are international differences also due to variations in tax treatments, accounting standards, cultural preferences and so on. And as a reminder, all metrics need to be calibrated with sector-specific data.

2.3 EXERCISES

2.1 For each transaction below, indicate whether it is a debit or credit and which account is affected:
  a. Purchase office supplies for cash
  b. Receive payment from a customer
  c. Pay off a bank loan
  d. Record depreciation on equipment

2.2 Illustrate the journal entry for buying a \$1,200 laptop using a credit card. Show both debit and credit accounts.

2.3 A company has assets of \$500,000 and liabilities of \$180,000. Calculate its equity. If liabilities increase by \$20,000, what happens to equity, assuming assets remain constant?

2.4 Using the sample income statement for A-One Engineering provided in the chapter, calculate:
  a. gross margin
  b. operating income
  c. net income

2.5 From the sample balance sheet for A-One Engineering, compute:
  (a) Working Capital
  (b) Debt-to-Equity Ratio
  (c) Percentage of assets tied up in equipment

2.6 Explain why a company can report a net loss on the income statement but still have positive cash flow from operations.

2.7 If inflation is 3%, what is the present value of $10,000 received 5 years from now?

2.8 A project costs $200,000 upfront and generates $80,000 annually for 3 years. Calculate NPV at a 6% discount rate.

2.9 In lieu of buying $20 worth of lottery tickets each week, you decide to invest that money in a stock market fund that is expected to provide an average 7% return. Assuming a discount rate of 2.5%, what is the lump sum payout equivalent of a winning lottery ticket that you win 5 years from now?

2.10 A project requires $1M upfront and generates $300,000/year for 5 years. Calculate NPV at a 7% discount rate and estimate IRR.

2.11 A new machine costs $100,000 (fixed), with $20/unit variable costs and $50/unit selling price. Calculate the break-even point.

2.12 Compare TCO for a $50,000 software license (5-year life, $5,000/year maintenance) versus a $12,000/year cloud subscription.

2.13 Suppose you buy a $40K car for 10% down and the rest is 0% interest, with monthly payments for 3 years. If the discount rate is 3%, what is the true cost of the car in today's dollars?

2.14 A solar installation costs $500,000 and returns $120,000 annually for 5 years. Estimate IRR (use trial and error or spreadsheet).

2.15 Compute the payback period for an investment of $150,000 that saves $50,000 annually.

2.16 Recalculate the payback period for the same investment above assuming a discount rate of 8%.

2.17 A metal part manufacturing company's balance sheet data sow the following:

Current Assets = $5.4M, Current Liabilities = $699K, Cash = $125K.

Compute:

a. The current ratio
b. The quick ratio
c. The cash ratio
d. Provide an opinion on the heath of this company

2.18 Compare the following investment alternatives:

Option A: Invest $250,000 in equipment saving $500,000 immediately.

Option B: Hire testers for $250,000, saving $750,000 in two years (discount rate = 10%).

Compute NPVs for both and recommend the better option.

2.19 Scenario: TechNova Inc., a mid-sized technology firm, is considering two initiatives to improve operational efficiency:

- Initiative 1: Upgrade servers at a cost of $400,000, expected to save $120,000 annually for 5 years.
- Initiative 2: Migrate to a cloud-based solution at a cost of $250,000, expected to save $90,000 annually for 4 years.

Assume a discount rate of 8%.

a. Calculate the NPV for each initiative.
b. Estimate the IRR for each initiative (use iterative approach or spreadsheet).
c. Compute the payback period for both initiatives.
d. Recommend which initiative TechNova should pursue, considering both financial metrics and qualitative factors (e.g., scalability and risk).

NOTES

1 An account that is paired with a regular asset account to reduce its recorded value on the balance sheet.

2 The sum of currents into a node in an electrical circuit equals the sum of the currents out.

REFERENCES

Drake, P. P., & Fabozzi, F. J. (2010). *The basics of finance: An introduction to financial markets, business finance, and portfolio management.* John Wiley & Sons.

Eqvista. (2025). *Debt-to-equity ratio by industry.* https://eqvista.com/debt-to-equity-ratio-by-industry/, accessed 12/18/2025.

Kenton, W. (2025). *Debt-to-EBITDA ratio explained: Definition, calculation, and significance.* Investopedia. https://www.investopedia.com/terms/d/debt_edbitda.asp, accessed 12/18/2025.

Withrow, S. (2025, June 18). *Performance benchmarks: Construction companies.* https://www.jmco.com/articles/construction/performance-benchmarks-construction-companies/

3 Understanding Business Structures and Finance

Engineers, scientists, and technologists are great at solving complex problems—whether it's optimizing a bridge's load-bearing capacity, refining a chemical process for higher yield, or architecting scalable software infrastructure. Yet many of these professionals launch into entrepreneurial ventures, join startups, or lead technical teams without fully grasping the "business system" that funds, steers, and sustains their work. This chapter bridges that gap by demystifying the foundational structures and financial frameworks that underlie every technical enterprise.

We begin with the nature of the business entity—sole proprietorships, LLCs, corporations, etc. and their international equivalents. We will explore how these forms affect personal liability, taxation, funding access, and operational flexibility, with practical implications for professionals navigating global markets. If you are planning to start your own business, this section is essential reading. But if you work for someone else, especially a small business, understanding these structures is also the key to understanding how they make major decisions.

Next, we examine fundamental accounting standards, starting with US GAAP and IFRS, their major differences, and the often-overlooked impacts of taxes, tariffs, and cost structures on project viability and profitability. We'll then dive into key financial ratios and metrics that serve as diagnostic tools for assessing company health, much like stress tests in engineering or control variables in scientific experiments.

Finally, we cover corporate reporting requirements, including how to read and contribute to the documents that communicate financial reality to stakeholders, investors, and regulators.

By understanding these concepts, you'll gain the ability to evaluate opportunities, mitigate risks, and align your technical innovations with sustainable business outcomes—transforming you from a specialist who builds things into a leader who builds profitable, resilient organizations.

But remember that you must consult appropriate professionals including attorneys and accountants before undertaking the start of your own business, in whatever form you may choose.

3.1 TYPES OF BUSINESS ENTITIES

The legal structure of the business entity is a critical aspect of EST ventures, as the structure affects personal liability, taxation, funding access, and operational flexibility. US Federal tax concepts are broadly consistent, though state entity, tax, and filing rules vary, influencing the choice of formation state.

We will largely focus on US business entities, though some of the similarities and differences in other countries will also be covered.

DOI: 10.1201/9781003733850-3

3.1.1 Business Structure in the United States

In the United States, the primary business configurations are

- sole proprietorships
- partnerships
- LLCs
- Corporations (S- and C-)

A summary of US business structures and features is shown in Table 3.1. Each of these entity types is discussed subsequently.

A *sole proprietorship* is the simplest form of business, and it is a popular structure for consultants and content creators. In a sole proprietorship, a single person is the owner or proprietor of the business entity. Sole proprietorships:

- are easy to start,
- have very modest paperwork filing requirements,
- are operationally agile, and
- have pass-through taxation for the owner.

Pass-through taxation means that the business does not pay income tax. Instead, profit/losses "pass through" to the owners' personal tax returns.

However, proprietorships have the disadvantage of *unlimited liability*, which means that if the business is sued for any performance issue or malfeasance, all the

TABLE 3.1
Summary of US Business Structure Features

US Business Entity	Key Features
Sole Proprietorship	• Unlimited liability • Pass-through tax • Easy setup • No funding flexibility
Partnership	• Shared liability • Pass-through tax • Moderate funding
Limited Liability Company (LLC)	• Limited liability • Pass-through tax • Flexible
S-Corporation	• Limited liability • Pass-through tax (ownership limits)
C-Corporation	• Limited liability • Double taxation • Best for scaling/initial public offering • VC's prefer

assets of the owner are at risk, including their home, personal property, and retirement accounts. Some of the personal risks can be mitigated by purchasing business and errors and omissions insurance. Because of these disadvantages, it can be difficult to raise capital for this kind of business. However, a sole proprietorship can be converted to a more sophisticated business structure as it grows.

A *partnership* is a collection of individuals who choose to pool resources and expertise for a competitive advantage. A partnership has the following advantages:

- It provides flexibility for ownership and management structure.
- Taxation is pass-through.
- It provides a platform for moderate funding potential from banks, venture capitalists, angels, and friends or family.
- Liability for performance or malfeasance is shared by the owners (the partners), which provides some investor protection.

This type of structure is popular for groups of consultants and professional services firms with diverse practices, such as engineering, fractional CTO or CIO, and boutique research collaboratives.

A *limited liability company (LLC)* is a hybrid business structure that provides the legal protections of a corporation with the flexibility of a partnership for ownership and management structure. As the name suggests, an LLC also provides some limitations on the personal liability of the owners. LLCs have pass-through tax liability, and they are an ideal business structure for small technical teams. While LLCs are operationally flexible, most institutional venture investors prefer a C-Corp for standardized equity, option plans, and financing instruments. An LLC is a US-specific structure, but it has close international analogs in "private limited liability companies."

Corporations are the most sophisticated form of business structure. In the United States, there are two forms of corporation: C and S.

The *S-Corporation* or *S-Corp* is a business structure with pass-through tax treatment and ownership limits that is suitable for small businesses, and sometimes medium-sized ones. The S-Corp is a uniquely American structure—most countries achieve pass-through via partnerships or specific private limited companies, without the exact shareholder limits. S-Corps can be converted to a C-Corp before institutional financing, which venture investors generally prefer. Another advantage of S-Corps is that there are fewer filing requirements than for a C-Corp. S-Corp owners are typically individuals (with specific allowances such as certain trusts and resident aliens). Corporations and partnerships cannot be shareholders. The number of partners in an S-Corp is limited to 100 [SBA].

A *C-Corporation* or *C-Corp* is a business structure that is the most sophisticated and most suitable for medium-to-large and complex businesses. A C-Corp is also commonly known as a "public/private limited company" or "corporation." A C-Corp has limited liability for shareholders, and there can be unlimited shareholders, including business entities and foreign nationals. *Double taxation* typically applies, which means that both the profits of the company and the earnings that are passed through to the shareholders are subject to applicable taxes. C-Corps are very attractive to investors because of the liability protections afforded. There are many more filing requirements for a C-Corp than for any other type of business configuration.

The C-Corp tends to be used for large and complex businesses, while the S-Corp tends to be used for small to medium-sized businesses. There are no official definitions for "small," "medium," or "large" business, but the US Small Business Administration (SBA) classifies manufacturing firms with 500 employees or fewer and non-manufacturing firms with annual receipts under $7.5 million as "small" [SBA].

The choice of business structure is entirely situational and depends on financial factors, legal and taxation considerations, and the business sector. For example, engineers, scientists, and technologists often favor LLCs for prototyping labs since their limited liability shields personal assets from intellectual property disputes. Scientists in biotech, on the other hand, lean toward C-Corps for attracting grants and investors.

Remember, US federal laws governing the structure, operations and taxation of businesses are consistent, but states, counties, and municipalities differ in how they treat business entities, and often influence the choice of state for incorporation. For instance, Delaware is a frequent choice for incorporation because of its well-developed, business-friendly laws and court system, and deep pool of business attorneys and related professionals.

3.1.2 International Business Structures

Across the globe, there are similar business structures to the various US forms. However, no perfect one-to-one equivalents exist. One feature to note—unlike the United States, many countries permit 100% foreign ownership in specific sectors, while others impose sectoral caps or require local directors.

Table 3.2 shows a comparison of common equivalent business structures, focusing on liability protection, taxation, and funding flexibility, and a brief discussion follows.

There are international equivalents to sole proprietorships (single-owner business with unlimited personal liability) and partnerships (multi-owner arrangements sharing profits, losses, and often liability). These structures vary by country in naming, formation requirements, liability rules, and tax treatment, and there is no exact equivalent due to local laws. Close analogs are common, however. For example, in the United Kingdom, Australia, and Canada, the "sole trader" designation is used, and in India, the "proprietorship" is common for small businesses.

International equivalents to partnerships are typically multi-owner structures that have unlimited joint liability. For instance, in the United Kingdom, an "Ordinary Partnership" has unlimited liability under a partnership agreement. Canada has a "General Partnership, via provincial registration." India has a "Partnership Firm" under the Partnership Act with unlimited liability unless designated as a limited liability partnership (LLP). Japan has a "Kumiai" or general partnership.

The US LLC, with its limited liability and flexible pass-through taxation, has different European equivalents or partial equivalents, including:

TABLE 3.2
US and International Equivalent Business Structures [Deloitte 2025]

US Structure	Close International Equivalents Features	Examples by Region/Country
Sole Proprietorship	Proprietorship, single owner	• India–Proprietorship • The United Kingdom/Australia/Canada-Sole Trader
Partnership	Multi-owner structures; general partnerships with unlimited joint liability	• Canada-General Partnership • India-Partnership Firm • Japan, general partnership • UK-Ordinary Partnership
LLC	Private limited company with limited liability (often pass-through or flexible tax)	• Australia—Proprietary Limited (Pty Ltd) • China—Limited Liability Company • France/Belgium/Luxembourg—SARL • Germany/Austria/Switzerland—GmbH • India—Private Limited Company • Italy—S.r.l. • Japan—Godo Kaisha (G.K.) • UK—Private Limited Company (Ltd)
S-Corp	No direct equivalent; closest are pass-through partnerships or private companies electing flow-through	• Australia—Proprietary Company (Pty Ltd) • Canada—Canadian Controlled Private Corporation (CCPC) • Germany—GmbH • Many countries use partnerships (e.g., UK LLP) for pass-through taxation
C-Corp	Public or private limited/share company	• Australia—Public Company Ltd • Canada—Corporation (federal/provincial) • China—Joint Stock Limited Company • France—Société Anonyme (S.A.) • Germany—Aktiengesellschaft (AG) • India—Public Limited Company • Japan—Kabushiki Kaisha (K.K.) • UK—Public Limited Company (PLC)

- Germany's GmbH,
- France's SARL and Italy's S.r.l.
- Japan's Godo Kaisha (G.K.),
- and China's LLC, which is like a partnership, but state-influenced.

Equivalents to the US S-Corp with its limited liability, pass-through taxation is achieved through partnerships, and with fewer restrictions such as limits on the number of owners. Examples include

- Australia's Proprietary Company (Pty Ltd),
- Canada's Canadian Controlled Private Corporation (CCPC), and
- Germany's GmbH.

In most countries, if the term "Private Limited" is appended to the name, then the company is an LLC analog.

Finally, analogs to the C-Corp with its limited liability, corporate taxation, scalability and public listing exist in all countries. For instance,

- Australia's Public Company Ltd,
- China's Joint Stock Limited Company, and
- India's Public Limited Company.

Typically, if the company name is appended with "Public Limited," then it is a US C-Corp equivalent.

For tech companies whose operations span more than one country, many choose to incorporate in Delaware (as a US C-Corp) for venture capital friendliness, then set up subsidiaries using local LLC equivalents.

Another approach is to use hybrid models, such as the EU's Societas Europaea (SE), a special type of public limited liability company established under EU law. This business entity type is designed specifically for pan-European operations, allowing a single company to operate seamlessly across multiple EU member states without needing separate subsidiaries in each country. But this approach is not as common as national forms (e.g., GmbH or SARL) due to complexity and cost.

The SE approach is used by large companies with significant EU presence wanting unified governance, branding, and tax planning. Formation requires at least two entities from different member states or a merger. As of this writing, more than 3,000+ SEs exist, mostly in engineering-heavy sectors like manufacturing and infrastructure [Council of EU].

All these approaches create potential risks such as double taxation in cross-border issues.

3.1.3 Banking Entities

It is important to set the context of banking and other financial institutions and their role in the EST business ecosystem. Banks are a type of financial institution that are essential for any business to function. Banks provide the primary mechanism for exchanging money between businesses, customers, and government entities.

There are nine types of financial institutions:

- central banks,
- commercial banks,
- credit unions,
- investment banks and companies,
- retail banks,
- savings and loan institutions,
- brokerage firms,
- insurance companies,
- and mortgage companies.

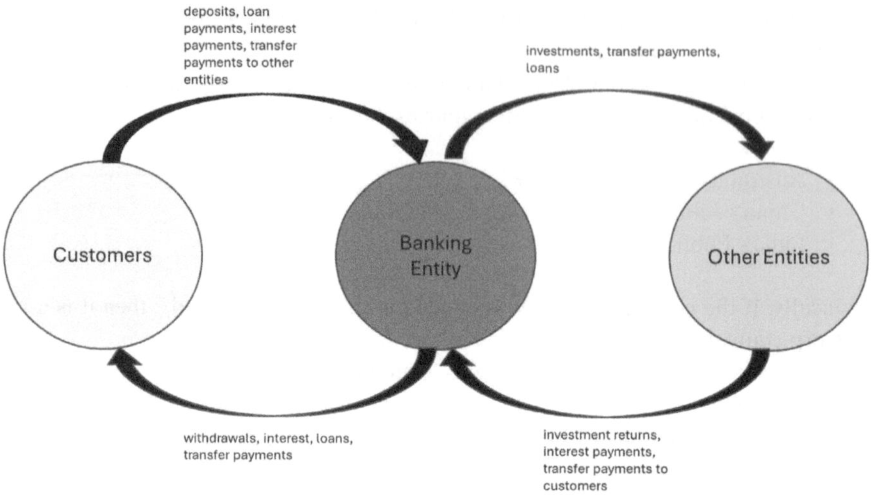

FIGURE 3.1 Banking interactions [Voas et al 2022].

For our purposes, we are mostly interested in central banks, investment banks and companies, and commercial banks.

Central banks, which were discussed in Chapter 2, interact directly with other financial institutions. For example, banks may act as financial intermediaries by accepting business deposits or by borrowing in the money markets. Banks then use those deposits and borrowed funds to make loans to businesses, individuals, and other entities or to purchase securities [International Monetary Fund 2017].

Figure 3.1 illustrates some monetary flows between the various participants in a financial system.

Other entities that use banks include individuals, merchants, service providers, governments, utilities, non-profit organizations, other banking entities, and others (e.g., consumers, investors, and businesses). There are many other kinds of transactions with various players including individuals and other legal entities, but we are only interested in the business-related interactions.

Financial sector institutions also serve as financial intermediaries by facilitating payments to and from businesses and other entities with which they interact via check payments and debit and credit transfers. The banking system supports the structure of all business, and it plays an important role in both facilitating and adding friction (in the form of added costs and time delays) to business transactions.

We will revisit some of these interactions and ways to reduce friction in Chapter 9 when open banking and financial technology firms are discussed.

3.2 ACCOUNTING STANDARDS

Accounting standards establish rules for recording and reporting financial data, ensuring transparency and comparability. Understanding that these are important is essential for EST professionals seeking funding, grants, or partnerships. Divergences from standards can impact R&D capitalization, revenue recognition, and tariff treatments, affecting everything from project budgets to investor confidence.

It takes an expert (i.e., accountants and financial experts) to deal with matters involving accounting standards, and you would not have any direct interaction with them. Therefore, the goal here is only to provide some familiarity with what the accounting standards are and how they are used.

3.2.1 The Financial Accounting Standards Board (FASB)

The *Financial Accounting Standards Board (FASB)* in the United States is an independent, non-governmental organization that is responsible for creating and administering the *Generally Accepted Accounting Principles (GAAP)* that are used by US companies. GAAP is a rules-based framework emphasizing detailed guidelines, historical cost, and conservatism.

The use of GAAP is mandatory for US public companies and widely used by private firms for credibility with investors. For EST companies, GAAP requires expensing most R&D as incurred (not capitalized, limiting balance sheet assets) but allows capitalization for certain development costs, aiding tech firms in accurate project tracking.

3.2.2 The International Financial Reporting Standards (IFRS)

In over 140 countries, a different set of rules—the *International Financial Reporting Standards (IFRS)*—is used. The IFRS are created and maintained by the *International Accounting Standards Board (IASB)*. The IFRS are a set of globally recognized accounting rules used to ensure that financial statements focus on economic substance and fair value. These rules promote global comparability, making them ideal for multinational EST collaborations. IFRS permits greater R&D capitalization during development phases, potentially boosting reported assets for science/tech firms.

3.2.3 Differences between GAAP and IFRS

GAAP and IFRS differ in specificity and flexibility, impacting EST reporting. GAAP's rules lead to stricter revenue recognition. Meanwhile, IFRS allows reversal of impairments (except goodwill) and prohibits last-in-first-out (LIFO) inventory, which affects hardware/semiconductor firms. Research and development under GAAP is mostly expensed, but IFRS allows capitalization post-feasibility, inflating biotech assets.

The US GAAP and IFRS became much more aligned ("converged") in how leases must be reported, after both adopted major new lease-accounting standards. Leases are similar post-convergence, but tariffs/customs duties are treated as expenses under both, with IFRS emphasizing fair value adjustments.

For global EST companies, hybrid reporting (e.g., GAAP for US operations, IFRS for EU) is common, but differences can add significantly to a company's compliance costs [Ross 2025].

3.3 TAXES AND TARIFFS

Taxes and tariffs play a significant role in the decisions that senior managers make for their businesses. Oftentimes, decisions are made because of tax or tariff implications that appear as mysterious to those who are not experts in tax and tariff policy as they apply to the business.

Taxes are mandatory financial charges imposed by governments. Business taxes refer to the taxes that companies must pay on their income, property, transactions, or operations. Taxes on businesses can include. corporate Income, payroll Taxes, sales, and property taxes.

A *value-added tax (VAT)* is a type of consumption tax applied incrementally at each stage of production and distribution. A VAT is collected throughout the supply chain. Each business charges VAT on its sales and can deduct the VAT it paid on its purchases. Thus, the tax burden falls on the end consumer. This approach is unlike a sales tax, which is only collected at the final point of sale to the consumer.

VATs are widely adopted in Europe, Asia, and many other regions, but the United States does not have a federal VAT and uses sales taxes instead.

Tariffs are another form of taxation on designated goods and services. Ancient kingdoms taxed goods entering ports to raise revenue and control trade. Throughout history, tariffs have been used as:

- revenue tools,
- for protection of domestic industries, and
- as levers of diplomacy or coercion.

In the 1700s–1800s, tariffs were the main source of government income worldwide.

A summary of tariff influences on technology over the centuries is given in Table 3.3.

The first major US tariff was enacted to fund the new federal government and protect early American manufacturing, and tariffs were used throughout the 19th and 20th centuries to protect US interests during industrial expansion. After World War II, worldwide reconstruction, a new era of cooperation, and the emergence of computing technologies led to reductions in many tariffs worldwide.

TABLE 3.3
Summary of Tariff Influences on Technology in Last 250 Years

Era	Tariff Intent	Impact on Technology
1789–1930	Revenue, protectionism	Insignificant
1947–1995	Global tariff reduction	Foundations for global tech trade
1996–2010	ITA eliminates many tech tariffs	Tech becomes cheap and global
2018–2025	US–China tariff conflict	Tech becomes a major target
2025+	US import tariffs	Broad impact on all tech goods

By the late 20th century, technology products became increasingly globalized and supply chains spread across Asia, Europe, and North America. The landmark Information Technology Agreement (ITA) of 1996, adopted by more than 80 countries, eliminated tariffs on computers, semiconductors, telecom equipment, and many electronics and created the foundation for cheap global tech. This is why, for decades, laptops, phones, and chips have been subject to minimal tariffs.

Today, technology products have become central to national security, economic competitiveness, and global supply chain dominance. Advanced nations use tariffs to protect their interests, sometimes leading to trade wars (e.g., the United States vs. China), while developing nations often maintain higher tariffs to protect emerging tech industries. Tariffs matter to technological businesses because they influence consumer prices, supply chain location decisions, semiconductor manufacturing strategy, and national security and export controls.

Taxes and tariffs directly reduce profit margins and influence entity structure choice, for example, C-Corps for US R&D credits, and place of incorporation, as corporate taxes vary globally. Tariffs, on steel, aluminum, copper, and semiconductors, raise input costs 10–30% for hardware and renewables, but these can sometimes be treated as operating expenses.

Failure to pay taxes can result in penalties, though businesses will use every maneuver to legally avoid paying taxes. Mitigation strategies include entity structuring (e.g., LLCs for pass-through) and transfer pricing for multinationals. Effective financial analysis should simulate tax and tariff impacts on the value of products for supply chain decisions.

Taxes and accounting treatments are a factor in project costing. Project cost structuring includes answering questions such as

- Is software a capital item or expense?
- Should we hire permanent versus contract employees?
- What are the tax implications for development investments and profits?

These complex issues require input from lawyers, accountants, and experts. But they also play a significant role in any financial activity and are often the key to understanding why a company makes certain financial decisions.

3.4 COST STRUCTURE AND RATES

Effective management of costs and bank borrowing separates surviving EST companies from thriving ones. In an environment of moderating inflation but persistent supply risks, maintaining a lean cost structure is a clear competitive advantage. Keeping expenses under control is a preferable, though not mutually exclusive, strategy to borrowing, raising private capital or winning government grants. We will discuss raising capital through various means in Chapter 7. For now, let's look at the ordinary costs of doing business—paying expenses and borrowing money.

3.4.1 Fixed versus Variable Expense

The costs of doing business can be divided into *fixed* and *variable expenses (or costs)*. Fixed expenses are independent of output, and include salaries, rent, software licenses, and depreciation on lab equipment. Variable expenses scale with activity, for example, materials, cloud computing usage, subcontractor fees, or prototype fabrication.

A characteristic of many technology and science firms is a high fixed-cost, low variable-cost model. That is, once R&D and talent are invested, additional units cost little to produce, leading to excellent gross margins (often 70–90%) and strong operating leverage as revenue scales. For example, software licenses, cloud-based services, and pharmaceutical products. Conversely, hardware-heavy companies carry higher variable costs from components and inventory, resulting in gross margins of 30–50%; for example, computers, robotics, electric vehicles, and aircraft. Meanwhile, consulting firms rely on billable hours (with utilization rates of 70–85% are ideal) plus reimbursable expenses.

Understanding this structure is crucial for pricing decisions—tech firms often use subscription models to recover fixed R&D expenses quickly. Engineers and technologists can influence this by optimizing utilization (e.g., reusable IP reduces future fixed expenses) or negotiating volume discounts on variable inputs.

3.4.2 Borrowing Rates

Having explored cost structure, we now examine how borrowing costs shape financial planning. Bank borrowing rates directly influence a business' profitability and cash flow. For many years, the London Interbank Offered Rate (LIBOR) was the main benchmark interest rate representing the average rate at which global banks lent to each other on a short-term, unsecured basis. LIBOR also served as the reference rate for pricing of corporate loans, credit lines, and commercial paper (capital raising) programs. LIBOR was phased out in 2023 because the interbank lending market shrank, leaving too few real transactions to support the benchmark, and some banks were caught manipulating submissions, undermining trust in the rate [US Securities and Exchange Commission 2023].

After LIBOR's demise, regulators advocated for benchmarks based on actual transactions, not bank estimates. These kinds of benchmarks are called *risk-free rates (RFRs),* and different ones are used around the world.

The Secured Overnight Financing Rate (SOFR) is used in the United States as the primary benchmark. Considered extremely robust because it reflects high-volume and secured lending, SOFR is now the standard reference for corporate loans, mortgages, derivatives, and floating-rate (variable-rate) notes in the United States.

Other countries use different benchmarks, for example, the Sterling Overnight Index Average (SONIA) in the United Kingdom, the Euro Short-Term Rate in the Eurozone, the Swiss Average Rate Overnight (SARON) in Switzerland and, the Tokyo Overnight Average Rate (TONA) in Japan.

So, when businesses borrow from banks, they pay some rate that is based on these benchmarks (the benchmark rate plus some delta), and these benchmarks change all the time. These costs can be significant. For example, technology startups face cost of capital challenges, such as early-stage equity that is expensive (causing dilution). In addition, debt rates can be high, ranging from 8–12% for established firms to 15–25% for high-risk ventures. Interest rates on equipment loans or lines of credit add pressure on capital-intensive projects like pilot plants or data centers.

Tariffs and supply-chain costs can further complicate pricing due to variable rates. US tariffs on electronics can inflate hardware costs, forcing firms to absorb these, pass them on to customers, or reshore operations.

3.5 KEY BUSINESS FINANCIAL RATIOS AND METRICS

Several financial ratios and metrics are important indicators that CFOs, boards, and investors, banks and other lenders use to monitor the ongoing health and creditworthiness of a business. The key 10 are:

1. *backlog/book-to-bill*
2. current ratio
3. customer acquisition cost (CAC) Payback
4. debt-to-equity
5. free cash flow (FCF)
6. gross margin
7. net burn/runway
8. operating cash flow ratio
9. revenue growth
10. rule of 40 score

Metrics #s 5, 6, 9 and 10 were introduced in Chapter 1 and #s 2, 4, 8 were introduced in Chapter 2. *Backlog (or book-to-bill)* represents work (products or services) that has been contracted but not yet delivered (nor billed). Backlog/book-to-bill is a leading indicator for future revenue, especially in the civil, aerospace, and biotechnology sectors.

Now let's look at the final two metrics, burn rate and customer acquisition cost, in more detail.

3.5.1 Burn Rate

The *burn rate* is a measure of how much cash a company is spending more than it earns each month—the rate at which it is burning through its cash reserves (Equation 3.1):

$$\textit{burn rate} = \textit{monthly cash spend}—\textit{monthly cash earn} \tag{3.1}$$

When spending exceeds earnings, the burn rate is positive. If earnings exceed spending, then the burn rate becomes negative—indicating that the company is profitable.

The *runway* is how many months the company can survive at the current burn rate before running out of cash, assuming no new revenue or funding (Equation 3.2):

$$\textit{runway (in months)} = \textit{cash in bank} \div \textit{burn rate} \tag{3.2}$$

These two metrics are crucial metrics for startups, growth-stage tech firms and even established engineering, science and technology firms during lean quarters.

EXAMPLE 3.1 CALCULATING BURN RATE AND RUNWAY

A tech startup has $12M cash in the bank, a monthly revenue stream of $800K and monthly expenses of $1.5M. Then its burn rate is:

$1,500,000–$800,000 = $700,000

Then the company's runway is

$12M ÷ $700,000 = 17.1 months

So, this company is "safe" for more than 12 months, but it needs to reduce its burn (reduce expenses or increase revenue) or raise more capital soon.

Managing the monthly burn/runway is the #1 survival metric for engineering, science, and technology firms.

3.5.2 Customer Acquisition Cost

As engineers, scientists, and technologists, we are focused on the creation, optimization, and maintenance of products and services within our fields of expertise. But we should also recognize that acquiring and retaining customers is one of the most important aspects of any business, and there are important metrics used to monitor customer acquisition and retention, particularly the *customer acquisition cost (CAC) payback period.* The CAC payback period is a compound metric derived from two sub-metrics—the customer acquisition cost (CAC) and the customer *monthly recurring revenue (MMR).* The CAC is calculated via Equation 3.3:

$$customer\,acquisition\,cost = \frac{sales\,and\,marketing\,expense}{number\,of\,new\,customers} \tag{3.3}$$

Sales and marketing expenses include marketing, sales staff, and any other direct costs related to customer acquisition.

The *monthly recurring revenue (MRR)* is the average monthly revenue expected from a particular customer or client. This can be simply estimated by taking the average of actual and anticipated revenue of a company over some time.

The *CAC payback is* the number of months needed by a company to recoup the initial costs incurred in the process of acquiring a new customer. The CAC payback period is computed as the total costs associated with acquiring new customers over a specific period divided by the number of customers acquired (Equation 3.4):

$$CAC\,payback\,period = \frac{CAC}{\left(\frac{MRR}{customer}\right)\cdot gross\,margin} \tag{3.4}$$

It is considered healthy for the CAC payback to be less than 12 months.

3.5.3 Targets for the 10 Key Metrics

Some targets and interpretations for the ten key metrics are given in Table 3.4. These ranges will differ depending on the sector and will change constantly. Therefore, consider these targets for discussion purposes only. These targets have been assembled from a wide variety of public financial reports.

TABLE 3.4
Targets and Alerting Thresholds for the Ten Key Metrics

Metric	Importance of the Metric	Target	Red Flag Threshold
Backlog/ book-to-bill	Leading indicator for future revenue (especially civil, aerospace, biotech). Ratio > 1.0 = growing pipeline.	1.1–1.5	<0.9 sustained
Current ratio	Can you pay the bills next year? Critical for bonding capacity (civil) and creditor confidence.	1.5–2.5	<1.2
Customer acquisition cost (CAC) payback	How many months will it take to recover sales and marketing spend?	<12 months	>24 months
Debt-to-equity	How leveraged is the company? Bonding companies and banks look here first before lending or issuing bonds.	<1.0 (infrastructure)/<0.5 (services)	>1.5
Free cash flow (FCF)	The capacity to pay salaries, dividends, debt, and new capital expenditures.	>15% of revenue (services)/>8% (civil/biotech)	<0 for 2 or more quarters
Gross margin	Shows pricing power and corporate efficiency.	65–90% (services)/25–35% (civil/highway)	<50% (services)/<22% (engineering)
Net burn/runway	Shows how many months of cash left at the current burn. The most important metric for startups and growth firms.	> 18 months (safe)/> 24 months (great)	<9 months
Operating cash flow ratio	Are operations generating cash, or is the company surviving on financing?	>1.2	<0.8
Revenue growth	Primary driver of valuation multiples. Investors pay for growth first, profitability second.	20–40%+ (services)/5–12% (civil/biotech)	<5% sustained
Rule of 40 score	Best health metric for tech/ services: Growth % + free cash flow margin %. Used by every venture capitalist and public investor.	≥40 (healthy)/≥50 (excellent)	<30 = danger zone

These metrics, together with their recommended ranges, provide a basis for discussing and advocating for various expenditures and actions in an engineering, science, or technology organization.

EXAMPLE 3.2 USING METRICS TO ACHIEVE FINANCIAL GOALS

Suppose you are seeking additional budget from management. Then you should focus on improving gross margin. Engineering teams can achieve this, for instance, by reducing unnecessary cloud expenditures and optimizing yield, which directly enhances profitability.

To navigate a downturn, prioritize increasing free cash flow. A practical approach is to shift from capital expenditures to operating expenses—for example, leveraging cloud services instead of purchasing physical servers. This strategy preserves liquidity and reduces upfront costs.

If you need to justify headcount, aim to improve the Rule of 40 score. Engineering can support this by automating workflows and accelerating delivery, thereby increasing efficiency and customer lifetime value relative to acquisition costs.

Finally, when it's time to terminate an underperforming project, look for indicators such as negative net present value (NPV) or a lifetime value (LTV) on a project or initiative that is less than three times the customer acquisition cost (CAC).

The point of these metrics is that they provide levers for rigorous financial analysis. You can compute these metrics and use them to provide clear and objective data to support any decision.

3.5.4 Financial Dashboards

The thresholds and ranges we have been discussing are used for alerting in dashboard applications and reports. Here are few examples of how to lay out some of these. First, consider Figure 3.2.

This dashboard includes the following metrics:

1. free cash flow (last 12 months + forecast)
2. rule of 40 score (trailing + next 12 months)
3. gross margin trend
4. revenue growth (year over year and quarter over quarter)
5. monthly burn/runway (in months)

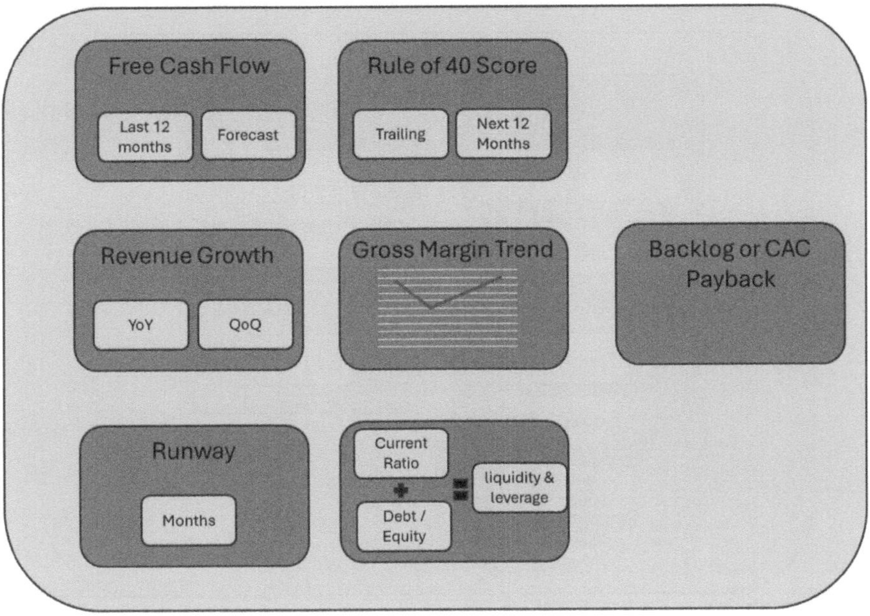

FIGURE 3.2 Key financial metrics dashboard.

6. current ratio + debt-to-equity (liquidity and leverage)
7. backlog or CAC payback.

Further customization by sector can lead to different dashboards.

EXAMPLE 3.3 SAMPLE MANAGEMENT DASHBOARD FOR A HIGHWAY ENGINEERING FIRM

For a highway engineering firm, consider including in the management dashboard:

- Gross and net profit margins
- Overhead as a percent of revenue
- Current and quick ratios
- Debt-to-equity and return on assets [JMCO 2025].

The dashboard is depicted in Figure 3.3.

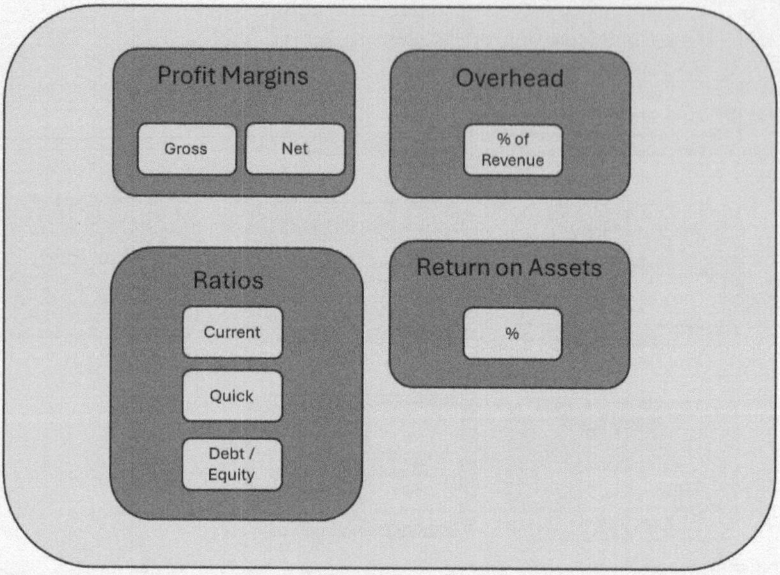

FIGURE 3.3 Simplified dashboard for highway engineering firm.

There are many other ways to organize dashboards and there are other, more complex and esoteric metrics that senior executives, financial analysts, investors, and government regulators include in their dashboards. Some of these will be discussed in later chapters.

3.6 CORPORATE REPORTING

Government regulators, business partners, investors (both individual and corporate) rely on accurate and truthful information to make significant decisions such as purchasing, selling, investing, partnering, taxing, and more. In all advanced countries, registered businesses of a certain type and structure (typically, public C-Corps and their equivalents) are required to file detailed reports of their activities and finances on a regular basis.

3.6.1 The US 10-K Report

The most important report for US public C-Corporation is 10-K report. This report is an annual filing required by the US Securities and Exchange Commission (SEC) for publicly traded companies. Private corporations do not file this form. The 10-K provides a comprehensive overview of the company's financial performance and operations for the fiscal year. It is also used to disclose risks to the company—a very important consideration for investors, customers, and other stakeholders.

3.6.1.1 Key sections of the US 10-K

The 10-K report is formally called the "Annual Report pursuant to Section 13 or 15(d) of the Securities Exchange Act of 1934," and it can be found at the SEC Electronic Data Gathering, Analysis, and Retrieval (EDGAR) site https://www.sec.gov/edgar by searching on the company name.

Table 3.5 lists the key sections of the 10-K report, a brief description of them, and an engineering analogy for each.

Looking at the table, the first section of a 10-K report lays out the company's core identity and operational blueprint.

Item 1: Business describes the organization's products, services, markets, and revenue model. Think of this as the system architecture diagram for a business—it defines the components (business units), their interactions (supply chains and distribution channels), and the overall design (strategic positioning). For engineers, scientists, and technologists, this is akin to understanding the architecture before diving into detailed design.

Item 1A, Risk Factors, highlights potential threats such as supply chain disruptions, regulatory changes, and competitive pressures. This section is the business equivalent of a Failure Mode and Effects Analysis (FMEA). Just as failure mode and effects analysis (FMEA) is used to identify failure points in a system to mitigate risk, companies disclose vulnerabilities to inform investors and stakeholders. These risks often dictate contingency planning and resource allocation.

Item 2, Properties, lists physical assets like factories, data centers, and mines. This is comparable to an infrastructure map in engineering—knowing where critical resources and facilities are located helps assess capacity, scalability, and exposure to geographic risks (e.g., natural disasters or geopolitical instability).

Item 7, Management's Discussion and Analysis (MD&A), provides management's perspective on trends, costs, and outlook. This section is like performance

TABLE 3.5
Sections of the 10-K Report, Brief Description and Engineering Analogy

Section	Description	Engineering Analogy
Part I		
Item 1: Business	Core operations, products, markets, revenue model	System architecture diagram
Item 1A: Risk Factors	Threats: supply chain, regulation, competition	Failure mode and effects analysis/Failure Modes
Item 2: Properties	Factories, data centers, mines	Infrastructure map
Part II		
Item 7: MD&A	Management's Discussion and Analysis—trends, costs, outlook	Performance profiling + root cause analysis
Item 8: Financial Statements	Balance sheet, income, cash flow	Bill of materials + power budget
Notes to Financials	Depreciation, R&D spend, leases	Component specifications
Part III & IV		
Executive comp, legal, exhibits	Bonuses, lawsuits, contracts	Access control + bug reports

profiling and root cause analysis in engineering. It doesn't just present numbers; it explains why performance metrics look the way they do and what corrective actions or optimizations are planned.

Item 8, Financial Statements, includes the balance sheet, income statement, and cash flow statement. These are the company's bill of materials and power budget—a detailed breakdown of resources (assets), flows (cash), and constraints (liabilities). Just as engineers track material and energy usage, financial statements track capital and operational efficiency.

Notes to Financials dive deeper into specifics like depreciation methods, R&D expenditures, and lease obligations. These are analogous to component specifications in engineering documentation—critical details that influence system behavior and lifecycle costs.

The final sections cover executive compensation, legal proceedings, and exhibits such as contracts. This is the business equivalent of access control and bug reports in engineering. Executive pay structures reflect incentive alignment, while legal disclosures highlight compliance issues or litigation risks—both essential for maintaining system integrity and stakeholder trust.

By framing these sections through engineering analogies, technical professionals can better grasp the logic behind corporate reporting. Just as a well-documented design ensures reliability and scalability, a well-structured 10-K provides transparency and confidence for investors, regulators, and partners.

3.6.1.2 Using the 10-K Document

The 10-K is a complex document, and it takes time and experience to understand it. But key Sections are structured like a modular spec sheet, and you can focus on these for actionable insights. For example, the disclosures (especially in Items 1 and 1A) act as constraint identifiers (think of the Theory of Constraints discussed in Chapter 1), as they quantify throughput targets, inventory risks, and operating expenses. And it is always very important to read the footnotes on all financial statements, as they can include very important amplifying or clarifying information.

Let's see a few examples of how an engineer, scientist, or technologist might use a 10-K in their own decision-making processes.

EXAMPLE 3.4 USING 10-K RISK DISCLOSURES TO GUIDE PRODUCT DESIGN DECISIONS

Suppose an engineering market analyst at a US auto parts supplier reads an automotive manufacturer's 10-K. He sees a discussion in Section 1A that the company plans to change the battery mounts for its vehicles. This discovery triggers a decision to initiate a new product design program to modular battery mounts.

EXAMPLE 3.5 USING MD&A INSIGHTS TO SHAPE A BID STRATEGY IN CIVIL ENGINEERING

A civil engineer at a small highway construction firm is considering whether to submit a new highway project bid. The client is a public, multinational infrastructure company that is the prime contractor for the project. Reviewing the client's latest 10-K, she spots in Item 7 (MD&A) a 16% increase in CapEx guidance for road maintenance and a risk factor in Item 1A about budget shortfalls due to inflation.

This data informs her decision to propose a cost-efficient modular design that reduces material use by 18%, aligning with the client's financial constraints—securing the $13M contract and boosting her firm's backlog.

EXAMPLE 3.6 USING R&D AND REVENUE SIGNALS IN A 10-K TO SHAPE BIOTECH STRATEGY

A biotechnologist researching gene therapies at a startup evaluates a potential partnership with a pharma giant. From the partner's 10-K Item 8 (financial statements), he notes a $290M R&D allocation increase and a 23% revenue growth from oncology products in Item 7.

Coupled with Item 1A risks on clinical trial delays, he decides to pivot his team's focus to faster-to-market oncology applications, accelerating their joint venture and unlocking $7M in milestone funding.

EXAMPLE 3.7 RESPONDING TO COMPETITOR COST PRESSURES IDENTIFIED IN A 10-K

A software technologist leading an AI team at a software services company considers expanding into edge computing. Analyzing a competitor's 10-K Item 7 reveals a 50% gross margin erosion from cloud costs and Item 1A warnings on data center shortages.

This data prompts her to prioritize on-premises hybrid solutions, reducing projected costs by 27% and positioning her firm to capture 16% more market share in the $21B edge sector.

These are just a few examples that show how the 10-K form is a valuable planning tool.

3.6.2 The US 10-Q Report

A 10-Q report is a quarterly filing with the US SEC that provides an update on the company's financial position during the fiscal year. This is formally called the Quarterly Report pursuant to Section 13 or 15(d) of the Securities Exchange Act of 1934. This report is less detailed than a 10-K but includes:

- Unaudited financial statements for the quarter.
- Management's discussion of results and trends.
- Updates on risk factors and significant events since the last filing.

The 10-Q helps investors track performance between annual reports and spot emerging issues early in the same way that the 10-K does. It can be used to inform you of some of the same kinds of decisions as were driven by the 10-K, only on a faster basis.

3.6.3 International Equivalents to the 10-K

The international equivalents to the US 10-K form provide cross-border signals for multinational engineering (e.g., supply chains spanning Asia/Europe). They often emphasize environmental, social, and governance (ESG) concerns more than US forms, influencing sustainable design.

There are required disclosures, like the US 10-K, for companies headquartered in other countries. Table 3.6 shows some of these along with the regulating authority and access point, and the focus of interest to engineers, scientists, and technologists. This table was compiled from information found at https://iriscarbon.com/.

TABLE 3.6
International Equivalents to the US 10-K Disclosure With Some Key Details [iris.com]

Country/Region	Equivalent Form	Key Regulator/Access	Engineer, Science, Technologist Focus
Canada	Annual Information Form (AIF) + MD&A	SEDAR (System for Electronic Document Analysis and Retrieval)	Resource extraction costs (e.g., mining tariffs for electrical components)
China	Annual Report (A-share)	CSRC (China Securities Regulatory Commission); CNINFO	Raw material hedging (e.g., lithium prices)
EU/UK	Annual Financial Report (under Transparency Directive)	ESMA (European Securities Markets Authority); Companies House (UK)	ESG risks (e.g., carbon taxes)
Japan	Yuho (Annual Securities Report)	EDINET (Electronic Disclosure for Investors' NETwork)	Supply chain resilience (e.g., earthquake risks)

Canadian companies file an Annual Information Form (AIF) along with Management's Discussion & Analysis (MD&A), accessible through SEDAR (System for Electronic Document Analysis and Retrieval). These documents provide detailed insights into resource extraction costs, which are critical for engineers working in sectors like mining, energy, and manufacturing.

For instance, tariffs on raw materials such as copper or rare-earth elements can significantly affect the cost structure of electrical components. Engineers and scientists can use this information to optimize designs for cost-efficiency or explore alternative materials to mitigate supply chain risks.

Chinese companies issue Annual Reports for A-share listings, regulated by the CSRC (China Securities Regulatory Commission) and published on CNINFO. These reports frequently address raw material hedging strategies, which are crucial for engineers in sectors like battery technology and electronics.

For example, fluctuations in lithium prices can impact chemical battery optimization and cost modeling. Engineers can leverage this data to plan for material substitutions, recycling strategies, or alternative chemistries to maintain competitiveness despite market volatility.

In the European Union and the United Kingdom, companies publish an Annual Financial Report as mandated by the Transparency Directive. These filings are overseen by regulators such as ESMA (European Securities and Markets Authority) and Companies House in the United Kingdom. For engineers, these reports often emphasize ESG (Environmental, Social, and Governance) risks, which can directly influence project design.

For example, carbon taxes and emissions targets may drive civil engineers to adopt low-emission materials or energy-efficient construction methods. Understanding these disclosures helps technical teams align with sustainability mandates and secure contracts in regions prioritizing green compliance.

In Japan, the Yuho report is the primary annual disclosure, regulated through EDINET (Electronic Disclosure for Investors' NETwork). These filings often highlight supply chain resilience, which is vital for engineers in earthquake-prone regions.

To illustrate, disclosures about geographic risks or infrastructure vulnerabilities can inform mechanical and civil engineers to design redundant systems, earthquake-resistant structures, or diversified sourcing strategies. This proactive approach reduces operational downtime and enhances safety in high-risk environments.

For multinational firms listed in the United States, Form 20-F is the standard annual filing made. These reports follow IFRS, which differ from US GAAP in areas such as asset depreciation and R&D capitalization. For engineers, these differences can influence lifecycle costing and project valuation.

For instance, IFRS often allows longer asset depreciation periods, which can make civil engineering projects appear more cost-effective over time.

Understanding these nuances helps engineers align project proposals with financial reporting expectations in global collaborations.

3.7 EXERCISES

3.1 Explain the differences between an LLC and a C-Corporation in terms of liability, taxation, and funding flexibility. Provide one scenario where each would be the preferred choice.

3.2 A group of engineers wants to start a consulting firm. Which business structure would you recommend and why? Consider liability, tax treatment, and funding needs.

3.3 Match the following US business structures with their closest international equivalents:
   a. LLC
   b. S-Corp
   c. Sole Proprietorship

   Provide reasoning for your choices.

3.4 Why is Delaware a popular choice for incorporation? Discuss at least three advantages.

3.5 List three major differences between GAAP and IFRS and explain how each difference might impact a technology company's financial reporting.

3.6 Under IFRS, R&D costs can be capitalized after feasibility is established. How does this affect the balance sheet compared to GAAP treatment?

3.7 Explain the difference between VAT and sales tax. Why does VAT create complexity for multinational engineering firms?

3.8 A US robotics company faces a 20% tariff on imported sensors. If sensors account for $500,000 of annual costs, calculate the new cost and suggest two mitigation strategies.

3.9 Give three examples of fixed costs and three examples of variable costs for a biotech firm. How does this cost structure influence pricing strategy?

3.10 A startup borrows $2M at SOFR + 8%. If SOFR is 5%, what is the annual interest expense? How would this affect cash flow planning?

3.11 A company has $10M in cash, a monthly revenue of $600K, and monthly expenses of $1.4M. Calculate burn rate and runway. Is the company in a safe zone?

3.12 If a company spends $120,000 on marketing and acquires 60 customers, what is the CAC? If each customer generates $500/month at 80% gross margin, compute the CAC payback period.

3.13 A firm has a current ratio of 1.1 and a debt-to-equity ratio of 1.8. What do these numbers indicate about liquidity and leverage? Suggest corrective actions.

3.14 A SaaS company has 25% revenue growth and 10% free cash flow margin. Calculate its Rule of 40 score. Is it considered healthy?

3.15 Design a simple dashboard for a civil engineering firm showing at least five key metrics. Explain why you chose these metrics.

3.16 Locate the "Risk Factors" section in a sample 10-K report for a technology company. Identify two risks that could directly impact engineering or R&D operations. Explain how an engineering team might mitigate these risks.

3.17 Using MD&A for Strategic Decisions MD&A section of a 10-K or international equivalent (e.g., AIF and Yuho). Summarize one trend or cost driver mentioned and propose an engineering or design decision that aligns with this trend (e.g., adopting modular designs to reduce material costs).

3.18 Compare the US 10-K with one international equivalent (e.g., EU Annual Financial Report or Japan's Yuho). Identify one key difference in disclosure focus (such as ESG reporting or supply chain resilience) and discuss how this difference could influence engineering project planning for a multinational firm.

REFERENCES

Deloitte. (2025). *A roadmap to comparing IFRS standards and US GAAP: Bridging the differences.* https://www.deloitte.com/us/en/services/audit-assurance/articles/us-aers-a-comparison-of-ifrs-standards-and-us-gaap-bridging-the-differences.html

JMCO. (2025). *Performance benchmarks: Construction companies.* https://www.jmco.com/articles/construction/performance-benchmarks-construction-companies/

International Monetary Fund. (2017). *Monetary and financial statistics manual and compilation guide* [Online]. https://www.imf.org/en/Data#manuals

Ross, S. (2025, April 18). *GAAP vs. IFRS: What's the difference?* Investopedia. https://www.investopedia.com/ask/answers/011315/what-difference-between-gaap-and-ifrs.asp, accessed 1/22/2026.

US Securities and Exchange Commission, Office of Investor Education and Advocacy. (2023). *What you need to know about the end of LIBOR—Investor bulletin.* https://www.investor.gov/introduction-investing/general-resources/news-alerts/alerts-bulletins/investor-bulletins/what-you-need-know-about-end-libor-investor-bulletin

Voas, J., Laplante, P., Kassab, M., Lu, S., Ostrovsky, R., & Kshetri, N. (2022). *Cybersecurity considerations for open banking technology and emerging standards* (No. NIST Internal or Interagency Report (NISTIR) 8389 (Withdrawn)). National Institute of Standards and Technology. https://doi.org/10.6028/NIST.IR.8389-draft

4 Business Models and Approaches

In engineering, science, and technology companies, innovation and flawless execution are important, but so too is structuring the business model to recover investments, price work profitably, secure funding when needed, and build alliances that accelerate success. This chapter demystifies the commercial side of technical innovation, equipping you with practical frameworks to turn R&D, prototypes, and projects into sustainable revenue streams.

We will start by exploring how companies position themselves strategically to recover their investments and beat the competition. Next, we contrast commercial pricing (direct client billing) with recovery through overhead rates (common in government-funded work), highlighting how multipliers and utilization drive profitability in consulting-heavy firms.

We then explore the difference between investment campaigns (long-term internal funding bets) and captures (targeted pursuits like grants or contracts), guiding allocation in resource-constrained environments. Finally, we will examine teaming versus partnering decisions and public–private partnerships, weighing short-term alliances for specific bids against deeper strategic relationships for shared intellectual property and risk.

By understanding these models, you will learn to align technical excellence with economic viability. This will ensure that your innovations not only solve problems but also generate the capital needed to scale them. In an era where high-tech ventures face intense competition and funding volatility, these approaches provide a blueprint for turning ideas into enduring businesses.

4.1 HOW TO RECOVER THE INVESTMENT: CHOOSING A COMPETITIVE MODEL

In the high-tech business world, where innovation drives markets and competition is fierce, Michael Porter's competitive strategy framework provides a timeless roadmap for achieving sustainable advantage. The concepts were first introduced in his seminal work *Competitive Strategy: Techniques for Analyzing Industries and Competitors* [Porter 1980], and they are still taught in business programs worldwide.

The main thrust of the work is to describe three generic competitive strategies:

- cost leadership,
- differentiation, and
- focus,

each tailored to different market dynamics and resource strengths [Porter 1980]. Let's look at the three strategies more closely.

DOI: 10.1201/9781003733850-4

4.1.1 Cost Leadership

Cost Leadership centers on becoming the lowest-cost producer in an industry while maintaining acceptable quality, often through scale, efficiency, or process innovation. For example, in computer hardware manufacturing, this could mean optimizing semiconductor fabrication to undercut competitors on price per chip, achieving economies of scale that deter new entrants. Scientists in pharmaceuticals, particularly in generic drug firms, apply this approach by streamlining R&D for bioequivalents, reducing costs to capture market share in commoditized segments. Technologists in cloud computing leverage automated infrastructure to offer services at razor-thin margins, forcing rivals to match or exit.

When Henry Ford revolutionized automotive engineering through the introduction of the assembly line in 1913, Ford Motor Company used cost leadership. Ford's innovation minimized production costs via economies of scale and standardized processes, allowing Ford to offer vehicles at below-industry prices while maintaining quality. The strategy captured massive market share, with the Model T selling over 15 million units by 1927. This established the company as a global leader, outpacing rivals focused on custom craftsmanship [Porter 1980].

As another example, civil engineering firm Bechtel Corporation used cost leadership for megaprojects like the Hoover Dam (1930s) and Channel Tunnel (1980s–1990s). By optimizing supply chains and large-scale efficiencies, Bechtel underbid competitors while delivering on time, securing dominance in infrastructure and becoming one of the world's largest private engineering firm by revenue ($17B+ in 2025) [CompaniesHistory.com 2025].

4.1.2 Differentiation

Differentiation is a competitive strategy that involves creating unique products or services that command a premium price, emphasizing innovation, branding, or superior performance.

For instance, Chrysler Corporation demonstrated differentiation in the mid-20th century by emphasizing luxury and engineering innovation, such as advanced suspension systems and powerful engines in models like the Chrysler 300 series (1955). By positioning vehicles as high-performance status symbols with unique features, Chrysler commanded premium prices and built brand loyalty, emerging as a market leader in the luxury segment despite economic challenges, with sales surging in the post-WWII era.

Note that differentiation is not an appropriate strategy in commodity industries. A *commodity* is a product with little or no differentiation. Structural steel, concrete, generic electrical components, bulk chemicals, and basic fasteners are all examples of commodities. Items like bearings, motors, and transformers, and adhesives are examples of partially differentiated commodities [Porter 1980].

4.1.3 Focus

Focus is a competitive strategy that narrows the scope to a specific segment, applying either cost leadership or differentiation within that niche. For example, Tesla Inc.

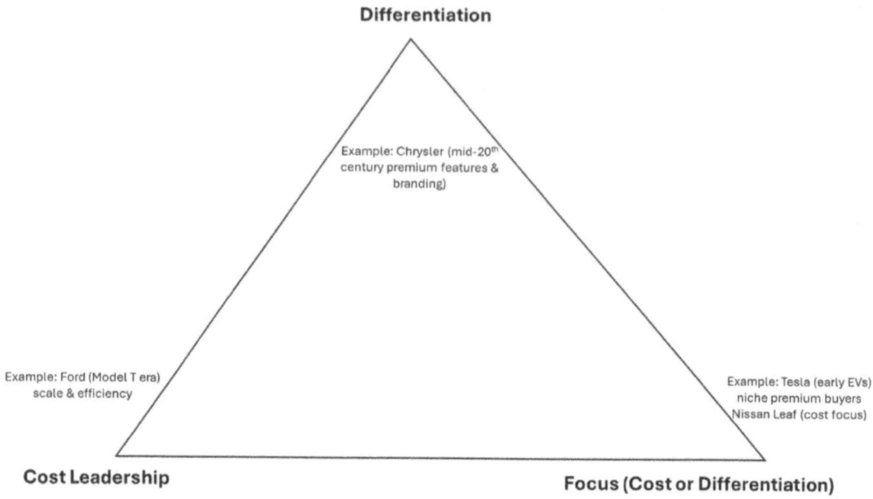

FIGURE 4.1 Porter's competitive strategies—automotive industry examples.

leveraged a differentiation focus on the electric vehicle industry starting in the early 2000s. Through proprietary battery technology and autonomous driving software, Tesla targeted eco-conscious premium buyers with high-priced, innovative models like the Roadster (2008) and Model S (2012). This focus on sustainability and performance disrupted traditional automakers, propelling Tesla briefly to a peak market cap exceeding $1 trillion by 2021 and leadership in the EV sector. In today's AI-driven landscape, a focused differentiation strategy often yields the highest returns for innovators.

Conversely, Nissan used cost focus in the hybrid/electric segment with the Leaf (2010), streamlining production to offer affordable, reliable electric vehicles for price-sensitive consumers. By emphasizing low operating costs and environmental benefits without luxury frills, Nissan captured early market share in the mass-market electric vehicle space, selling over 500,000 units by 2020 and establishing itself as a leader in accessible green technology.

Porter's model emphasizes that success isn't about being the best at everything but about choosing a deliberate path to outperform the competition (Figure 4.1).

For EST professionals, this means aligning technical expertise with strategic positioning to create value that competitors can't easily replicate.

If you are an entrepreneur, by adopting one of Porter's approaches, you can turn inventions into defensible businesses, avoiding the "stuck in the middle" trap of mediocre performance [Porter 1980].

4.1.4 Porter's Five Forces

A precursor to his competitive strategy work, Porter's "Five Forces," introduced in his 1979 Harvard Business Review article, is a framework for analyzing industry attractiveness and competitive intensity. This framework can help EST professionals assess market dynamics to inform strategy, such as entering a new field or defending against rivals [Porter 1979].

Porter motivates his competitive strategies by introducing five forces:

- rivalry among existing competitors,
- bargaining power of suppliers,
- bargaining power of buyers,
- threat of new entrants,
- and threat of substitute products.

All of these are interrelated in that four are forcing functions for competition/rivalry (see Figure 4.2). Understanding these forces helps professionals to analyze competitive industries.

In Porter's theory, high forces signal tough markets, while low forces indicate opportunities, motivating the competitive strategy choice. Let's discuss these forces using the ubiquitous example of the electric vehicle (EV) industry.

Rivalry among existing competitors: Intense competition and price wars lower profits. In the EV world, intense competition among established players drives rapid innovation but compresses margins due to price wars and performance races.

Bargaining power of suppliers: Powerful suppliers can raise prices or reduce quality. In the EV world, suppliers of rare-earth minerals, and other raw materials such as lithium and cobalt, hold significant power because these materials are scarce and critical for battery production. Price hikes (e.g., for semiconductors) or supply disruptions can impact manufacturers.

Bargaining power of buyers: Strong buyers (e.g., governments in defense technology) demand concessions. For example, in civil engineering, state departments of transportation hold power over highway contractors, squeezing margins. For electric vehicles, large automakers purchasing batteries in bulk

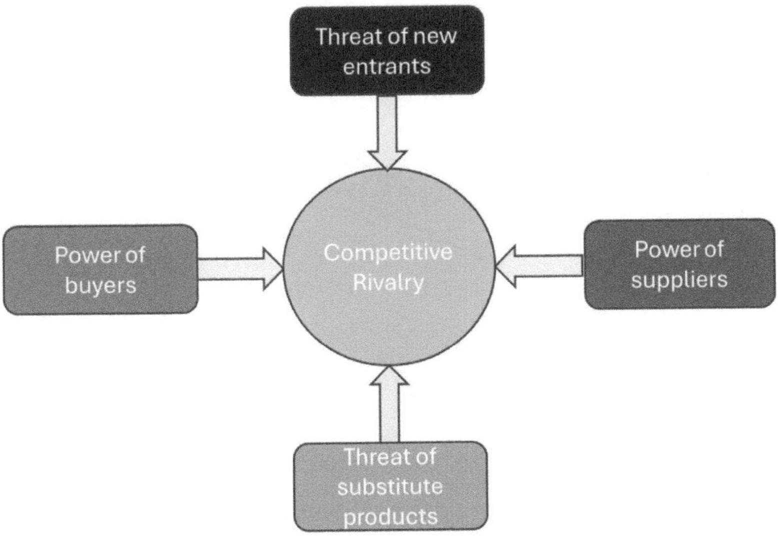

FIGURE 4.2 Porter's 5 forces [Porter 1979].

exert pressure on battery suppliers for lower prices and better performance. In defense technology, government buyers have even greater leverage.

Threat of new entrants: Barriers like capital requirements, patents, or economies of scale deter newcomers. In biotech, high R&D costs and FDA approvals create strong barriers, protecting large, established companies. In the case of the electric vehicle industry, high capital requirements for battery plants, specialized R&D, and compliance with environmental regulations create strong barriers. Patents on battery chemistry and economies of scale also deter newcomers.

Threat of substitute products: Alternatives erode demand (e.g., electric vehicles substituting gas cars). In renewables, solar panels threaten traditional grid infrastructure. For EVs alternative energy storage technologies (e.g., hydrogen fuel cells or solid-state batteries) pose a substitution threat to conventional lithium-ion batteries.

The electric vehicle example illustrates how engineering and technology firms must analyze industry dynamics before investing heavily, as high rivalry and supplier power can erode profitability even in fast-growing sectors. It's the same for startup companies or established companies that want to enter new markets. A thorough analysis of the extant forces is critical before too much capital is invested.

4.1.5 Christensen's Theory of Disruptive Innovation

Clayton Christensen's Disruptive Innovation, first outlined in a 1995 Harvard Business Review article and his 1997 book *The Innovator's Dilemma*, explains how simpler, cheaper technologies initially underperform but eventually upend established markets by targeting overlooked segments. It contrasts with sustaining innovation—incremental improvements for high-end customers (Figure 4.3).

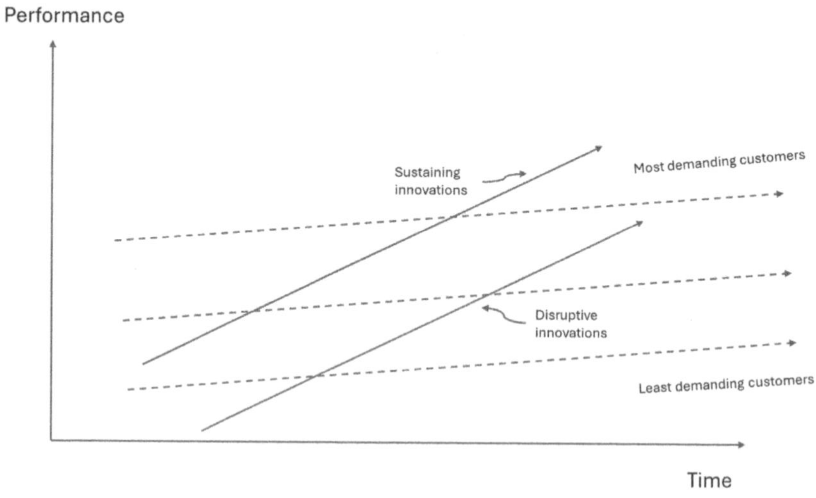

FIGURE 4.3 Disruptive innovation, performance over time.

Source: adapted from [Christensen 1997]

Disruptions come in two forms: low-end and new market. Low end means end starting cheap for underserved users, for example, budget smartphones overtaking premium ones. New market involves creating demand where none existed, for example, personal computers for non-experts [Christensen 1997].

Christensen's theory highlights how simpler, often "inferior" technologies start in overlooked markets and eventually displace incumbents The theory is illustrated by the following examples.

EXAMPLE 4.1 PERSONAL COMPUTERS DISRUPTING MAINFRAMES/MINICOMPUTERS (1970S–1980S):

Mainframe giants like IBM dominated with expensive, powerful computers for enterprises. Personal computers began as "toys" for hobbyists. These new entries to the market were cheaper, less powerful, but affordable to small companies and individuals.

Personal computers improved rapidly, advancing so that they replaced minicomputers and even mainframes for many tasks. By the 1990s, personal computers dominated the workplace, disrupting the $100B+ mainframe/minicomputer industry.

EXAMPLE 4.2 DIGITAL PHOTOGRAPHY DISRUPTING FILM (1990S–2000S)

For most of the 20th century and into the early 21st century, Kodak led chemical film photography with high-quality, profitable products. Kodak invented the digital camera in the mid-1970s but chose to remain focused on chemical film products.

Then, in the early 1980s, digital cameras from companies such as Sony and Canon emerged, albeit as low-end products with poor resolution, expensive memory, and no print capability. They targeted casual users who valued convenience over quality. As sensors improved and costs fell, digital cameras overtook film, and numerous players in the digital camera industry won. Kodak filed for bankruptcy in 2012 [Christensen 1997].

EXAMPLE 4.3 STREAMING DISRUPTING CABLE TV & DVDS (2000S–2020S)

In the early 2000s, Netflix started mailing DVDs to subscription customers, causing low-end disruption to major video chains, such as Blockbuster, and mom-and-pop stores.

Then low-quality streaming emerged, targeting those who were tired of the in-store experience. As bandwidth grew, services such as Netflix and Disney+ offered on-demand convenience, displacing cable bundles. As a result, US pay-tv subscribers dropped from 100M (2014 peak) to ~70M (2025) [CableCompare 2026].

These examples show the pattern. Disruptors enter underserved/low-margin segments with "good enough" solutions, improve relentlessly, and climb the value chain. This approach to new market entries often catches incumbents off-guard.

For EST professionals, Christensen's theory is a warning against ignoring "inferior" technology. Incumbents often dismiss disruptors until too late, as they focus on profitable core customers. The theory urges vigilance and strategic pivots.

4.2 PRICING MODELS

Pricing models are the backbone of financial strategy. Setting a price for high-tech products and services determine is about how value is translated into revenue and how risk is shared between clients and vendors. In high-tech and engineering environments, where uncertainty and complexity are common, pricing approaches must balance transparency, adaptability, and performance incentives. Selecting the right model is not just a contractual detail. It influences project flexibility, cost predictability, and profitability.

EST companies can directly charge customers for products or services. Alternatively, the vendor can use recovery through overhead rates, which is the most common form of pricing in government contracts or cost-reimbursable projects where indirect costs are bundled into multipliers.

The choice impacts cash flow predictability, profit potential, and risk exposure. Commercial pricing offers upside from efficiency but demands market savvy, while overhead recovery provides stability but caps margins via audits. We will discuss overhead recovery in Section 4.3.

In this section, we explore the most widely used models including time and materials and fixed price to cost-plus, unit pricing, and innovative structures like performance-based and shared savings. We will cover the following pricing strategies:

- cost plus,
- fixed price,
- performance-based,
- retainer,
- shared savings,
- time and materials,
- and unit price

highlighting their features, advantages, disadvantages, and best-fit scenarios. Understanding these models enables professionals to align technical execution with financial sustainability, ensuring projects deliver both engineering excellence and economic value.

4.2.1 Cost-Plus

Cost-plus pricing, also known as *cost-reimbursable*, is a pricing model where the client agrees to reimburse the vendor for all allowable project costs along with an additional fee. Allowable project costs typically include labor, materials, and overhead.

The "additional fee" can take different forms, including

- a fixed amount (*cost-plus-fixed-fee*),
- a performance-based incentive (*cost-plus-incentive-fee*), or
- an award tied to specific criteria (*cost-plus-award-fee*).

For example, a defense contractor might be reimbursed $1 million in costs for a prototype, receive a 10% fee ($100,000), and earn an additional $50,000 for meeting performance targets.

Cost-plus pricing emphasizes transparency, as costs are tracked and documented through receipts and detailed reports, and it aligns incentives by rewarding efficiency and performance. Its flexibility makes it ideal for projects with evolving requirements or high uncertainty, such as research-intensive initiatives or complex defense contracts. Vendors benefit from reduced financial risk since costs are covered, while clients gain adaptability for scope changes.

The cost-plus model also carries drawbacks. For example, clients face cost uncertainty and administrative burdens due to extensive tracking and audits, and vendors may lack strong motivation to control costs in pure cost-plus arrangements.

4.2.2 Fixed Price

The *fixed price* model is a pricing approach where the client agrees to pay a predetermined amount for the entire project or a specific deliverable, regardless of the actual time or resources consumed. This amount is established upfront based on a clearly defined scope of work, which provides cost certainty for the client. Typically, payments are structured around milestones, ensuring progress is tied to financial commitments.

For instance, a residential construction company might agree to build a house for $300,000, with payments tied to milestones such as foundation, framing, and completion.

One of the key advantages of this model is that it simplifies billing and budgeting, as clients know the total cost from the outset. It also places accountability on the vendor, encouraging efficiency since the vendor bears the risk of cost overruns.

However, this approach comes with challenges. Vendors face higher financial risk if actual costs exceed the agreed price, and the rigidity of scope can lead to disputes or additional charges when changes occur. To mitigate risk, vendors may inflate bids to account for uncertainties.

Fixed price contracts are best suited for projects with well-defined requirements, such as construction or website development, government contracts requiring strict budget control, and short-term projects with predictable deliverables.

4.2.3 Performance-Based (Incentive or Value-Based)

The *performance-based pricing* model, also known as *incentive-based pricing* or *value-based pricing*, ties payment to achieving specific performance metrics, outcomes, or value delivered. Rather than paying solely for effort or time, this approach rewards results, such as cost savings, revenue increases, or meeting project milestones. Value-based pricing often does not require measurable operational metrics.

Typically, the structure includes a base fee supplemented by bonuses for meeting or exceeding agreed targets, which aligns the interests of both the client and the vendor around shared objectives. For example, a vendor might receive a base payment of $100,000 for implementing a software system, with an additional $20,000 bonus if the system achieves 99% uptime during its first year.

This model offers several advantages. It encourages high-quality results and efficiency by focusing on outcomes rather than inputs, and it promotes risk sharing since both parties are motivated to achieve project goals. Additionally, it aligns costs with tangible benefits, making it a value-driven approach.

Performance-based pricing is best suited for projects with measurable outcomes, such as system uptime, user adoption, or cost savings. It is commonly used in construction projects with performance bonuses for early completion and in consulting engagements tied to business results like increased sales.

Performance-based pricing also presents challenges. Operationally, it requires clear, measurable metrics and mutual agreement on performance criteria, which can be complex to define. Some outcomes, such as customer satisfaction, are difficult to quantify, and vendors face the risk of reduced payment if targets are not met. Performance-based models are often combined with other pricing approaches, such as fixed-price contracts with added incentives, to balance predictability and motivation.

4.2.4 Retainer

A *retainer* pricing model involves the client paying a fixed recurring fee, such as a monthly payment, to secure ongoing access to the vendor's services or resources. Unlike project-based arrangements, this model does not necessarily tie each payment to a specific deliverable.

Retainers provide predictable revenue for vendors and stable costs for clients, making them particularly useful for ongoing support or consulting engagements. They often include a set number of hours or services within each billing period, offering flexibility for ad-hoc work while fostering long-term collaboration. For example, a civil engineering firm might charge a $5,000 monthly retainer for up to 40 hours of design and engineering work.

In addition to "engineer on call" services, retainers are frequently used for ongoing consulting, legal, or marketing services, IT support or maintenance contracts, and creative agencies providing regular services like social media management.

The advantages of this approach include financial predictability, adaptability to changing needs, and stronger client–vendor relationships. However, retainer-based work carries certain disadvantages, such as the risk of underutilization if clients do not fully use the services, ambiguity regarding scope that can lead to disputes, and limited suitability for projects requiring clearly defined outputs.

4.2.5 Shared Savings

Shared savings (sometimes called *gainsharing*) is a pricing model designed to align the interests of clients and vendors by sharing the financial benefits of cost reductions achieved through efficiencies or innovative practices. Under this approach, both parties agree to split savings below a predetermined target or baseline, often as a percentage, which incentivizes vendors to operate efficiently and discourages unnecessary client-driven changes.

The advantages of shared savings include

- motivating vendors to identify cost-saving measures,
- fostering innovation through creative solutions, and
- creating a win-win scenario for both sides.

However, the model also presents challenges, such as the complexity of establishing accurate baselines, the administrative burden of tracking costs, and the potential for disputes over savings calculations or attribution.

Implementing this model requires clear benchmarks and robust cost-tracking mechanisms to ensure transparency and fairness. It is commonly used in cost-plus or collaborative contracts where flexibility and trust are essential. For instance, a contractor who reduces construction costs by $100,000 through efficient material use might split the savings equally with the client, resulting in $50,000 for each party.

Shared savings is best suited for projects with measurable cost-saving opportunities, such as energy efficiency retrofits, supply chain optimization, or large infrastructure initiatives.

4.2.6 Time and Materials

The *time and materials (T&M)* pricing model is based on billing the client for the actual time spent and materials used during the project. Under this approach, rates are typically set on an hourly or daily basis, and material costs are billed at actual or marked-up rates. For example, a software development project might bill developers at $100 per hour and add $5,000 for server costs, invoicing monthly based on actual hours and expenses.

The T&M model offers flexibility, as it easily accommodates scope changes or evolving requirements, making it ideal for projects where specifications are unclear or iterative work is expected. Clients benefit from transparency because they receive detailed breakdowns of labor and material costs, while vendors face lower financial risk since payment is guaranteed for work performed regardless of project outcomes.

However, T&M also carries disadvantages. Clients bear the risk of cost overruns if the project takes longer than anticipated, and close monitoring is required to avoid inefficiencies. Disputes may arise over what constitutes billable work. T&M is best suited for agile or research-driven projects, consulting engagements, maintenance contracts, or any initiative with unpredictable durations or resource needs.

4.2.7 Unit Price

The *unit price* model is a pricing approach where the client pays a fixed rate for each unit of work or deliverable, such as per square foot, per item produced, or per line of code. This method is based on measurable units of output, meaning the total cost depends on the quantity delivered. It is commonly used in contracts involving repetitive tasks, where work can be easily quantified.

To illustrate, consider the following unit pricing items.

- Construction projects (e.g., $100 per square foot for paving).
- Manufacturing or production contracts (e.g., $50 per widget produced).
- Service contracts with measurable outputs (e.g., translation at $0.10 per word).

The model offers several advantages:

- It scales well because costs align with the volume of work completed.
- It provides transparency since costs are easy to verify based on units delivered.
- It offers flexibility by accommodating changes in quantity without renegotiating rates.

However, there are drawbacks. Estimating total costs can be challenging because they depend on accurate quantity forecasts, scope creep can occur if additional units are added unexpectedly, and there is a risk that vendors may prioritize quantity over quality to maximize earnings.

Unit pricing is best suited for projects with clearly measurable outputs, such as construction projects priced per square foot, manufacturing or production contracts priced per item, and service agreements like translation billed per word.

4.2.8 Choosing the Right Pricing Model

Selecting the right pricing model requires balancing project clarity, risk tolerance, and stakeholder alignment to ensure financial and technical success. Let's look at some ways to choose the pricing model.

4.2.8.1 Factors That Determine Pricing

When choosing a pricing model, several critical factors must be considered to ensure alignment between project requirements and financial strategy. These include:

- project clarity,
- risk tolerance,
- stakeholder alignment,
- robust monitoring, and
- control mechanisms.

Project clarity means that the scope is well-defined. Fixed-price contracts are most effective when the scope and deliverables are well-defined, whereas time-and-materials (T&M) or cost-plus models are better suited for projects with evolving or uncertain specifications.

Risk tolerance also plays a major role. For example, fixed-price arrangements transfer most of the financial risk to the vendor, while T&M and cost-plus models shift that burden to the client. The type of project further influences the decision; agile initiatives typically favor T&M or retainer-based models for flexibility, while predictive projects such as construction lean toward fixed-price or unit-price structures for cost certainty.

Stakeholder alignment is another key consideration, as performance-based or shared-savings models can effectively align interests, provided that clear, measurable metrics and mutual trust are established. Contractual clarity is also essential to avoid disputes, particularly in flexible models like T&M or cost-plus, where scope changes are common.

Finally, robust monitoring and control mechanisms are necessary for models that involve variable costs, such as T&M and cost-plus, which require detailed cost tracking and adherence to frameworks like PMBOK's Control Costs process [PMBOK], including tools such as earned value management (see Chapter 6).

Table 4.1 shows the relative differences between models along the dimensions of client or vendor risk, flexibility, best uses, and a summary example.

Next, let's see how to choose the pricing model based on these factors.

4.2.8.2 Pricing Model Decision Diagram

Figure 4.4 provides a decision diagram for choosing a pricing model for some sort of development or construction project from the vendor's perspective. Retainer-based models are used for services and, therefore, are not shown in the figure.

In Figure 4.4, note that shared savings can be used in almost all cases. A decision diagram from the client's perspective is left as an exercise.

An example of using the decision-making process shown in Figure 4.4 follows.

TABLE 4.1
Pricing Model Comparison

Model	Client Risk	Vendor Risk	Flexibility	Best For	Example
Cost-plus	High (cost uncertainty)	Low (costs covered)	High (scope changes)	Complex, uncertain projects	Prototype: $1M costs + 10% fee
Fixed price	Low (fixed cost)	High (cost overruns)	Low (rigid scope)	Well-defined projects	House build $300k, milestone pay
Performance-based	Variable (depending on outcome)	Variable	Moderate	Outcome-driven projects	$20k bonus @ 99% uptime
Retainer	Moderate (underutilization)	Low (guaranteed payment)	High (ad-hoc work)	Ongoing services, support	$5k/month for 40 hours of engineering services
Shared savings	Moderate	Moderate	High	Large infrastructure projects with cost-saving potential	$400M highway construction/shared surplus upon completion
Time and materials	High (cost overruns)	Low (paid for work)	High (scope changes)	Agile, unclear scope projects	Development @ $120/hour, not-to-exceed $200k
Unit price	Moderate (quantity-based)	Moderate (quantity risk)	Moderate (quantity changes)	Repetitive, measurable tasks	$50/component; 10k units

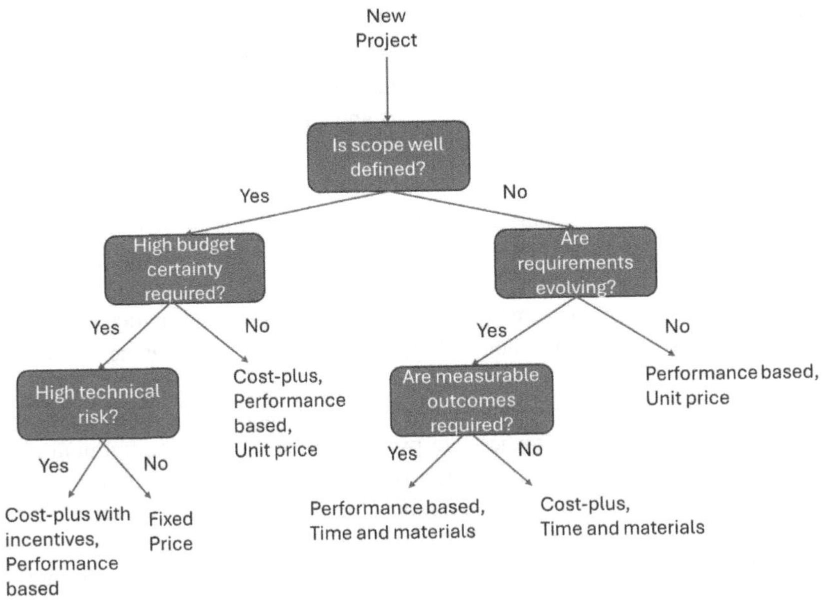

FIGURE 4.4 Choosing a pricing model—vendor's perspective.

EXAMPLE 4.4 SELECTING PRICING MODELS FOR A PRODUCTION LINE SOFTWARE CUSTOMIZATION PROJECT

An industrial manufacturer needs some customization of its production line system to be done by a software firm. Here are different pricing strategies the software company can use based on different assumptions.

Cost-plus: The vendor is reimbursed for costs (e.g., $100,000 in labor/materials) plus a 15% fee ($15,000). Use if this project is considered innovative or high-risk.

Fixed price: The software company quotes $150,000 for the entire system, with payments at milestones (e.g., design, development, and deployment). Use if the scope is well-defined.

Performance-based: The software company is paid $100,000, with a $25,000 bonus if the software system increases sales by 10%. Use if the outcomes are written in the contract.

Retainer: The vendor charges $10,000/month for ongoing development and support. Use only if this has long-term maintenance revenues.

T&M: The software company charges $120/hour for developers, with monthly billing and a not-to-exceed price of $200,000. Use if requirements are unclear and iterations are expected.

Unit price: The software company charges $5,000 per module completed (e.g., 10 modules = $50,000). Use if this project can be cast as modular development with clear deliverables.

Agile projects often favor flexible models like T&M or retainers, while predictive projects such as construction lean toward fixed-price or unit-price structures for cost certainty. Performance-based and shared-savings models can align interests effectively but demand clear, measurable metrics and trust between parties.

4.2.8.3 Other Pricing Determination Strategies

A company can also use A/B offers to help refine customer needs that determine pricing. An *A/B offer* is a pricing or marketing experiment where two different versions of an offer are presented to separate groups of customers to compare perceived value. Conventionally, the versions are labeled "A" and "B." The goal is to determine which offer generates better results, such as higher conversion rates, revenue, or customer engagement.

For instance, in commoditized work like basic CAD, small price hikes can lose customers, while in specialized niches like proprietary biotech assays, premium pricing tends to stick. Testing strategies such as A/B offers or surveys can help refine pricing decisions; for example, a 10% increase might reduce some software-as-a-service conversions by 15% in elastic markets but still boost revenue if high-value users remain engaged. A market or price is elastic if a small change in price changes demand significantly.

The Project Management Institute's Project Management Body of Knowledge Guide offers the following advice for choosing the right pricing model:

- *Plan Cost Management*: Choose a pricing model during this process to align with the cost management plan.
- *Estimate Costs*: T&M and Cost-Plus rely heavily on accurate cost estimation (e.g., bottom-up estimating).
- *Determine Budget*: Fixed Price and Unit Price models require a firm cost baseline; T&M and Cost-Plus may include contingency reserves.
- *Control Costs*: All models benefit from monitoring (e.g., EVM for Fixed Price and cost tracking for T&M) to manage variances and ensure budget adherence [PMBOK].

Regardless of the model, contractual clarity and robust monitoring are essential to prevent disputes and manage variable costs. For example, Earned Value Management is often used for cost control.

4.3 RECOVERY USING OVERHEAD RATES

Overhead rates are indirect cost multipliers applied to direct labor or project costs to recover expenses that are not directly billable to a specific project. These costs typically include items such as facilities, utilities, IT systems, insurance, administrative staff, and other general business expenses necessary to keep the organization running. In EST, the overhead rate pricing strategy is a balance between technical value and financial sustainability.

In practice, overhead rates are expressed as a percentage of direct labor costs. For example, overhead rates of 100–180% to cover office space, equipment, and administrative support are common. General and administrative (G&A) costs, such as executive salaries and bidding expenses, are often recovered at the rate of 10–25%. Combined, these rates form a *wrap rate*, which is often 2.5–3.8 times the base salary of an engineer, scientist or technologist in government or cost-reimbursable contracts.

Here is an example for an engineer working on a US government project. Suppose the direct labor cost is $100,000, the overhead rate is overhead rate: 150% (1.50 multiplier), the G&A rate is 15% (0.15 multiplier), and the negotiated fee is 10%. Then, to compute the price, we have:

Step 1: Apply Overhead and G&A

$$\text{Subtotal} = \$100,000 \times (1 + 1.50 + 0.15) = \$100,000 \times 2.65 = \$265,000$$

Step 2: Add Fee

$$\text{Total Price} = \$265,000 + (\$265,000 \times 0.10) = \$265,000 + \$26,500 = \$291,500$$

Therefore, the total billed amount would be $291,500 for $100,000 in direct labor per year for that person under these rates.

Overhead rates are always subject to audit (e.g., by the US Defense Contract Audit Agency in US federal work). While overhead recovery provides predictable cost coverage, it can restrict profit margins and create challenges if utilization drops below expected levels.

4.4 INVESTMENT CAMPAIGNS VERSUS CAPTURES

EST companies often entertain investment campaigns and captures. *Investment campaigns* are internal, longer-horizon bets to develop differentiated capabilities (e.g., proprietary tools, IP, and platforms). *Captures* are externally funded opportunities to be pursued such as contracts, grants, and bids to bring in near-term revenue. Captures stabilize cash flow but often pay at commodity margins unless the work is highly specialized.

4.4.1 Investment Campaigns

Investment campaigns produce proprietary assets e.g., (algorithms, battery chemistries, and simulation models) that raise switching costs and block entrants, that is, they create a "moat." A *moat* is a defensible advantage that competitors can't easily copy. Tesla's battery R&D is a classic illustration of compounding capability that improves cost/performance over time. These initiatives typically consume cash today for returns later.

Differentiated capabilities allow for price on value rather than cost, aligning with the goal of choosing competitive models that avoid "stuck in the middle." Campaigns typically follow long-horizon NPV/IRR logic: cash outflows now (e.g., for R&D, tooling, and talent) with inflows later (e.g., premium pricing, licensing, and lower unit costs). Investment campaigns should be evaluated for long-term value.

For example, suppose a company wishes to invest in an R&D project to build a proprietary optimization engine for energy-efficient production lines. The time horizon is 18–24 months, at a cost of $2.5M. The desired result of the R&D project is 15–20% IRR based on premium licensing at a lower delivery cost. The strategic goal is differentiation, in this case, the ability to command value-based or performance-based fees later.

The disadvantage of investment campaigns is that they create a high burn rate, raise the risk of technical failure, and tie up capital. That is, they drain cash with a delayed ROI.

4.4.2 Captures

A *capture* is a near-term, externally funded opportunity that a company pursues to bring in revenue. The goal of pursuing captures is to win work as quickly as possible to improve cash flow and keep teams utilized. Captures are typically pursued through contracts, grants, or competitive bids.

In comparison to investment campaigns, captures promise near-term cash with lower uncertainty. Wins like government contracts or client projects translate into revenue now, improving utilization and covering overhead/wrap rates.

EXAMPLE 4.5 STRUCTURING CAPTURE STRATEGIES IN AN INDUSTRIAL COMPANY

A high-performance industrial motor company decides to pursue three fixed-scope retrofit contracts this year ($1.2M each). Management chooses to add performance bonuses tied to kWh reduction and shared savings clauses to avoid commodity margin erosion.

The outcome of the incentive program yields ready cash to fund the campaign, at the same time increasing margins via incentives/gainsharing.

Because many captures are bid competitively, the work can be commoditized (standard deliverables and tight specifications), pushing the company toward cost competition and slimmer fees unless performance-based structures or unique IP are embedded in the product. Therefore, you should evaluate captures with shorter payback and strategic fit, emphasizing execution discipline and margin protection.

4.4.3 Suggested Portfolio Mix

Technology companies typically allocate via portfolio: 60–70% captures for stability, 30–40% campaigns for differentiation [O'Reilly] (see Figure 4.5). This approach balances cash generation with long-term moat building.

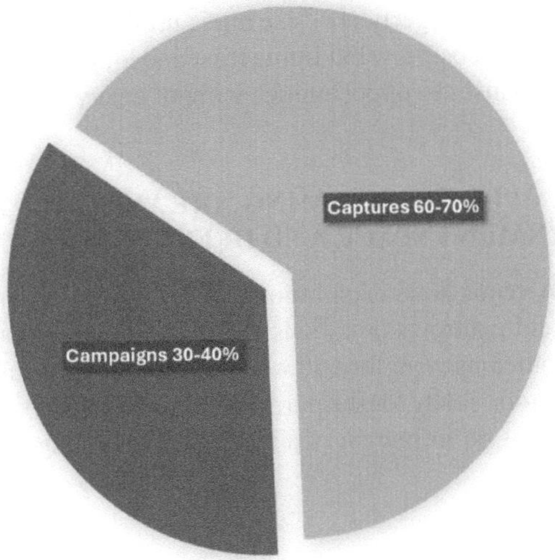

FIGURE 4.5 Portfolio allocation: captures versus investment campaigns.

When considering investing in a new technology versus using a capture strategy, here's a brief checklist that might help make the decision.

1. Is the opportunity strategic IP (campaign) or revenue now (capture)?
2. What's the evaluation horizon (NPV/IRR vs. payback)?
3. Which pricing model best preserves margin (fixed, T&M, performance-based, shared savings, and unit)?
4. Do wrap rates and controls (EVM/PMBOK) protect profitability?
5. Portfolio balance: Are we near 60–70% captures/30–40% campaigns?

Investment/capture campaigns should be formulated with long-horizon NPV/IRR goals to balance shorter payback and strategic fit.

4.5 TEAMING VERSUS PARTNERING DECISIONS

Oftentimes, it is advantageous for a company to join forces with another company and/or with government entities in some large-scale endeavor. Let's look at three strategies for pursuing these collaborations: teaming, partnering, and public–private partnerships.

4.5.1 Teaming

Teaming with other companies forms temporary alliances for specific opportunities (e.g., prime-sub on a government bid). Teaming is a short-term, bid-specific arrangement often used to pursue a single contract or project. It offers speed and flexibility, with shared risk but limited reward because the alliance dissolves after the bid or project ends. Intellectual property (IP) sharing is minimal, and governance is loose, which keeps transaction costs low but limits trust and long-term synergy. This model is ideal for quick responses to opportunities without heavy commitments.

EXAMPLE 4.6 TEAMING ARRANGEMENTS IN GOVERNMENT SPACE AND DEFENSE SOLICITATION

Several NASA Artemis lunar exploration and US Space Force missile-warning and command-and-control solicitations offered teaming options. That is, temporary contractor teams were formed where one firm served as prime and others as subcontractors, solely for the purpose of bidding and executing a specific program. There was no formation of a joint venture or long-term partnership.

Teaming operates as the fast-response counterpart to the more deliberate approaches covered earlier in the chapter. Because it demands little upfront investment and avoids long-term commitments, teaming aligns naturally with capture-driven pursuits where speed, compliance, and utilization matter more than differentiation or moat building. It also echoes the dynamics of cost-plus and T&M pricing models, where flexibility and rapid mobilization often outweigh the pursuit of premium margins. As a result, teaming gives EST firms a practical mechanism to compete quickly in crowded markets while reserving deeper investments and deeper partnerships for opportunities better suited to long-horizon campaigns.

4.5.2 Partnering

Partnering with other companies creates deeper, ongoing relationships, and it may involve strategic collaboration or legal joint ventures, depending on how the relationship is structured. Partnering, in the form of a *joint ventures*, equity stakes, or *co-development,* represents a long-term strategic relationship, for sustained advantage.

Partnering requires higher upfront investment, but promises shared long-run value through deeper integration and co-created IP. Governance is tighter to protect intellectual property and manage joint decision-making. This model suits situations where firms seek sustained advantage, shared innovation, and market expansion beyond a single project.

Both these collaborative forms have risk—teaming can leak capabilities, while partnering locks you into a situation. Teaming and partnering also involve working with external entities, and financial and cultural factors significantly influence the choice between the two.

From a financial perspective, teaming has lower upfront costs, shared resources, but less control. While partnering requires a higher initial investment (for instance, in joint ventures, there are significant setup costs), it offers long-term cost savings and shared expertise.

The potential for culture clashes must also be considered. For example, corporate differences can lead to misaligned goals or clashing management styles that can disrupt collaboration. Operationally, a startup may prioritize innovation, while a large firm emphasizes compliance.

There may be international differences in time zones, language barriers, and regulatory differences (e.g., EU vs. US standards) that require clear contracts and communication protocols. For example, suppose a US engineering firm partners with a German manufacturer. Cultural differences (such as German precision vs. US flexibility) require joint training and aligned incentives for success.

4.5.3 Public–Private Partnerships

A *public–private partnership (PPP)* is a hybrid partnering model where government and private entities co-finance, build, and operate infrastructure. A PPP is a large-scale collaboration between government and private entities, typically spanning 20–30 years. These involve significant risk allocation. That is, private partners assume construction and operational risks, while public entities may cover demand risk. Control is exercised through detailed contracts and layered guarantees, ensuring compliance and performance over decades.

PPPs are typical in highways, hospitals, sports stadiums, and renewable-energy projects. In these situations, the private side handles design/build/operations while the public partner provides such resources as land/permits and revenue sharing, for example, through tolls, availability payments or use fees.

PPPs are best for infrastructure or complex projects requiring long-term financing and lifecycle management.

In EST situations, PPPs excel for large-scale projects but require rigorous risk allocation analysis. For example, the private partner can see risks in construction delays, while the public partner may see demand shortfalls. In the United States, many sports stadiums have been built using PPPs only to see financial shortfalls due to construction delays or significant operating losses due to overestimated demand for the facility [Akintoye].

Table 4.2 summarizes the three collaboration models discussed in this section across four dimensions: time horizon, risk/reward, and intellectual property (IP) control.

TABLE 4.2
Collaboration Options: Teaming versus Partnering versus PPP [Akintoye]

Model	Time Horizon	Risk/Reward	IP/Control
Teaming	Short-term (bid-specific)	Shared risk; limited reward	Low IP sharing; loose alliance
Partnering (joint venture/ co-development)	Long-term	Higher upfront; shared long-run value	Deeper IP sharing; tighter governance
PPP (Design-Build-Operate)	20–30 years	Private assumes construction/operations risk; public may cover demand	Contractual control; layered guarantees

The choice among these three models depends on project scope, time horizon, and strategic objectives. Teaming works for short-term bids, partnering for sustained innovation and shared IP, and PPPs for long-term infrastructure requiring robust governance and risk sharing. But a PPP is ideal when private innovation meets public scale.

4.6 BONDING FOR PUBLIC AND PRIVATE INVESTMENTS

Beyond core pricing and partnership models, EST companies must navigate additional mechanisms that secure performance, mitigate risks for clients, and unlock nondilutive capital, particularly in public-sector or high-stakes private projects.

Performance and *payment bonds* are insurance-like guarantees required on many construction, infrastructure, and engineering projects to protect owners from contractor default. A *surety* (*bonding*) company or agency issues the bond, promising to complete the work or pay damages if the contractor fails. In public projects (e.g., highways and bridges), bonds are often required by law, such as the US Miller Act for certain federal contracts.

Private owners and Public–Private Partnerships increasingly mandate bonding for risk transfer. For example, a $500M PPP toll road might require a 100% performance bond plus payment bonds for subcontractors.

Bond amounts typically equal 100% of contract value for performance and 50–100% for payment. Premiums range 1–3% of the bond amount, but these can be lower for companies in a strong financial position (see Table 4.3) [PMI].

For EST firms, poor liquidity (e.g., as measured by a low current ratio) or high leverage can disqualify the company from bidding on bonded work, limiting access to the governmental infrastructure markets (see Table 4.3). Other qualifications depend on the contractor's working capital, net worth, backlog, and credit (often using credit rating agencies such as Dun & Bradstreet or surety pre-qualification). Altman's Z-score (see Chapter 5) is also used as a bid qualifying factor by some governmental agencies.

Overall, bonding companies underwrite projects based on the "three Cs": Character (reputation), Capacity (experience/track record), and Capital (based on financial metrics such as the Z-score). Major global surety companies include Travelers, Liberty Mutual, Chubb, Zurich, and CNA.

Companies that bid on bonded work need to build relationships early and have strong financials, as these yield lower premiums and higher bonding limits. Poor metrics can result in denial or high collateral commitments before the bond is issued [Akintoye].

TABLE 4.3
Bonding Requirements (Public/PPP Projects) [PMI]

Bond Type	Typical Amount	Notes
Performance Bond	≈ 100% of contract value	Ensures completion if contractor defaults
Payment Bond	50–100% of contract value	Protects subcontractor payments
Premiums	~1–3% of the bond amount	Lower for strong financials; surety qualification required

4.7 EXERCISES

4.1 Select three technology companies (e.g., a cloud provider, EV manufacturer, or biotech firm). For each, identify whether their primary competitive strategy is cost leadership, differentiation, or focus. Justify your classification with two concrete features of the firm's offering or operations.

4.2 Choose an engineering or technology industry (e.g., aerospace components, medical devices, and AI software). Analyze the industry using Porter's Five Forces and rank the forces from strongest to weakest. Briefly explain how this structure affects profitability.

4.3 Identify one disruptive innovation and one sustaining innovation (not mentioned in the book) from the past 20 years in engineering or technology. Explain how incumbents responded and whether the response succeeded or failed.

4.4 For each of the following projects, recommend the most appropriate pricing model and explain your rationale:
a. Early-stage R&D prototype
b. Software system with fixed regulatory requirements
c. Long-term engineering support services
d. Energy-efficiency retrofit with measurable savings

4.5 Using Table 4.1, compare fixed-price and time-and-materials contracts. Identify:
a. Who bears cost risk
b. Who benefits from efficiency
c. What behaviors each model incentivizes

4.6 An engineering firm bills $120,000 in direct labor for a government contract. Overhead is 140%, G&A is 18%, and the negotiated fee is 8%.
a. Compute the total billed amount
b. Identify which components are subject to audit

4.7 Design three measurable performance metrics for a performance-based pricing contract in:
a. A construction project
b. A software deployment
c. A research collaboration
Explain why each metric is objective and defensible.

4.8 Redraw Figure 4.4 from the client's perspective, showing how risk tolerance, budget certainty, and scope clarity influence the choice of pricing model.

4.9 In a shared-savings contract, a baseline annual operating cost is $2.0M. After improvements, costs drop to $1.7M. The agreed sharing split is 60% client/40% vendor.
a. Calculate the savings and payout
b. Discuss one potential dispute that could arise in this model

4.10 A firm is considering raising prices for a specialized engineering service by 12%. Design a simple A/B pricing experiment, specifying:
a. Target customer groups
b. Success metrics
c. How elasticity would be inferred

4.11 Classify the following as investment campaigns or captures, and explain why:
   a. Developing a proprietary AI optimization engine
   b. Bidding on a competitively priced infrastructure retrofit
   c. Re-engineering an existing client's system under contract

4.12 A company currently earns 90% of revenue from captures and 10% from investment campaigns.
   a. Identify two risks of this imbalance
   b. Propose a two-year roadmap to move toward a 70/30 portfolio

4.13 An engineering firm wants to enter the offshore wind market but lacks turbine IP. Compare teaming and partnering as entry strategies and recommend one, considering finance, IP, and culture.

4.14 In a toll-road PPP:
   a. Identify one risk best borne by the private partner
   b. Identify one risk best borne by the public partner
   c. Explain how poor risk allocation can lead to project failure

4.15 A contractor is denied a performance bond due to weak financial metrics. Using the three Cs (character, capacity, capital):
   a. Identify two likely red flags
   b. Recommend concrete steps the firm could take over 12–18 months to become bond-eligible

REFERENCES

CableCompare. (2026, February 11). *U.S. cable TV subscribers 2025: Ongoing decline & cord-cutting trends*. CableCompare.

Christensen, C. M. (1997). *The innovator's dilemma: When new technologies cause great firms to fail*. Boston, MA: Harvard Business School Press.

CompaniesHistory.com. (2025, December 26). *Bechtel*, accessed 2/18/2026.

Porter, M. E. (1979). How competitive forces shape strategy. *Harvard Business Review, 57*(2), 137–145.

Porter, M. E. (1980). *Competitive strategy: Techniques for analyzing industries and competitors*. New York, NY: Free Press.

5 Risk and Opportunity Management

In the relentless pursuit of innovation that defines engineering, science, and technology fields, risk is not an adversary to avoid but a fundamental force to harness. As Peter Bernstein eloquently captures in his seminal work *Against the Gods: The Remarkable Story of Risk (1996)*, humanity's progress stems from our evolving ability to quantify, mitigate, and even profit from uncertainty [Bernstein].

In EST, managing uncertainty means viewing risks as opportunities for strategic foresight and resilient design. Whether technical failures in a satellite launch or market disruptions in biotech, risks are ever-present in every engineering, scientific, and technological field.

A recurring theme of this chapter and the book is the timeless adage "cash is king," which underscores liquidity's role as the ultimate buffer against risk. In volatile business landscapes, ample cash reserves enable pivots during downturns, funding for breakthroughs, and the agility to capitalize on fleeting opportunities like government grants or emerging tech trends.

We'll explore personal and organizational risk profiles, drawing on tools like the Balanced Scorecard and quantitative metrics to turn risky situations into competitive advantages, ensuring that your technical expertise translates into sustainable success.

5.1 DEFINING AND IDENTIFYING RISKS

A *risk* is any event or condition that could interfere with objectives. More formally, ISO/IEC/IEEE 16085 provides the standard risk language used across modern engineering:

> "Risk is the effect of uncertainty on objectives."

The standard explicitly recognizes that the effects of risk may be negative or positive and that risk is evaluated relative to objectives, not just specific hazards or failures [ISO 2021]. Positive consequences presented by embracing risk include:

- higher rewards for the activity than expected,
- the establishment of new markets or business relationships, and
- significant increase in company reputation.[1]

Our interests going forward are mostly in managing the downside consequences of risk.

DOI: 10.1201/9781003733850-5

5.1.1 Fundamental Risks

In EST companies, the fundamental types of risk are:

- geopolitical risk,
- market risk,
- disruption risk, and
- bubbles.

Geopolitical risk is due to the complex relationships between competing and cooperating nations, for example, US-China trade tensions disrupting semiconductor supply chains. *Market risk is* a function of volatile supply and demand in a particular market, for instance, oil price swings impacting chemical engineering firms' feedstock costs. *Disruption risk* is due to a paradigm-shifting change, for instance, technical or logistical innovations rendering current technology obsolete. Henry Ford's implementation of production line technology disrupted and revolutionized the automobile industry. A *bubble* is *an* overhyped market leading to a crash, such as the 2022 crypto winter, wiping more than 70% of market value from blockchain ventures.

5.1.2 Risks and Unknown Unknowns[2]

On a snowy January 13, 1982, an Air Florida Boeing 737 jet departed from Washington, DC's National Airport and crashed into the 14th Street Bridge less than a mile from the airport. Of the 79 people on board the plane, only 5 survived. The plane also struck seven vehicles on the bridge, killing four motorists and injuring another four.

The unfortunate passengers on the plane should have known that there was a risk of the plane crashing, no matter how small that risk may have been. In some cases, the passengers may have even taken out special crash insurance. But what were the expectations of the motorists in the vehicles on the bridge who were killed or injured? Surely, the thought of an airplane crashing into the bridge never crossed their mind. For the motorists, therefore, this crash represented an unknown but manifested uncertainty [Laplante and Kassab 2022]. The possibility of unknown unknowns, like this unfortunate disaster, highlights the problems in risk identification.

One form of unknown unknown is referred to as a "black swan" event. A *black swan* is a rare and unpredictable outlier event with extreme impact, extensively covered in Taleb's groundbreaking book [Taleb 2010]. The term is drawn from the discovery of black swans in Australia in 1697, upsetting a supposed certainty that all swans are white.

Taleb argues that even rigorous fields like stock options trading, science and engineering, people attribute success to talent when it's really the result of randomness. A black swan event has three key characteristics:

- unpredictability—nothing in the past suggests its possibility (it is an "outlier" or "unknown unknown").
- massive impact—it carries extreme consequences, reshaping history, markets, societies, or personal lives.
- retrospective explainability—after the event, experts create narratives that make the event seem predictable (the "narrative fallacy") [Taleb 2010].

Some pundits cite the outbreak of World War I, the 2008 financial crisis, Google's dominance, and the COVID-19 pandemic as black swan events. But just as risk can manifest in both negative and positive ways, there are rare, positive black swan events too, for example, the ubiquity of personal computers and the internet. These events highlight how history is often driven by outliers rather than predictable trends [Laplante 2026].

While black swan events, by definition, can't be predicted, EST leaders need to position their organizations and moderate their predictions to be able to react to black swan events.

5.1.3 Gartner Hype Cycle

The *Gartner Hype Cycle* is a graphical model (Figure 5.1) mapping technology maturity through five phases: Technology Trigger (the initial buzz), Peak of Inflated Expectations (or hype bubble), Trough of Disillusionment (post-crash realism), Slope of Enlightenment (practical adoption), and Plateau of Productivity (mainstream stability).

For EST professionals, this visualization helps contextualize and explain risk situations and their resulting success and failure.

The Technology Trigger is the phase where a breakthrough, proof-of-concept, or research result sparks initial attention, long before mature products or proven business value exist.

To illustrate, in the late 2010s, Generative AI, with breakthroughs such as transformer architectures and early large language models, was a clear technology trigger.

The Peak of Inflated Expectations (or hype bubble) occurs when early successes are extrapolated into unrealistic promises. In EST, we often see these as fantasized expectations, limited pilots or exaggerated demos. For example, in the early 2020s, overinvesting in the metaverse hype, with little to no results, represented a technology that had reached a peak of inflated expectations.

The Trough of Disillusionment follows when inflated expectations collide with reality. Projects fail, budgets are cut, vendors consolidate or disappear, and public

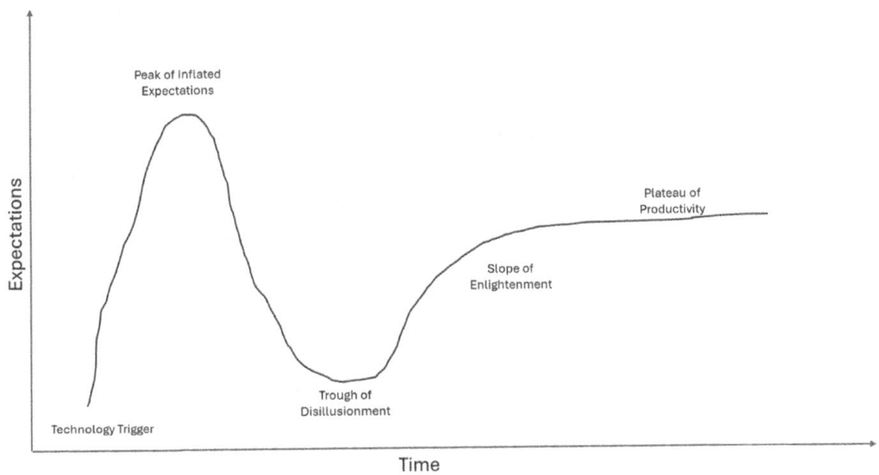

FIGURE 5.1 Gartner hype cycle [Gartner, Inc. 2012].

sentiment turns sharply negative. For missing opportunities in the trough, consider AI's 2010s disillusionment before the 2020s boom.

The Slope of Enlightenment begins after hype collapses and practical understanding emerges. Organizations learn what works and what does not. For instance, cloud computing experienced a clear slope of enlightenment—after early skepticism and failures, firms learned how to use cloud services effectively for scalability, DevOps, disaster recovery, and cost control.

Finally, the ubiquitous nature of Relational Database Management Systems represents a technology at the Plateau of Productivity. Performance, tooling, and reliability are well understood; innovation is incremental rather than disruptive. Organizations no longer ask, "should we use it?" but rather "which version and configuration?" [Gartner 2012].

EXAMPLE 5.1 THE DIESEL ENGINE AND THE GARTNER HYPE CURVE

The Diesel engine's early history is fascinating [Brunt 2023], and its development also traces the Gartner Hype Cycle exquisitely. Its Technology Trigger came in the 1890s when Rudolf Diesel introduced a radically efficient combustion concept that promised far greater thermal efficiency than steam or early gasoline engines. It could burn many types of fuels—crude oil, coal tar, even vegetable and nut oils—and produced no smoke, making it an excellent choice for military ships.

This new technology generated enormous interest from industrialists and military planners who saw the potential for a safer, more economical power source and a military engine that was more efficient than steam, more stable than petrol, with more easily accessed and transported fuels. That excitement soon crested into a Peak of Inflated Expectations, as early adopters imagined the Diesel engine quickly transforming transportation, manufacturing, shipping, and naval warfare. But the first prototypes were unreliable, difficult to manufacture, and prone to catastrophic failures.

As limitations became clear, the technology entered a Trough of Disillusionment. Early Diesel engines were heavy, expensive, and unsuited to many practical applications. But steady engineering improvements, like better fuel injection, materials, safety, and manufacturability, gradually restored confidence. This period corresponds to the Slope of Enlightenment, where industry learned how to apply Diesel power effectively, especially in marine, rail, and stationary applications.

By the mid-20th century, the Diesel engine reached the Plateau of Productivity, becoming the dominant global workhorse for transportation and power generation due to its reliability, fuel efficiency, and long service life. The Diesel engine represents a classic example of a breakthrough technology that matured exactly along the Gartner curve.

5.1.4 Company Risk Profile

Just like people, companies also vary in their willingness to take on risk, and this risk profile affects decision-making. For example, startup companies may be more willing to enter risky endeavors because they have less to lose than a large, well-established company. However, a larger company that is failing may need to take on more risk than it has historically to survive.

You can evaluate the risk profile of an engineering, science, or technology (EST) company using a structured framework that integrates quantitative and qualitative methods across financial, operational, regulatory, and market dimensions, while ensuring alignment with strategic goals like innovation and sustainability.

Possible frameworks include:

- NIST's Risk Management Framework (RMF),
- ISO 31000

The NIST RMF emphasizes integrating security and privacy risks throughout the system lifecycle for tech-heavy firms [NIST 2018], while [ISO 2018] is a global standard for risk management that promotes a process of identification, analysis, evaluation, treatment, and monitoring tailored to EST contexts like R&D projects.

EXAMPLE 5.2 EVALUATING A COMPANY RISK PROFILE USING STRUCTURED FRAMEWORKS

At a renewable-energy tech startup, you could use the NIST RMF to categorize risks by preparing an inventory of assets (e.g., proprietary algorithms and hardware prototypes), then assessing threats like cyber vulnerabilities or supply chain disruptions using semi-quantitative scoring (e.g., likelihood times impact on a 1–5 scale).

For small to medium-sized companies, you could complement this type of risk analysis with financial metrics like the Altman *Z*-score to gauge bankruptcy risk or the Piotroski *F*-score for operational health. We will look at these shortly.

Finally, you can monitor certain metrics via dashboards including leading indicators like cash burn rate and mitigate them through diversification. Then you can review the main cost drivers quarterly to adapt to emerging risks like AI ethics regulations or supply chain geopolitics. This holistic approach not only quantifies risks but also turns them into opportunities, like pivoting to sustainable tech for grant funding.

5.1.5 Your Personal Risk Profile

We continuously live with personal risk, for example, getting hit by a car when crossing a street, or having our identity stolen through some business transaction. Your personal attitude about risk, whether you are eager to embrace it or play it more safely and try to avoid it, greatly influences your decisions and behavior in both the workplace and at home.

Your personal risk tolerance in your career refers to how much uncertainty, volatility, or potential downside you're willing to accept in pursuit of upside rewards, for example, higher pay, equity, or more exciting work. Your risk tolerance is shaped by:

- personal factors (e.g., age, family responsibilities, and financial cushion),
- personality (i.e., thrill-seeker vs. stability-seeker), and
- market realities (e.g., layoffs in Big Tech vs. steady government contracts).

It drives every big decision that you will make during your career. For example,

- switching jobs,
- beginning a side hustle,
- using startup equity versus cash for some business endeavor, and
- choosing when to exercise stock options.

High-risk tolerance might mean working in startups or buying volatile stocks; low tolerance favors stable roles with predictable income. Aggressive risk-takers tend to thrive in early-stage ventures, while conservatives excel in regulated fields like defense. In those business domains where cycles boom and bust (e.g., AI hype vs. crypto winters), mismatched tolerance can lead to burnout or regret. For example, taking an 80% equity in a startup company that you join, when you need cash for a mortgage. Assessing your risk profile honestly (via quizzes or reflection) helps optimize decisions.

To help you assess your personal risk profile more rigorously, Table 5.1 provides a brief, eight-question survey instrument. This short-form instrument was synthesized from several more comprehensive surveys found online. It is not validated and is intended for educational use only. It aligns conceptually with established investor risk-tolerance assessments but is not a substitute for a full psychometric evaluation.

To take the test, answer each question with an ordinal number from 1 to 5, where 1 indicates a very conservative sentiment and 5 indicates a very aggressive sentiment. Then add up the score column total, which will be in the range of 8 to 40. Then assign yourself a risk identification according to the following rules:

40–32	Aggressive	You might be a startup founder or angel investor.
31–24	Moderately Aggressive	You might work for one of the "Magnificent 7."
23–16	Balanced	You might be a typical senior engineer, with a 60/40 stocks/bonds portfolio and a 6-month emergency fund.
15–10	Conservative	You prefer cash, bonds, and real estate, and hate volatility.
<10	Ultra-Conservative	You probably like government bonds and keep a lot of cash on hand.

TABLE 5.1
How to Identify Your Personal Risk Profile

#	Question	1 (Very Conservative)	5 (Very Aggressive)	Your Score (1–5)
1	If your net worth dropped 30% tomorrow, how would you feel?	Panic, sell everything	Stay calm—this is a buying opportunity	
2	What % of income could you lose for 12 months without lifestyle change?	<10%	>50%	
3	Equity versus cash comp: $200k cash or $600k restricted stock (4-year vest, 50% chance $0)	Take cash	Take restricted stock	
4	Job stability: BigTech $450k compensation; stable versus startup; $300k + equity moonshot	BigTech	Startup	
5	Emergency fund size	12+ months expenses	<3 months	
6	Investment style	100% bonds/ index funds	70%+ individual stocks, options, crypto	
7	Age/family situation	45+, kids, mortgage	25–35, single/dual-income-no-kids, no debt	
8	"I'd rather miss upside than lose principal"	Strongly agree	Strongly disagree	

Compare your actual portfolio/job choices to your score, and if there is a mismatch, you may be over-stressed and/or undercompensated. You can also use your awareness of your risk profile when negotiating. For example, "My profile is balanced, so I'd take 50% equity if vested over 3 years." Or "My risk profile is 19, therefore I need $350k cash compensation, not $500k in restricted stock." Or "I'm 34 years old and happy to take 80% equity."

Over time, certain events and experiences may change your risk tolerance, so you should reassess yourself every 2–3 years or after major life changes (e.g., marriage, recession, and job loss). In EST domains, high tolerance often yields outsized rewards, but there are higher failure rates; therefore, it may be wise to diversify your earnings through, for example, side gigs or passive income sources to hedge against risk.

Finally, note that your personal risk profile as a budget manager at work may not be the same as your risk profile with your own money. For example, you may be risk-averse with either your personal or business finances, but risk tolerant in the other; the correlation depends on many factors that are difficult to capture. Just be aware that any person can be a high-risk taker in one aspect of life, and a low-risk taker in another.

5.2 BANKRUPTCY AND DISTRESS PREDICTION METRICS

Financial health is the foundation of any organization's ability to manage risk and seize opportunities. While strategic foresight and operational excellence are critical, they cannot compensate for a weak balance sheet or deteriorating liquidity. Fortunately, there are metrics that help engineers, technologists, and managers assess the likelihood of financial distress before the situation becomes catastrophic.

Among these, the Altman Z-score remains one of the most widely used predictors of bankruptcy risk, offering a composite measure based on profitability, leverage, liquidity, and efficiency ratios. Complementing this, the Piotroski *F*-score provides a simplified yet powerful framework for evaluating financial strength and improvement trends, particularly useful for private firms and value investors. Together, these metrics enable decision-makers to answer two critical questions: "Are we financially stable enough to win major contracts and fund innovation?" or "Are we strengthening or weakening over time?" By understanding these metrics, professionals can transform raw financial data into actionable insights, ensuring resilience in volatile markets and positioning their organizations for sustainable success.

5.2.1 Altman *Z*-Score

The Altman Z-score, introduced in 1968, remains one of the most trusted tools for assessing bankruptcy risk. The metric combines multiple financial ratios such as liquidity, profitability, leverage, and efficiency into a single composite score that signals whether a firm is in a safe, gray, or distress zone. By applying the Z-score, decision-makers can quickly gauge financial stability, anticipate bonding capacity issues, and take proactive measures to avoid catastrophic failure. It can also be used for corporate self-assessment, and evaluation of potential partners or merger and acquisition targets [Altman 1968].

5.2.1.1 Versions of the *Z*-Score

The original form of the Z-score for public manufacturing firms is given in Equation 5.1:

$$
\begin{aligned}
Z = 1.2\cdot\frac{\textit{Working Capital}}{\textit{Total Assets}} + 1.4\cdot\frac{\textit{Retained Earnings}}{\textit{Total Assets}} + 3.3\cdot\frac{\textit{EBIT}}{\textit{Total Assets}} \\
+ 0.6\cdot\frac{\textit{Market Value of Equity}}{\textit{Book Debt}} + 1.0\cdot\frac{\textit{Sales}}{\textit{Total Assets}}
\end{aligned}
\tag{5.1}
$$

The calculation yields a real number that is interpreted as follows: $Z > 2.99$, then the company is in the "Safe Zone," that is, there is a very low chance of bankruptcy.

If Z is in the range 1.81–2.99, then the company is in the "Grey Zone," which means that senior leaders need to watch the company closely.

If $Z < 1.81$, then the company is in the "Distress Zone," meaning the company is at a high risk of bankruptcy.

Altman later evolved his scoring system for privately held manufacturing/construction firms, forming the Altman Z'-score, as shown in Equation 5.2 [Altman 2012]:

$$Z' = 0.717 \cdot \frac{Working\,Capital}{Total\,Assets} + 0.847 \cdot \frac{Retained\,Earnings}{Total\,Assets} + 3.107 \cdot \frac{EBIT}{Total\,Assets} + 0.420 \cdot \frac{Book\,Equity}{Book\,Debt} + 0.998 \cdot \frac{Sales}{Total\,Assets} \tag{5.2}$$

The interpretation of the metric also changes as follows: Safe Zone: $Z' > 2.90$, Grey Zone: $1.23 \leq Z' \leq 2.90$, Distress Zone: $Z' < 1.23$.

Finally, Altman derived a Z" score for non-manufacturing sectors, which is highly relevant for EST enterprises [Altman 2012]. The Z" metric is given in Equation 5.3:

$$Z'' = 6.56 \cdot \frac{Working\,Capital}{Total\,Assets} + 3.26 \cdot \frac{Retained\,Earnings}{Total\,Assets} + 6.72 \cdot \frac{EBIT}{Total\,Assets} + 1.05 \cdot \frac{Book\,Equity}{Book\,Debt} \tag{5.3}$$

The zones are again tuned for the new situation as:

- Safe Zone: $Z'' > 2.60$,
- Grey Zone: $1.10 \leq Z'' \leq 2.60$,
- Distress Zone: $Z'' < 1.10$.

Again, these zones differ by sector.

5.2.1.2 Example Calculation

Here is an extended example showing the calculation of the Z"-score. Consider a mid-sized highway engineering firm, like A-One Engineering, introduced in Chapter 1. Its first quarter 2026 numbers are summarized below:

Item (from Balance Sheet + P&L)	Amount	Formula Input (TA = Total Assets = $1,781,000)
A. Working Capital	$495,000	Current Assets–Current Liabilities
B. Retained Earnings	$564,000	Cumulative profits kept in the business
C. EBIT (Operating Income before interest & tax)	–$2,000	
D. Book Equity	$1,064,000	Total Equity
E. Book Debt	$390,000	Current + Long-term loans
F. Revenue (annualized from Q1)	$2,020,000	Q1 revenue × 4

Then we compute the component terms of Z" from Equation 5.3 as

Term	Calculation	Result
6.56 × (Working Capital/TA)	6.56 × (495k/1,781k)	1.83
3.26 × (Retained Earnings/TA)	3.26 × (564k/1,781k)	1.03
6.72 × (EBIT/TA)	6.72 × (–2k/1,781k)	–0.01
1.05 × (Equity/Total Debt)	1.05 × (1,064k/390k)	2.86

Summing the terms yields

$$Z'' = 6.56\times(\$495k / \$1,781k) + 3.26\times(\$564k / \$1,781k) + 6.72 \times(-\$2k / \$1,781k) + 1.05\times(\$1,064k / \$390k) \approx 5.71$$

Using the zone interpretations for Z" ≈ 5.71 implies that the company is well within the Safe Zone.

The firm is financially solid despite a tiny quarterly loss since it has strong working capital, and low debt dominates the score.

5.2.1.3 Interpreting and Using the *Z*-Score

Different EST sectors will have somewhat different desirable Z"-scores. You can see the range of differences in Table 5.2, which was compiled from a number of sites. The rightmost column in the table provides a rationale for the difference in the metric.

Because it requires detailed ratio analysis and sometimes market value data, the Z-score is most used by lenders, bonding companies, and government agencies to assess short-term survival and qualification for large contracts. It is particularly relevant for asset-heavy industries such as civil, mechanical, and electrical engineering, where liquidity and leverage are critical factors. Table 5.3 lists some of the "rules of thumb" used by these decision makers based on Z-scores.

More generally, decision makers will look for the following red flags:

Z' dropping below 2.6

Bonding companies get nervous; state DOTs may disqualify the company from prequalification lists.

A sudden 1.0+ point drop in one year

This is usually caused by taking on too much debt for equipment or losing a major client.

Z-scores >10

These typify software as a service or cash-rich biotech companies; traditional engineering firms almost never exceed 8.0.

TABLE 5.2
Altman *Z*-Score Ranges by Sector

Firm Type	Typical *Z*″-Score Range (Healthy)	Grey Zone	Distress Zone	Rationale
Civil/Highway Engineering	3.2–6.0	2.0–3.2	<2.0	Small firms often have 3.3–4.0 during growth quarters
Mechanical & Industrial Engineering Consultancies	3.5–7.0	2.2–3.5	<2.2	Heavy CapEx years can dip to 3.5
Electrical/Power Engineering	3.0–5.5	1.8–3.0	<1.8	Project-backlog volatility common
Chemical & Process Engineering	2.8–5.0	1.8–2.8	<1.8	Oil price swings cause big swings
Aerospace & Defense Engineering	2.5–4.5	1.6–2.5	<1.6	Long payment terms hurt working capital
Environmental Engineering/ Consulting	3.5–7.5	2.3–3.5	<2.3	Grant-funded work creates very high scores
Biotech/Life-Sciences R&D	4.0–12+	2.5–4.0	<2.5	Cash-rich biotechs routinely >8; cash-burn phase can drop to 3–5
Software Engineering/SaaS	5.0–15+	3.0–5.0	<3.0	Very high sales/assets ratio
Hardware/Semiconductor Design & Engineering	3.5–8.0	2.2–3.5	<2.2	R&D spend keeps scores moderate
Robotics/Automation Engineering	4.0–10.0	2.5–4.0	<2.5	High fixed-asset efficiency
Renewable-Energy Engineering (solar, wind)	2.8–5.5	1.8–2.8	<1.8	Project-finance heavy, long receivables

TABLE 5.3
Z-Score Rules of Thumb for Decision Makers

Industry Cluster	Minimum "Safe" *Z*″ to Win Big Contracts	Typical Score When Bidding on $100M+ Projects
Traditional engineering (civil, mechanical, electrical, chemical)	≥3.2	3.5–5.5
Environmental/renewables engineering	≥3.0	3.5–6.0
Pure tech/software engineering	≥4.0	6.0–12+
Biotech & deep-tech R&D	≥3.5 (cash runway >24 months)	5.0–15+

5.2.2 Piotroski *F*-Score

While the Altman *Z*-score is a powerful tool for predicting bankruptcy risk, it requires multiple ratios and sometimes market value data, which can be challenging for private firms. To complement this, the Piotroski *F*-score, introduced in 2000, offers a simpler yet highly effective framework for assessing financial strength and improvement trends.

Designed originally for value investors, the *F*-score uses nine binary tests derived from a company's financial statements to produce a score between 0 and 9. Each "yes" answer earns one point, making the metric intuitive and easy to apply even without advanced analytics.

The nine test questions and their evaluation criteria are given in as a scoresheet in Table 5.4 [Piotroski 2000].

The highest *F*-score a company can have is 9, and the lowest is 0. A score of 8–9 indicates excellent financial health, 6–7 suggests solid strength, and an *F*-score of 3 or less signals weakness or potential distress. This simplicity makes the *F*-score particularly useful for engineers, technologists, and managers evaluating acquisition targets, strategic partners, or their own firm's trajectory [Piotroski 2000].

TABLE 5.4
Questions for the Piotroski *F*-Score

Test	Evaluation Rule	Points False (0 points)/True (1 point)
1. Positive ROA this year?	Net Income > 0	
2. Positive operating cash flow this year?	Cash from Operations > 0	
3. ROA improved?	ROA this year > ROA last year	
4. Cash flow > Net Income?	CFO > Net Income	
5. Long-term debt ratio decreased or stayed the same?	Total Debt/Assets decline or flat	
6. Current ratio improved? (better liquidity)	Current Assets/Current Liabilities Increase	
7. No new shares issued? (no dilution)	Average shares this year ≤ last year	
8. Gross margin improved?	Gross Margin % Increase	
9. Asset turnover improved? (better efficiency)	Revenue/Average Total Assets Increase	

EXAMPLE 5.3 APPLYING THE PIOTROSKI *F*-SCORE TO EVALUATE FINANCIAL HEALTH

Suppose the senior managers of A-One Engineering complete Table 5.4, yielding the following results shown in Table 5.5.

TABLE 5.5
F-Score Evaluation for A-One Engineering

Test	Evaluation Rule 2026 Result for A-One Engineering	Points False (0 points)/True (1 point)
1. Positive ROA this year?	Small loss	0
2. Positive operating cash flow this year?	Strong operating cash	1
3. ROA improved?	From small profit to small loss	0
4. Cash flow > Net Income?	Cash Flow > Net Income	1
5. Long-term debt ratio decreased or stayed the same?	Debt increased slightly	0
6. Current ratio improved? (better liquidity)	Slight improvement	1
7. No new shares issued? (no dilution)	No dilution	1
8. Gross margin improved?	Tiny improvement	1
9. Asset turnover improved? (better efficiency)	Asset turnover up	1

Adding up the last column, we see that the *F*-score = 6 for A-One Engineering. The interpretation is that A-One is financially healthy and quietly improving despite the tiny accounting loss—exactly the kind of boring, cash-generating engineering firm that the board members and value investors love.

Metrics like the *Z*- and *F*-scores give managers a concise, data-driven snapshot of whether the firm is quietly strengthening or drifting toward trouble. These metrics represent an early-warning system as essential as any fault sensor in a critical system.

5.2.3 Altman *Z*-Score versus Piotroski *F*-Score

To wrap up this section, let's briefly compare the two risk metrics just introduced. The Altman *Z*-score is primarily a bankruptcy prediction model. In contrast, the Piotroski *F*-score focuses on financial strength and improvement trends rather than imminent failure.

Unlike the *Z*-score, the *F*-score does not require market value inputs, making it ideal for private firms and early-stage ventures. Investors and managers often use the *F*-score to identify undervalued companies or to gauge whether a firm's fundamentals are strengthening over time. This makes it particularly useful in technology, biotech, and R&D-heavy sectors, where book value often understates true potential.

In short, the Z-score answers the question: *"Are we at risk of bankruptcy or losing bonding capacity?"* while the F-score asks: *"Are we getting stronger and creating long-term value?"* Together, they provide a holistic view. One metric is focused on survival and compliance, the other on growth and resilience. For engineering leaders, using both metrics in tandem ensures not only that the firm can weather financial storms but also that it is positioned for sustainable success.

Use the Z-score for survival and qualification. Use the F-score for growth and valuation conversations. Both take less than 15 minutes to derive in a spreadsheet once you have the financial numbers in hand.

5.3 RISK AND OPPORTUNITY MANAGEMENT

Effective risk and opportunity management lies at the heart of strategic decision-making in EST organizations. While earlier sections of this chapter focused on identifying risks and quantifying financial distress, this section shifts toward the *active management* of uncertainty. That is, it addresses the question of how individuals and companies can deliberately shape their portfolios, allocate resources, and design metrics to improve outcomes. The goal is not simply to avoid negative events but to position projects and organizations so they can withstand disruptions and capitalize on emerging opportunities.

Whether a firm is evaluating competing R&D initiatives, deciding which engineering features to prioritize, or balancing long-term innovation against short-term constraints, leaders must navigate limited budgets, competing objectives, and shifting external conditions. Tools such as diversification, the profitability index, and the balanced scorecard provide structured approaches for making these complex decisions.

Each method offers a different lens. Diversification reduces exposure to avoidable risks. Profitability index helps rank competing investments, and the balanced scorecard ensures alignment between technical activities and organizational strategy.

Taken together, these frameworks provide a comprehensive toolkit for transforming uncertainty into strategic advantage. By applying them thoughtfully, engineers and managers can build resilient project portfolios, avoid common pitfalls such as suboptimization, and ensure that both risks and opportunities are managed in a way that supports long-term organizational success. This section illustrates how these tools work in practice and how they can be adapted to the unique challenges of EST domains.

5.3.1 Diversification

Diversification is one of the most fundamental tools for managing risk in both personal finance and organizational strategy. By spreading exposure across multiple assets, markets, or product lines, investors and companies reduce their vulnerability to *unsystematic risk*. That is, the kind of risk tied to the fortunes of a single firm, technology, or sector. In finance, modern portfolio theory shows that combining assets with imperfectly correlated returns can dramatically reduce volatility without sacrificing expected return. In practical terms, this means that a well-constructed portfolio of equities, typically on the order of 20 to 30 stocks, can nearly eliminate firm-specific risk while maintaining strong performance potential [Fisher and Lorie 1970].

For EST organizations, diversification works in much the same way. Companies can diversify their revenue and risk exposure by:

- expanding their product mix,
- customer base,
- geographic footprint, and
- R&D or grant funding sources.

A renewable-energy firm, for example, may balance its portfolio across solar, wind, and grid-storage technologies, shielding itself from policy changes or commodity-price swings that might impact one area more than another. Similarly, a software consultancy might diversify across industries such as healthcare, finance, and energy to avoid the cyclicality of any one vertical. This approach not only reduces vulnerability to downturns but also enables organizations to capture upside in emerging markets or technologies, turning uncertainty into a strategic opportunity.

It is important to understand, however, that diversification cannot eliminate *systematic risk*—the broad economic forces that affect all firms and markets. Recessions, supply-chain disruptions, or geopolitical events will still impact both individuals and organizations regardless of how well-diversified they are. What diversification can do is minimize avoidable, uncompensated risks so that decision-makers are only exposed to the uncertainties that come with the possibility of higher returns. In practice, this means engineers, managers, and investors should diversify where possible. But they also need to remain mindful of the macroeconomic forces that cannot be hedged away. This ensures that the risks they retain are purposeful, strategic, and aligned with long-term goals.

5.3.2 PROFITABILITY INDEX[3]

When organizations face constrained budgets and competing investment opportunities, they need a systematic way to determine which projects offer the greatest return on risk. The *profitability index* (PI) provides exactly this kind of guidance by expressing the ratio of a project's net present value (NPV) to its investment cost.

In essence, PI serves as a "bang for the buck" measure that helps rank opportunities when resources are limited, making it especially valuable in engineering, R&D, and technology portfolio decisions where capital must be allocated carefully.

The profitability index (PI) is the NPV of the investment or activity divided by the cost of the investment, *I*:

$$\mathrm{PI} = \mathrm{NPV} / \mathrm{I} \tag{5.4}$$

A profitability index > 1 indicates a worthwhile investment.

EXAMPLE 5.4 EVALUATING A PROCESS IMPROVEMENT INITIATIVE USING THE PROFITABILITY INDEX

Suppose a process improvement initiative will cost \$50,000 and is estimated to yield \$100,000 savings in two years. Assuming the discount rate is 3%, is this initiative worth undertaking?

To see, first, we compute the future value of the initiative.

$$PV = \$100,000 / 1.03^2 \approx \$94,274$$

Then, plugging the PV of the initiative into Equation 5.4, we have

$$PI = \$94,274 \ / \ \$50,000 \approx 1.885$$

We see that the profitability index is greater than 1, therefore this initiative is worth undertaking.

One of the drawbacks of the profitability index as a metric for making resource allocation decisions is that it can lead to a suboptimization of the result. To see this effect, consider the following situation. An engineering manager is faced with some hard budget-cutting decisions and needs to remove some features from the proposed system. To reach this decision, an analysis is prepared on the basis of the profitability index shown in Table 5.6.

Suppose the capital budget is \$500K. The PI ranking technique would cause the manager to pick A and B first, leaving inadequate resources for C. Therefore, D will be chosen, leaving the overall NPV at \$610K. However, using an integer programming approach will lead to a better decision (based on maximum NPV) in that features A and C would be selected, yielding an expected NPV of \$620K.

TABLE 5.6
A Collection of Prospective Software Features, Their Project Cost, NPV, and Profitability Index

Feature	Projected Cost (in \$10s of thousands)	NPV (in \$10s of thousands)	PI
A	200	260	1.3
B	100	130	1.3
C	300	360	1.20
D	200	220	1.1

EXAMPLE 5.5 SUBOPTIMIZATION USING PROFITABILITY INDEX

Here is one more example of how choosing by PI can suboptimize. The director of a program office must decide on which of the four projects shown in Table 5.7 to fund with a $400,000 R&D budget:

TABLE 5.7
Four Projects to Fund

Project	Investment	NPV	PI
A	100,000	130,000	1.3
B	100,000	125,000	1.25
C	300,000	360,000	1.20
D	200,000	220,000	1.1

Ranking by profitability index selects A and B, then D (total NPV: $475,000). An integer programming approach, however, would select A and C (total NPV: $490,000).

It's clear from these examples that you should use the profitability index in conjunction with NPV to balance the returns on a project portfolio to avoid suboptimization.

5.3.3 Balanced Scorecard

The *Balanced Scorecard (BSC)* method, originally developed by Kaplan and Norton in 1992 as a performance measurement framework. This is a powerful tool for engineers, scientists, and technologists. BSC helps to manage project risks and opportunities by translating strategic objectives into measurable KPIs across four interconnected perspectives: financial, customer, internal business processes, and learning & growth.

Financial objectives answer the question "To financially succeed how should we appear to our shareholders?" and include considerations of cost efficiency and ROI. Customer "To achieve our goals how should we appear to our customers?" and considers issues such as stakeholder satisfaction and market needs. Internal business processes address the question "What business processes must we excel at?" and address such topics as innovation and operational efficiency. Learning & growth answers the question "How will we sustain our vision through change and improvement?" and incorporates concerns such as team skills and knowledge adaptation. A strategy map for Balanced Scorecard is shown in Figure 5.2.

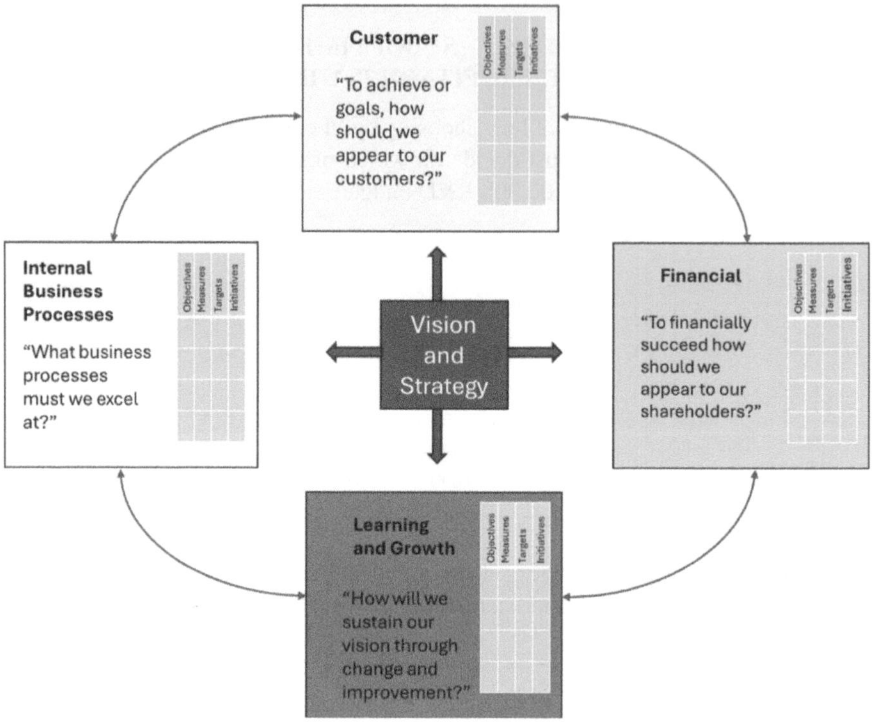

FIGURE 5.2 Balance scorecard strategy map.

The map highlights causal relationships among perspectives, used to drive and monitor implementation while controlling sociotechnical requirements [Kaplan and Norton 1992].

To apply Balanced Scorecard in technical projects, we start by mapping project goals to these perspectives, identifying risks (e.g., budget overruns as a financial risk or technical failures as an internal process risk) and opportunities (e.g., scalable innovations as a learning and growth opportunity). The basic steps are as follows:

1. **Assessment**: Develop BSC plan, strategic elements, and change management.
2. **Strategy**: Define customer value, strategic themes, and results.
3. **Objectives**: Outline strategy action components.
4. **Strategy Map**: Establish cause-and-effect links.
5. **Performance Measures**: Set measures, targets, and baselines.
6. **Initiatives**: Identify strategic projects.
7. **Automation**: Implement software, reporting, and knowledge sharing.
8. **Cascade**: Align unit and individual scorecards.
9. **Evaluation**: Review strategy results and revise strategies.

Table 5.8 provides a template that is useful in conducting these steps:

TABLE 5.8
Balanced Scorecard Template—Engineer/Scientist/Technologist Edition (One-Page Version [Kaplan and Norton 1992]

Perspective	Objective (What we want)	KPI/Metric	Target (2026)	Risk if we miss	Opportunity if we exceed	Leading Indicator (early warning)	Owner
Financial	Deliver >3× ROI on project spend	Project NPV @ 12%/Fully loaded cost	>3.0×	Budget cut/ cancellation	Extra headcount & CapEx	Monthly burn versus value created	Project Lead
	Keep gross margin impact positive	Annualized $ saved or revenue added	+$4.2M	Margin compression	Bonus refresh	Cloud/material cost trend	Tech Lead
Customer/ Stakeholder	Hit all regulatory & client milestones on time	% milestones delivered on or early	100%	Contract penalty	Follow-on contracts	Stakeholder NPS/ regulator feedback	PM/Lead Eng
	Achieve adoption/ performance spec	System availability or spec compliance	99.9% or ±2%	Reputation damage	Reference customer	Beta-user satisfaction score	Product/Eng
Internal Processes	Remove the #1 constraint every quarter	Throughput of the constraining process	+40% versus baseline	Schedule slip	Faster time-to-market	Bottleneck utilization %	Process Owner
	Zero critical safety/ compliance incidents	Number of Level 1 incidents	0	Project shutdown	Industry award	Near-miss rate	Safety/ Quality
Learning & Growth	Team can independently run next-gen tech by EoY	% team certified or shipped with new tech	80%	Obsolescence	New patent/ product line	Training hours + internal contributions	People Lead
	Capture & reuse knowledge	# reusable assets/lessons added to library	≥12	Repeated mistakes	20–30% productivity gain	Post-mortem completion rate	Knowledge Owner

EXAMPLE 5.6 APPLYING THE BALANCED SCORECARD TO A RENEWABLE-ENERGY ENGINEERING PROJECT

An engineer leading a renewable-energy project might set KPIs like "reduce material costs by 15%" (financial, mitigating supply chain risks), "achieve 95% on-time delivery to clients" (customer, capturing market expansion opportunities), "streamline prototype testing cycles" (internal processes, addressing efficiency risks), and "train team on new simulation software" (learning & growth, seizing tech advancement opportunities).

BSC's strength lies in its holistic view, allowing cross-functional teams to cascade strategy into actionable dashboards, conduct regular reviews to adjust for emerging risks (e.g., via Monte Carlo simulations for uncertainty), and capitalize on opportunities through balanced metrics that prevent over-focus on one area like cost at the expense of innovation.

Scientists in R&D can use BSC to monitor biotech experiments by tracking grant funding ROI (financial), regulatory compliance rates (customer/stakeholder), lab process error reductions (internal), and patent filings per quarter (learning & growth), enabling proactive risk mitigation like reallocating resources from underperforming areas. Technologists in software development might apply it to AI projects by measuring deployment costs (financial risk), user adoption metrics (customer opportunity), code quality benchmarks (internal risk), and skill-building workshops (growth opportunity).

The Balanced Scorecard approach shifts focus from siloed metrics (e.g., on-time delivery) to holistic strategy execution, often yielding 15–30% gains in efficiency and alignment. It has been widely adopted in engineering contexts to align technical projects with strategic goals, balancing financial metrics, risk, and reward with operational, customer, and learning perspectives [Kaplan and Norton 1992].

5.3.4 Financial Rewards

Incentive systems like financial rewards, performance ratings, or behavioral penalties are powerful tools for shaping decisions in EST organizations. However, even well-intentioned incentives can trigger unintended consequences when they interact with human psychology, organizational culture, or complex socio-technical processes. This phenomenon reflects a broader principle from economics. People respond to incentives, but not always in the way designers expect.

Poorly structured incentives may distort behavior, encourage short-termism, suppress innovation, or even degrade quality, safety, or morale. These truths need to be considered when managing risk and reward in EST organizations.

5.3.4.1 Incentives/Disincentives and Unintended Consequences

A classic illustration of unintended consequences is "The Parable of the Broken Window," in which a shopkeeper is forced to repair a window his son has broken. Superficially, the repair appears to "stimulate" local economic activity by employing a glazier. However, as Bastiat argues, this reasoning ignores the *opportunity cost*, that is, the goods, investments, or improvements the shopkeeper could have purchased instead. What looks like economic gain is, in fact, a net loss once hidden costs, lost time, and demoralization are considered.

Unintended consequences often ensue from *sunk costs*—unrecoverable expenditures on an uncompleted project. Continuing a failing project because "we already invested so much" is a classic unintended decision error.

Another manifestation of unintended consequences applies directly to modern engineering and technology organizations, where compliance requirements, reporting frameworks, or incentive-linked metrics may force teams to divert resources away from valuable innovation or customer-facing work toward activities that satisfy the metric rather than the mission.

EXAMPLE 5.7 INCENTIVE MISALIGNMENT: CUSTOMER-DEFECT CLOSURE METRICS AND UNINTENDED CONSEQUENCES

This real-world example comes from my own consulting experience. A semiconductor software company tied annual merit raises to a customer-satisfaction metric: the time required to close critical software defects. Engineers could "close" issues by fixing them, downgrading severity, attributing problems to user error, or providing temporary workarounds. While the reported resolution times improved dramatically, customer satisfaction declined.

Support engineers, reacting rationally to the incentive structure, began gaming the system to protect their compensation, even though doing so undermined product quality and their own professional pride. Instead of encouraging better software, the incentive pushed employees toward metric manipulation and away from meaningful fixes.

This pattern is not unique. Similar distortions occurred in multiple US police departments in the late 1990s, where crime statistics were artificially improved by downgrading offenses to lesser categories. The underlying driver was the same. Namely, a measurement-based reward system that unintentionally incentivized data manipulation rather than genuine improvement in public safety [Laplante 2010].

EXAMPLE 5.8 BEHAVIORAL RESPONSES TO INCENTIVES: THE ISRAELI DAYCARE LATE-PICKUP CASE

This case comes from behavioral economics: an Israeli day-care center introduced small fines to discourage late pickups. Counterintuitively, late pickups *increased*. The fine transformed lateness from a moral violation, which may have caused guilt or concern for overburdening the teachers, into a simple, affordable transaction.

Once commodified, the behavior persisted even after the fine was removed. Parents no longer viewed lateness as socially costly; the fine reframed it as a paid option. This case highlights a broader principle: financial incentives can crowd out intrinsic motivations, such as professionalism, responsibility, and empathy.

The deeper lesson for EST leaders is that incentive design must consider second-order and psychological effects, not just the direct behaviors being targeted. An incentive system may fail if:

- The metric is easier to manipulate than the underlying behavior is to improve.
- The incentive rewards short-term gains but ignores long-term consequences.
- The reward crowds out intrinsic motivations such as craftsmanship, safety culture, or teamwork.
- The system becomes adversarial, encouraging compliance theater instead of meaningful progress.

Well-designed incentives, by contrast, reinforce desired behaviors without encouraging perverse ones. Effective systems typically:

- Use multiple metrics, reducing the risk that optimizing one harms others.
- Emphasize leading indicators over lagging ones when possible.
- Reward learning, transparency, and long-term performance, not quick wins.
- Include qualitative assessments, especially peer or customer feedback.
- Revisit and recalibrate incentives regularly based on unintended behaviors observed in the field.

As engineering and technology projects grow more complex, for example, with distributed teams, multi-layered compliance frameworks, and AI-driven workflows, the risk of poorly designed incentives also rises. Leaders must remain vigilant for early signals of gaming, burnout, and mismatched priorities. Ultimately, incentive structures should align with mission, ethics, and long-term value creation. This ensures that teams are encouraged to excel in ways that genuinely benefit the organization, the customer, and society, rather than serving the metric.

5.3.4.2 Unlimited Personal Time Off and Unintended Consequences

Here is an extended example of the unintended consequences of seemingly harmless rewards.

EXAMPLE 5.9 UNINTENDED CONSEQUENCES OF UNLIMITED PTO POLICIES IN ENGINEERING ORGANIZATIONS

Unlimited personal time off (PTO) is often marketed as a progressive, employee-empowering benefit. In theory, it gives workers complete flexibility to take leave whenever necessary without tracking hours or "burning" days.

However, decades of organizational-behavior research and real-world HR experience show that unlimited PTO frequently reduces time off rather than increasing it. This makes it a prime example of an unintended consequence. That is, a policy designed to promote work–life balance ends up creating subtle psychological and organizational pressures that diminish it.

In many EST organizations, the culture already values productivity, speed, and "crunch periods." This is especially true in high-performance environments like aerospace, software development, biotech R&D, and semiconductor engineering. When a company introduces unlimited PTO without explicitly encouraging usage, employees may interpret the policy as:

- Ambiguous—There is no clear target for "acceptable" time off.
- Performance-sensitive—Employees worry that taking time off signals low commitment.
- Socially constrained—Nobody wants to be the person taking more vacation than peers.
- Manager-dependent—In inconsistent organizational cultures, some managers support PTO while others implicitly punish it.

Paradoxically, a fixed PTO bank can empower employees because they have a claim to it. Unlimited PTO removes this protection. Without the anchoring of "20 days a year" or "accrued hours," employees tend to internalize responsibility for the organization's burden, leading to:

- Fewer vacations taken overall, often 20–40% less than with traditional PTO structures.
- Work while sick because it is unclear whether sick days count.
- Hidden burnout since no system tracks unused time or overwork.
- Reduced transparency, as managers cannot monitor trends.

In high-intensity EST environments, these effects can produce safety and quality risks, especially in engineering fields where fatigue has measurable impacts on error rates, decision quality, and design-review rigor.

Leaders often underestimate the psychological mechanisms at play:

- Employees with high conscientiousness self-suppress PTO.
- Project-driven organizations create social pressure to "not delay the sprint."

- R&D workflows often make individuals feel irreplaceable.
- Early-career researchers fear appearing unserious or uncommitted.

The result is a structural mismatch between intention and behavior. Unlimited PTO succeeds only when paired with:

1. Explicit minimum PTO expectations (e.g., "take at least 4 weeks per year").
2. Managerial accountability for ensuring employees use time off.
3. Cultural modeling, where senior engineers and leaders visibly take PTO.
4. Workload planning, so leave does not create bottlenecks that punish the employee.

Without such safeguards, "unlimited" PTO becomes a psychological disincentive, not a benefit.

5.3.4.3 Reverse Auctions

A reverse auction is a behavioral economic mechanism in which participants bid downward, not upward, to win an item at the lowest unique bid. While originally designed for consumer engagement and entertainment, they have been commercialized on certain consumer shopping platforms. The underlying principles have powerful implications for EST organizations as they distinctly highlight the notions of risk and reward.

First, let's discuss the closely related "dollar auction." In this exercise, a professor offers to conduct an auction for a $100 bill. Bids start at $1 and continue in increments of $1. The highest bidder wins the right to buy the bill at the final price. However, the second-highest bidder is required to pay the amount of their final bid to the professor. The auction proceeds with bids slowly climbing to the low $10s, then $20s. Then, students who are very risk-averse drop out of the bidding. Still, the bidding continues into the $40s, $50s and $60s as those with moderate risk still see a bargain. As the bidding comes close to the $80s and $90s, many drop out, but some stay in for fear of being obliged to pay their last bid. Bidding continues and exceeds $100, even $200, as those who wish to avoid the loss stay in the game. In this thought exercise, the quest to minimize the maximum loss propels the bidding to a surprising end. No one wants to be the second-highest bidder, but eventually someone blinks and the auction ends [Shubik 1971].

Reverse auctions are like the dollar auction in that risk mitigation is a primary factor. A reverse auction differs from a traditional auction in three ways:

- Participants compete to offer the lowest price rather than the highest.
- Winning typically requires uniqueness—you must offer the lowest bid not matched by any other participant.
- Bids often carry small transaction costs or penalties that shape behavior.

This structure creates a complex mix of risk, strategy, and psychology. Participants must:

- Predict competitor behavior.
- Avoid clustering around common "psychological" numbers.
- Manage diminishing returns as repeated bids cost time or money.

The consumer auction house DealDash and the energy brokerage firm FreeMarkets are notable examples of firms using some form of reverse auction.

Though it may seem unusual at first glance, reverse-auction mechanisms have several meaningful applications in EST environments, particularly in procurement, resource allocation, project bidding, and internal competition for constrained resources [Chen 2025]. For example, in large engineering projects like civil construction, aerospace supply chains and semiconductor tooling, procurement often involves bidding. Traditional low-bid contract awards can create perverse incentives, including corner-cutting and unsustainable pricing.

A reverse auction can improve outcomes by:

- Rewarding unique value propositions, not just rock-bottom prices.
- Encouraging vendors to differentiate on innovation, risk-reduction strategies or lifecycle cost advantages rather than just cost-cutting.
- Preventing clustering around common price points that artificially inflate competition.

Both the myth of unlimited PTO and the reverse auction illustrate a central principle of this chapter. That is, systems that involve human decision-making rarely behave linearly. Even when designed rationally, metrics, incentives, and mechanisms interact with psychology, culture, and hidden feedback loops that can produce surprising outcomes.

Understanding these dynamics allows EST leaders to design systems that align human behavior with organizational objectives, reducing risk while unlocking new opportunities.

5.3.4.4 Stock Options and Equity

In EST startups, it is very common to offer *stock options* or equity compensation to align contributor incentives with long-term enterprise value. Both offer an opportunity for the employee to own a piece of the company. The main difference between the two is that the option does not need to be exercised, eliminating downside risk except for the risk of non-participation in future bounties.

The two most common employee stock option types in the United States are *incentive stock options and nonqualified/non-statutory stock options*. Both option types grant the right (but not the obligation) to buy shares at a fixed *exercise/strike* price, but they differ materially in tax treatment and eligibility.

Those offering stock options must model dilution and portfolio-level trade-offs. *Stock dilution* is the reduction in existing shareholders' ownership percentage, which occurs when a company issues new shares.

EXAMPLE 5.10 MODELING STOCK DILUTION FOR AN EARLY ENGINEER IN A STARTUP FUNDING ROUND

A robotics-software startup currently has 8,000,000 shares outstanding. The company raises a new funding round that issues 3,000,000 new shares to investors. At the same time, the company increases its option pool by 1,500,000 shares. An early engineer owns 80,000 shares before the round. Let's compute the new total number of shares.

Existing shares: 8,000,000
New investor shares: +3,000,000
New option pool shares: +1,500,000
New total shares = 8,000,000 + 3,000,000 + 1,500,000 = 12,500,000.
The engineer still owns 80,000 shares, but the denominator is now larger.

$$\text{Ownership }\% = \frac{80{,}000}{12{,}500{,}000} = 0.0064 = 0.64\%$$

Ownership dilution occurs because the total number of shares increased (from 8M to 12.5M), causing the engineer's percentage to fall from:

$$\text{Old ownership} = \frac{80{,}000}{8{,}000{,}000} = 1.0\%$$

to:

$$\text{New ownership} = 0.64\%$$

even though the engineer still owns the same number of shares.

Value dilution refers to whether the economic value of the stock shares decreases. Dilution does not always reduce the value of stock. For example, if the funding round increases the company's cash runway, strengthens growth, or improves valuation, the engineer's smaller slice may still be worth more in absolute dollars.

One way the company could offset dilution for employees is by providing *refresh grants.* That is, new stock awards are given to existing employees after a financing round. This helps maintain long-term incentives and partially restores their ownership percentage.

Beyond options, many EST organizations use *restricted stock* or *restricted stock units (RSUs).* Unlike options, RSUs generally require no purchase price; value is delivered when units *vest* and are taxed as ordinary income at vesting, with subsequent gains or losses taxed when shares are sold.

Vesting design governs retention risk and incentive timing. *Retention risk* refers to the possibility that employees will leave the company before their equity incentives (such as stock options or RSUs) fully vest, undermining the organization's ability to keep key talent. The prevalent model in startups is four-year vesting with a one-year cliff (nothing vests until month 12, then monthly/quarterly thereafter).

Some awards include *acceleration.* A *single-trigger acceleration* vests equity upon a change of control; a *double-trigger acceleration* requires both a change of control and a qualifying termination (e.g., without cause) to accelerate. This approach is generally favored by investors and acquirers to preserve post-deal retention.

For private companies, compliance and pricing hinge on *valuation,* that is, an independent appraisal of the fair market value of common stock that sets the minimum strike price for options. Corporate leaders who provide equity, stock, and stock options use them to ensure that incentives for contributors (earning, retaining, realizing equity) match the organization's long-term value creation [Brealey et al 2023].

It is beyond the scope of this text to discuss the nuances and personal tax implications of these various forms of incentives. If you are offered such incentives, you should consult with an accountant or licensed financial professional. The interest here is only in raising awareness that any form of reward or incentive can change behavior or have unintended consequences.

5.4 EXERCISES

5.1 In ~250–400 words, contrast known risks, unknown knowns, and unknown unknowns in an EST project of your choice (e.g., grid-scale battery deployment, medical device firmware, or a robotics line retrofit). Identify at least two monitoring tactics per category that would surface early warning signals.

5.2 Select one historical or plausible black swan that could affect your domain (e.g., supply-chain embargo on a critical rare earth, zero-day cyber exploit in a common PLC). Create a two-column plan:

a. *Robustness*: actions that reduce fragility (cash runway, inventory buffers, design redundancy).

b. *Optionality*: actions that preserve upside if the shock opens new markets (licensing strategy, pivot triggers, pre-approved "fast spend" limits).

5.3 Using Equation 5.3 (Z"), compute Z" for the following firm (TA = Total Assets):

Item	Amount
Working Capital	$1,200,000
Retained Earnings	$2,450,000
EBIT	$180,000
Book Equity	$3,000,000
Book Debt	$1,050,000
Total Assets (TA)	$5,900,000

a. Calculate each term and the final Z'.

b. Classify the firm (Safe/Grey/Distress) using the chapter's thresholds.

c. Briefly explain which component is driving the result.

5.4 Explain when you would choose each version of Altman's model (public manufacturing, private manufacturing/construction, non-manufacturing/EST). Then, for a mid-size SaaS firm, describe why Z" is more appropriate than Z or Z' and which inputs are most likely to fluctuate quarter-to-quarter.

5.5 A private engineering consultancy reports: positive net income; cash flow > 0; return on assets improved year over year; cash flow> net income;

leverage ratio increased; current ratio improved; no new shares; gross margin fell; asset turnover improved.

a. Compute the *F*-score.
b. Interpret in two to three sentences how a lender or acquirer might read this signal in conjunction with *Z'* from Exercise 3.

5.6 You manage a $20M annual budget across five programs: (A) EV power electronics, (B) wind blade materials, (C) grid-edge AI, (D) wastewater monitoring IoT, (E) hydrogen balance-of-plant.

a. Propose a diversified allocation (percentages + rationale).
b. Identify systematic risks you cannot eliminate and unsystematic risks you can reduce through portfolio design.
c. Specify leading indicators (2 per program) to track.

5.7 Your project management office has $1.0M to deploy in FY26. Evaluate the following features:

Feature	Cost ($k)	NPV ($k)
A	300	420
B	200	250
C	500	560
D	300	315

a. Compute the profitability index (PI) for each.
b. If you pick purely by PI until the budget is exhausted, which set do you choose?
c. Allocate the investment budget by inspection to maximize total NPV under $1.0M and compare choices.

5.8 For a laboratory automation project, create a one-page balanced scorecard with:

a. 2 objectives per perspective (Financial, Customer/Stakeholder, Internal Process, Learning & Growth),
b. 1 KPI + 2026 target per objective,
c. 1 leading indicator per objective,
d. 1 owner per objective.

Show cause-and-effect arrows (textual description is fine).

5.9 Analyze a metric you could link to a bonus in your organization (e.g., "critical bug closure time").

a. List three gaming vectors.
b. Propose a balanced metric set that closes loopholes while preserving motivation.
c. Add a review cadence and "kill-switch" conditions to replace the metric if distortions appear.

5.10 Draft a two-page policy outline that makes "unlimited PTO" work: minimum annual floor, manager accountability, coverage planning, reporting for underuse, and cultural modeling. Include three safety/quality guardrails for teams operating in regulated or high-risk engineering environments.

5.11 You need to allocate HPC cluster hours across six teams next quarter. Design a lowest-unique-bid–style internal mechanism where teams "bid down" runtime, priority, flexibility, or scope to win earlier access.
   a. Define eligibility, uniqueness, and floors (quality/safety).
   b. Explain how you will detect & deter collusion and publish post-mortems to sustain trust.
   c. Specify three metrics to evaluate the mechanism after one quarter (e.g., utilization, wait time variance, and defect rework).

5.12 A startup has 10,000,000 shares outstanding. It raises a new round issuing 2,500,000 new shares and simultaneously expands the option pool by 1,000,000 shares (created pre-money).
   a. Compute the new total shares and the ownership % for an early employee who owned 100,000 shares before the round.
   b. Briefly discuss the difference between ownership dilution and value dilution and one way the company might offset dilution for employees.

5.13 Create a decision matrix (bulleted, not a table) for a 200-person EST firm choosing between ISOs, NSOs, and RSUs for mid-career engineers. Address: eligibility, taxation timing, AMT risk, cash needs at exercise, retention effects, and ease of administration. Conclude with your recommended mix and why.

5.14 Your board believes "cash is king" and wants to cut all non-cash incentives to simplify. In 600–800 words, argue for a balanced risk-and-reward system that includes cash buffers *and* carefully designed equity and non-cash incentives. Use examples from earlier exercises (*Z*", *F*-score health signals, BSC leading indicators, reverse-auction mechanisms, and PTO policy) to show how the pieces work together to reduce downside risk while preserving upside opportunities.

NOTES

1 The accounting term for this is "*goodwill.*"
2 Portions of this section are adapted from [Laplante and Kassab 2022] with permission.
3 Portions of this section are adapted from [Laplante and Kassab 2022] with permission.

REFERENCES

Altman, E. I. (1968). Financial ratios, discriminant analysis and the prediction of corporate bankruptcy. *Journal of Finance*, *23*(4), 589–609.

Altman, E. I. (2012). *Altman Z-Score Plus: Enhancements to assess global credit risk in public and private companies across manufacturing and non-manufacturing industries.* NYU Stern School of Business. https://www.stern.nyu.edu/experience-stern/faculty-research/altman-launches-zscore-plus [stern.nyu.edu]

Bernstein, P. L. (1996). *Against the gods: The remarkable story of risk.* John Wiley & Sons.

Brealey, R. A., Myers, S. C., Allen, F., & Edmans, A. (2023). *Principles of corporate finance* (13th ed.). McGraw-Hill Education.

Brunt, D. (2023). *The mysterious case of Rudolf Diesel: Genius, power, and deception on the eve of World War I.* Simon and Schuster.

Chen, J. (2025). *Reverse auction*. Investopedia. https://www.investopedia.com/terms/r/reverse-auction.asp, accessed 1/22/2026.

Fisher, L., & Lorie, J. H. (1970). Some studies of variability of returns on investments in common stocks. *The Journal of Business*, *43*(2), 99–134. https://doi.org/10.1086/295259

Gartner, Inc. (2012). *Hype cycle for emerging technologies, 2012*. Gartner Research. http://www.gartner.com/technology/research/methodologies/hype-cycle.jsp

International Organization for Standardization. (2018). *ISO 31000:2018: Risk management—Guidelines* (2nd ed.). ISO. https://www.iso.org/standard/65694.html

International Organization for Standardization, International Electrotechnical Commission, & Institute of Electrical and Electronics Engineers. (2021). *ISO/IEC/IEEE 16085:2021—Systems and software engineering—Life cycle processes—Risk management*. IEEE Standards Association. https://standards.ieee.org/standard/16085-2020.html.

Kaplan, R. S., & Norton, D. P. (1992). The balanced scorecard: Measures that drive performance. *Harvard Business Review*, *70*(1), 71–79.

Laplante, P. A. (2010). Nexialism and the law of unintended consequences. *IT Professional*, *12*(4), 58–60.

Laplante, P. A. (2026). Of black swans and manure piles. *IEEE Reliability Magazine*, *3*(1), 1–5.

Laplante, P. A., & Kassab, M. (2022). *Requirements engineering for software and systems* (4th ed.). Taylor & Francis.

National Institute of Standards and Technology. (2018). *Risk management framework for information systems and organizations: A system life cycle approach for security and privacy* (NIST Special Publication 800-37 Revision 2). U.S. Department of Commerce. https://doi.org/10.6028/NIST.SP.800-37r2

Piotroski, J. D. (2000). Value investing: The use of historical financial statement information to separate winners from losers. *Journal of Accounting Research*, *38*(Supplement: Studies on Accounting Information and the Economics of the Firm), 1–41.

Shubik, M. (1971). The dollar auction game: A paradox in noncooperative behavior and escalation. *Journal of Conflict Resolution*, *15*(1), 109–111.

Taleb, N. N. (2010). *The black swan: The impact of the highly improbable* (2nd ed.). Random House Trade Paperbacks.

6 Project Cost Estimation and Budgeting

American Hall of Fame baseball player, Yogi Berra, famously stated "It's tough to make predictions, especially about the future." While this quote was unintentionally funny, it is essentially true. Estimating project costs is a prediction of the future, and it is hard to do, and often inaccurate because of

- unpredictable events,
- unknown risks,
- or unforeseen factors.

Effective cost estimation and financial analysis are critical for successful project management, particularly in engineering, technical and IT-driven initiatives. Many high-profile engineering and infrastructure projects have experienced massive cost overruns due to poor planning, scope creep, and inadequate risk management. Examples include the Sydney Opera House (~1,300% cost overrun), the Channel Tunnel (~80%), the Scottish Parliament Building (~980%), Berlin Brandenburg Airport (~257%), and the UK's National Programme for IT (~61%) [Flyvbjerg 2014]. These failures highlight the importance of accurate cost estimation, stakeholder engagement, and phased implementation. Tools like Earned Value Management and rigorous requirements engineering can help prevent similar issues in engineering projects at all scales.

This chapter introduces the financial tools and frameworks necessary to evaluate, budget, and make informed decisions for all kinds of projects. The aim here is to focus on the financial aspects of and not the management and control of the project. The latter aspects can be found in any project management text.

Throughout this chapter, an exemplar project, a small, 5 MW solar power farm, is used to illustrate many of the concepts. The cost estimates, WBS details, and contingency calculations for the solar farm are hypothetical, but are designed to be realistic and consistent with industry norms for small-scale solar projects.

By understanding cost estimation, financial metrics, and strategic decision-making models, engineers can align technical objectives with financial realities, ensuring project success and organizational profitability.

6.1 COST ESTIMATION

Cost estimation involves forecasting the expenses associated with labor, materials, equipment, overhead, and contingencies for some project. *Budgeting* translates these estimates into a financial plan to guide project execution. Cost estimation is

DOI: 10.1201/9781003733850-6

a cornerstone of project finance, enabling engineers to forecast expenses accurately and develop realistic budgets.

There are five fundamental steps in cost estimation:

1. define scope,
2. break down work,
3. estimate costs,
4. include contingencies, and
5. validate estimates.

The next sections explore each of these in detail.

6.1.1 Define Scope

Effective cost estimation and control begin with a well-defined and managed project scope. Defining and maintaining a robust project scope is a significant undertaking and as previously noted, there are many examples of spectacularly failed projects that suffered from failed scope management, along with other problems.

A project's scope can be visualized with a context diagram, use case diagram, rich picture (from Soft Systems Methodology), block diagram or other representation. The important aspect is that the visualization shows system boundaries and interactions with other systems, people, and the environment. The context diagram is important for two reasons. First, it provides a basis for dialog about the system for engineers and stakeholders throughout the project lifecycle. Second, the context diagram helps combat scope creep, that is, the unchecked growth of functional requirements beyond the intent of the system. The context diagram can also assist in requirements agreement and analysis when discussing ambiguous, complex, and missing requirements. An example of a context diagram for a solar wind farm is shown in Figure 6.1.

For a further discussion of stakeholder identification and managing project scope, see Laplante and Kassab (2022). For our purposes, let's assume that rigorous requirements engineering, stakeholder and scope definition have occurred.

6.1.2 Breakdown Work

A *work breakdown structure (WBS)* is a hierarchical decomposition of project functionality, and is the most common approach used for costing, project management, and process tracking. The WBS decomposes the project into manageable tasks, facilitating cost estimation, resource allocation, and responsibility assignment, but here our focus is only on using the WBS to estimate cost.

The WBS can be organized by deliverables or project phases. Some organizations accept activity-based WBS, but clarity and consistency within each project are key. Whichever approach, the WBS should conform to the following rules.

1. Use terms like tasks, sub-tasks, work packages, phases, entries, or activities to classify WBS components.

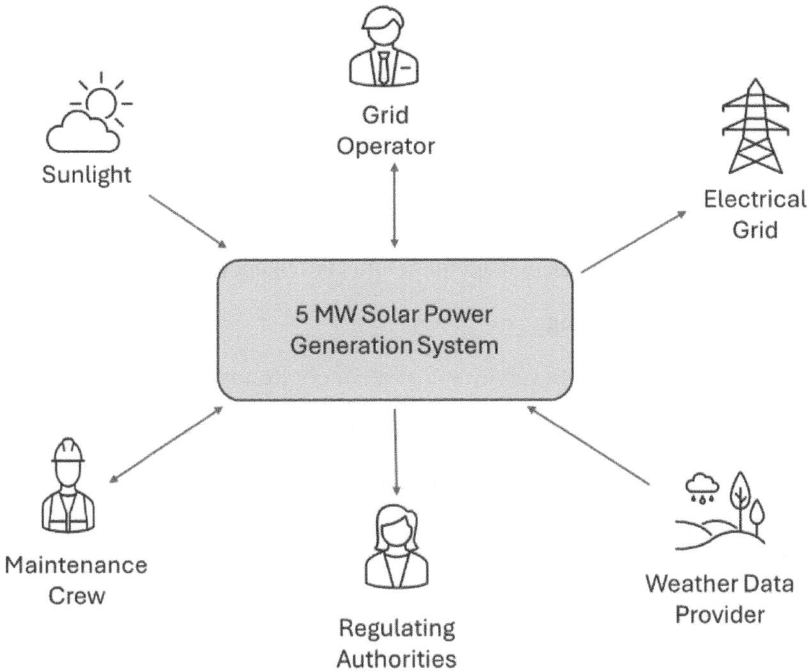

FIGURE 6.1 A simple context diagram for the solar power farm.

2. Components should be defined with durations between several days and a few months to avoid micromanagement or overly large, unmanageable tasks.
3. Do not mix activities and components in the WBS. That is, it should be organized by one or the other but not both.

There are many feasible WBS for any non-trivial project, for example, organizing by phase versus component, changing the order of project phases, varying the decomposition granularity, and more. Figure 6.1 illustrates one possible WBS for the 5 MW solar farm project.

This WBS is activity based, showing achievable tasks, that is, completion of the task can be substantiated, for example, by delivery, audit or inspection. This WBS can be decomposed further into three, four, five or more levels depending on the size and complexity of the project. For example, task 4.3 "Install Solar Panels" can be further decomposed into subordinate tasks such as task 4.3.1 "Install hardware," task 4.3.2 "Install electrical," and task 4.3.3 "Connect to master panel."

The hierarchy shown in Figure 6.2 can also be visualized as an n-ary tree. Figure 6.3 shows a tree equivalent to the WBS of Figure 6.2, with lower levels also depicted. Only the task numbers are shown for brevity. Popular project planning tools provide such visualizations, and each node can be annotated with resources, timelines, and dependencies.

5 MW Solar Farm Project

1. **Project Management**
    1.1 Develop project plan and schedule
    1.2 Coordinate stakeholders (e.g., client, contractors, regulators)
    1.3 Track budget and control costs
    1.4 Perform risk management and contingency planning
2. **Design and Engineering**
    2.1 Conduct site survey and assessment (topography, solar irradiance)
    2.2 Perform system design (panel layout, electrical schematics)
    2.3 Obtain permits and regulatory approvals
    2.4 Conduct final design review
3. **Procurement**
    3.1 Procure solar panels
    3.2 Procure inverters and mounting structures
    3.3 Procure electrical components (cables, transformers)
    3.4 Coordinate logistics and delivery
4. **Construction and Installation**
    4.1 Prepare site (clearing, grading)
    4.2 Install foundation and mounting structure
    4.3 Install solar panels
    4.4 Setup electrical system (wiring, inverters, grid connection)
5. **Testing and Commissioning**
    5.1 Test system (panel output, electrical integrity)
    5.2 Synchronize grid connection
    5.3 Conduct final inspection and certification
    5.4 Handover to client

FIGURE 6.2 Work breakdown structure for the solar power farm project.

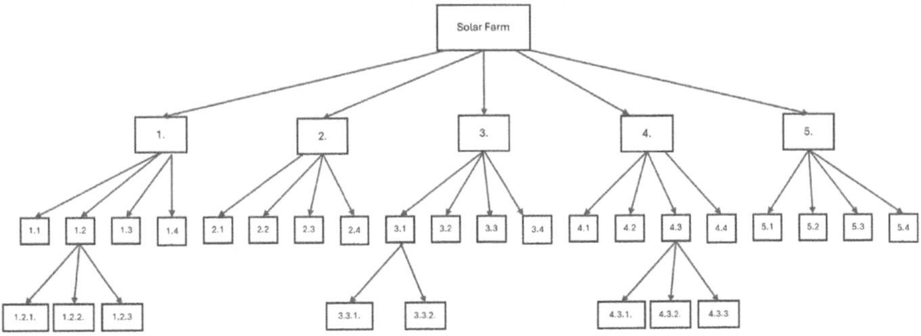

FIGURE 6.3 WBS (task numbers only) for the solar power farm depicted as an n-ary tree.

Any level in the WBS can be used to develop work packages. *A work package* serves as the smallest unit of cost estimation, management planning and control. A work package should include:

- WBS number and name.
- Brief task description.
- Estimated duration.
- Resources needed.
- Predecessor and successor tasks.
- Work products to be produced.
- Work products under version control (baselined).
- Risk factors.
- Objective acceptance criteria.

Higher-level activities aggregate work packages for subordinate tasks and activities and can be used to estimate cost in a bottom-up fashion, which will be discussed later. Figure 6.4 depicts a template for a generic work package.

Figure 6.5 depicts a hypothetical work package for one task, "procure solar panels" for the 5 MW solar farm project.

This is an example of a relatively high-level work package, but if you have good experience and historical data from other similar projects, it may not be necessary to decompose this activity further. Drilling down deeper in the WBS can give higher fidelity estimates, but it also takes more time and skill and can introduce significant noise and variability in the estimates.

The WBS is the foundation of project planning, used to identify work packages for schedule and cost estimation and to support tools like the critical path method (CPM) and the program evaluation and review technique (PERT). Brief descriptions only for creating a WBS and work packages were provided here for the purposes of setting up the cost-estimation process. For a more detailed discussion on these topics, you should consult an appropriate project management guide.

WBS Number and Name
Brief Task Description
Estimated Duration Resources Needed Predecessor and Successor Tasks
Work Products to Be Produced Work Products Under Version Control (Baselined) Risk Factors
Objective Acceptance Criteria

FIGURE 6.4 A template for a work package.

4.2 Install foundation and mounting structure
Description: Install foundations and mounting structure as per specification
Duration: 4 weeks **Resources Needed:** Concrete crew, concrete, cranes **Predecessor(s):** 4.2, **Successor(s):** 4.3
Work Products to Be Produced: As per specification. **Work Products Under Version Control (Baselined):** Not Applicable **Risks:** Weather, material delivery
Objective Acceptance Criteria: County and State building inspector approval

FIGURE 6.5 A work package for the 5 MW solar farm project.

6.1.3 Estimate Costs

There are three basic estimation techniques—*analogous estimation*, *parametric estimation*, and *bottom-up estimation*. Analogous estimation is based on the actual cost of similar projects. Parametric estimation uses some statistical models to determine cost. Bottom-up estimation uses the WBS by summing the detailed task costs for each task from the bottom up.

This section dives deeper into each technique, describing the methodology, applications, advantages and limitations of each, while using the solar farm system for examples.

6.1.3.1 Analogous Estimation

Analogous estimation uses historical data and experience from similar past projects to estimate costs for a new project. It relies on expert judgment to adjust for differences in scope, scale, or complexity. The methodology is as follows:

1. Identify a completed project with similar characteristics (e.g., size, technology, and complexity).
2. Extract cost data from the historical project (e.g., total cost, labor hours, and material expenses).
3. Adjust for differences using qualitative factors such as inflation, technological advancements, or site-specific conditions.
4. Apply the adjusted cost as the estimate for the new project.

Analogous estimation is great for early-stage estimates when detailed plans are unavailable (e.g., feasibility studies), for projects with well-documented historical data, such as building a new warehouse like the previous one, or in the case where a company focuses on the same kinds of projects all the time (e.g., large-scale irrigation system installation). This estimation technique is commonly used in construction, civil engineering, and repetitive manufacturing projects.

For example, say your engineering firm is asked to design a 5 MW solar farm. A similar 4 MW solar farm built two years ago cost \$5 million. Adjusting for inflation (5%) and increased panel efficiency (10% cost reduction), the estimate for the new solar farm would be:

$$\textit{estimated cost} = \$5{,}000{,}000 \times (1 + 0.05) \times (1 - 0.10) \approx \$4{,}725{,}000$$

This rough estimate guides early budgeting but requires refinement as details emerge.

Analogous estimation has the advantage of being fast and cost-effective, and it requires minimal data. It is also a useful technique in the early phases of project conception when details are sparse. Analogous estimation leverages organizational experience and lessons learned and helps to grow that experiential knowledge over time.

There are some shortcomings to this estimation technique, however. For example, the accuracy of the estimation depends on the similarity of projects; dissimilarities lead to errors. The subjective adjustments that are always necessary in non-identical projects can introduce bias errors. Finally, analogous estimation is far less reliable for novel or complex projects with no close analogs.

6.1.3.2 Parametric Estimation

Parametric estimation uses mathematical models to estimate costs (and sometimes project duration) based on certain project variables. This technique is great for projects with quantifiable metrics, such as software development (cost per line of code), construction (cost per square meter), or energy projects (cost per MW).

Simple linear interpolation and extrapolation models can be developed from historical data in-house. Other models can incorporate statistical tools (e.g., regression analysis) for more precision. More complex mathematical models are usually needed for non-trivial projects, but such models and tools, are often available from industry standards group, vendors, or researchers. For example, the COSYSMO (COnstructive SYstem engineering MOdel), which is intended for cost and effort estimation of mixed hardware/software systems, is based on a set of size drivers, cost drivers, and team characteristics [see Laplante and Kassab 2022].

This chapter does not cover the development of a parametric model, as these are specialized to each industry and application. Let's just assume that a parametric model is available for your situation.

For parametric estimation, the following steps are followed:

1. Identify the appropriate mathematical model for the project type.
2. Identify key project input parameters (e.g., square footage, lines of code, and labor hours) required by the model.
3. Adjust the project's parameter values for factors such as location or complexity.
4. Input the parameters into the model and generate the estimate.
5. Validate the output either through a "sanity check" by using industry benchmarks or by using another estimation method such as analogous estimation.

For example, for a highway construction project, historical data shows a cost of $2 million per mile for similar roads. The new project involves 10 miles, with a 10% cost increase due to urban terrain complexity; therefore, the estimated cost is:

$$\textit{estimated cost} = 10 \times \$2{,}000{,}000 \times (1 + 0.10) = \$22{,}000{,}000$$

This is a simple linear cost model. But for highway construction, there are many other factors to consider, for example, time of year, market price of materials, availability of construction crews and equipment and more. These factors could increase or reduce the cost estimate.

It is also quite important to choose the correct cost estimation model for each situation. For example, an electrical contractor has a standard price to install a single receptacle, switch, or light fixture. But if the contractor is asked to estimate the cost to wire a new residential home, they would use the square footage and a cost multiplier to estimate the cost (e.g., $5 per square foot).

Parametric estimation is more accurate than analogous estimation when reliable data and models exist. Parametric estimation is also quite useful for project management because it can be applied mid-project for progress monitoring and control.

But parametric estimation requires accurate and validated models, and even with these, poor data quality reduces the accuracy of the estimates. Using the wrong model, for example, if a complex project has linear cost relationships, can lead to faulty predictions. The technique is also sensitive to parameter misestimation (e.g., underestimating labor hours).

6.1.3.3 Bottom-Up Estimation

Bottom-up estimation estimates total project cost by summing the work package costs of the WBS tree from the lowest levels to the successively higher ones. That is, the cost of a work package at any level is equal to the cost of the sum of its child nodes in the tree. Applying this principle recursively from the lowest leaves in the tree to their parents yields the overall project cost. The cost of any subtree of the WBS can also be estimated this way. Typically, bottom-up estimation is warranted in complex engineering projects such as bridge construction, aerospace manufacturing, or large-scale IT implementations.

The procedure for bottom-up estimation is as follows:

1. Develop a WBS to decompose the project into tasks.
2. Estimate costs for each task at the lowest (leaf) level of the WBS tree, including labor, materials, equipment, and overhead, using quotes, vendor data, or expert input.
3. Working from the leaf nodes up, calculate the task costs next level up (the parent) by summing the estimated costs of the subordinate (child) work packages.
4. Continue summing up the tree this way until the root node is reached, which yields the total project cost estimate.

EXAMPLE 6.1 BOTTOM-UP COST ESTIMATION USING A WORK BREAKDOWN STRUCTURE

Consider the tree structure WBS in Figure 6.3 for the solar farm project. Let's assume that cost estimates are available for each work package at the lowest levels or leaves of the tree, that is, tasks 1.1, 1.2.1, 1.2.2, 1.2.3, 1.3, 2.1, etc. This cost information is based on quotes, historical data, or parametric models. Now summing the cost estimates at this gives us the estimate for the costs for the parent node. That is,

cost(task 1.2.)= cost(task 1.2.1) + cost(task 1.2.2) + cost (task 1.2.3)

cost(task 3.3.) = cost(task 3.3.1)+cost(3.3.2)

cost(task 4.3.)= cost(task 4.3.1) + cost(task 4.3.2) + cost (task 4.3.3)

Now continuing up the tree, and plugging in some simulated numbers, we calculate

cost (task 1)= cost(task 1.1)+cost(task 1.2)+cost(task 1.3)+cost(task 1.4)
= $50,000

cost (task 2)= cost(task 2.1)+cost(task 2.2)+cost(task 2.3)+cost(task 2.4)
= $100,000

cost (task 3)= cost(task 3.1)+cost(task 3.2)+cost(task 3.3)+cost(task 3.4)
= $3,000,000

cost (task 4)= cost(task 4.1)+cost(task 4.2)+cost(task 4.3)+cost(task 4.4)
= $1,500,000

cost (task 5)= cost(task 5.1)+cost(task 5.2)+cost(task 5.3)+cost(task 5.4)
= $50,000

Summing these up yields the total estimated cost of the project:

Total Cost = cost (task 1) + cost (task 2) + cost (task 3)+ cost (task 4) + cost (task 5) = $4,700,000. This estimate is somewhat higher than the $4M previously determined and so senior managers will need to go back and revisit those estimates and assumptions and reconcile them with this estimate.

Bottom-up estimation is well-suited for projects where the scope and tasks are well-defined. It is widely used for final budgets and contract bidding and is also useful for detailed planning. It is very accurate due to granular cost breakdowns, and it provides a high level of transparency, allowing stakeholders to review and adjust specific task estimates. It also supports cost control by tracking expenses at the task level.

Bottom-up estimation is time and labor-intensive in that detailed cost information must be collected or calculated. And the accuracy of the method depends on a complete scope definition—an incomplete WBS will lead to critical omissions. Finally, bottom-up estimation may miss systemic risks not captured at the task level.

6.1.3.4 Combining Cost Estimate Techniques

Whenever possible, it is best to use at least two of the three estimation techniques for each project at an appropriate stage of maturity. Typically, you would use analogous estimation for early feasibility, parametric estimation for mid-stage refinement, and bottom-up for final budgets. For example, the solar farm project might start with analogous estimates based on past projects, then be refined with parametric models (cost per MW), and finalized with detailed, bottom-up task costing.

A summary of the three basic cost estimation techniques is given in Table 6.1.

Whichever technique or techniques you choose, use commercial or free estimation tools as much as practical to ease the computational effort and reduce errors. As the

TABLE 6.1
Cost Estimation Technique Summary

Technique	Recommended For	Accuracy	Effort Required	Tools
Analogous	Early stages, limited data, similar projects	Low–Moderate	Low	Historical records, expert judgment
Parametric	Mid-stage, quantifiable metrics available	Moderate–High	Medium	Statistical models, cost databases
Bottom-Up	Detailed planning, well-defined scope	High	High	WBS, project management software

project unfolds, use Earned Value Management (EVM), to track actual costs against estimates, and revise estimates along the way. To account for risks, you should apply contingency estimates (e.g., 10–20% for early scope definition) and use Monte Carlo simulations for probabilistic cost modeling, especially for complex projects.

Finally, even if you use two or three estimation methods, you should always validate these estimates by cross-checking with industry benchmarks, vendor quotes, or expert reviews. The next sections further explore contingencies, validation techniques and EVM.

6.1.4 Accounting for Contingencies

Contingencies are reserve funds or time buffers added to cost estimates to account for uncertainties, risks, and unforeseen events that could increase project costs. Whichever cost estimation techniques are used, you should always condition the estimates with contingencies. Accounting for contingencies is critical for engineering projects, especially where complexities like weather, supply chain disruptions, or design changes can lead to cost overruns. For example, Berlin Brandenburg Airport lacked adequate contingencies for technical risks (e.g., fire safety systems), contributing to a €7.2B overrun [Flyvbjerg 2014]. A risk-based contingency analysis could have flagged issues early, saving costs.

Contingencies are generally represented as a percentage or fixed amount added to a base cost estimate to cover:

- known risks—these are identified uncertainties with a measurable probability (e.g., weather delays, labor shortages),
- unknown risks—these are unforeseen issues not captured in the initial estimate (e.g., regulatory changes, equipment failure), and
- scope creep—an incremental change in project scope.

Contingencies differ from *management reserves*, which are separate funds for major, unanticipated scope changes controlled by senior management. Contingencies are embedded in the project budget and must be recognized, understood, and managed by the project team.

Accounting for contingencies is important because they buffer against risks that could affect financial metrics like NPV or IRR. Recognizing contingencies also prevents cost overruns. Realistic budgets with contingencies align with cost–benefit trade-offs and build stakeholder confidence and trust. Finally, in some industries (e.g., construction), regulatory compliance mandates that contingency reserves be accounted for (e.g., 10–20% per PMBOK Guide [PMBOK 2021])

6.1.4.1 Determining Contingencies

Let's look at four methods for determining contingencies:

- a percentage-based approach,
- a risk-based approach,
- Monte Carlo simulation, and
- historical data analysis.

In the *percentage-based approach*, you add a fixed percentage (e.g., 10–20%) to the base estimate, based on project complexity and historical data. This method is often used with early-stage estimates (analogous or parametric) or on projects with moderate uncertainty. For example, for a \$4M base estimate, a 15% contingency adds \$600,000, yielding a \$4.6M total budget.

In the *risk-based contingency approach*, you use an expected monetary value (EMV) of each risk by assigning probabilities and cost impacts to those risks. That is, for some risk, the EMV given by Equation 6.1:

$$EMV(risk) = probability(risk) \cdot cost\ impact(risk). \tag{6.1}$$

The set of risks and their probabilities of occurrence for some project is called a *risk register*. We sum the EMVs for each risk to compute the contingency reserve (budget) for that project, yielding Equation 6.2:

$$contingency = \sum EMV(risk_i) = \sum probability(risk_i) \cdot cost\ impact\ (risk_i) \tag{6.2}$$

For example, for the solar power farm, the project risks include a 20% chance of \$200,000 weather delays and a 30% chance of \$100,000 material price hikes. Then:

$$contingency = (0.2 \times \$200{,}000) + (0.3 \times \$100{,}000) = \$40{,}000 + \$30{,}000 = \$70{,}000$$

It is appropriate to use the risk-based contingency approach with detailed projects that have a well-defined WBS.

Monte Carlo simulation uses statistical modeling to simulate thousands of cost scenarios based on a risk distribution (e.g., triangular for task costs). We set the contingency at a confidence level (e.g., 80% likelihood costs won't exceed budget), run the simulation and report the results. For example, a simulation shows an 80% chance that costs for the solar farm project won't exceed \$4.5M for a \$4M base estimate, so the contingency is \$500,000.

Monte Carlo simulation estimation is appropriate for complex projects that have high uncertainty. Simulation tools such as @RISK or Oracle Crystal Ball are often used to conduct the simulation. These simulation results can be used to reduce time and cost overrun risks and as justification to stakeholders for certain risk mitigation decisions that may be taken.

Finally, in *historical data analysis,* we use data from similar projects (analogous estimation) to set contingency based on past overruns. For example, past solar projects averaged 12% overruns, so a \$4M project adds \$480,000 contingency. This approach is advisable with projects with reliable historical records (e.g., repeat construction projects).

6.1.4.2 Integration into Cost Estimation

As discussed in the introduction to Section 6.1, contingencies are applied in the fourth step of the five-step estimation process, that is, after calculating the base estimate using estimation techniques previously discussed, and before the validation step.

Contingencies are included in the project budget but tracked separately to monitor usage (e.g., via Earned Value Management). Contingencies are allocated to WBS work and adjusted as risks materialize or are mitigated.

Let's look at some practical examples by applying contingencies to the 5 MW Solar Panel Installation Project WBS shown in Figure 6.2, with a base estimate of $4,000,000.

6.1.4.2.1 Scenario 6.1 Percentage-Based Contingency

Here, we use bottom-up estimation, assuming moderate complexity for the solar power project. Industry standards suggest a 15% contingency for solar projects. Therefore:

$$\textit{contingency} = \$4,000,000 \times 0.15 = \$600,000$$

$$\textit{total budget} = \$4,000,000 + \$600,000 = \$4,600,000$$

From a practical standpoint, we allocate the $600,000 contingency budget across WBS tasks (e.g., $300,000 to construction, $200,000 to procurement) to cover delays or price fluctuations. The actual distribution of this contingency across all tasks is based on experience and the judgment of the project manager.

6.1.4.2.2 Scenario 6. Risk-Based (EMV) Contingency

In this scenario, the project manager identifies the following risks for the solar power plant project:

25% chance of $150,000 delay due to weather
20% chance of $100,000 material cost increase
10% chance of $50,000 permitting delays

The EMV contingency is then:

$$\begin{aligned}\text{EMV} &= (0.25 \times \$150,000) + (0.2 \times \$100,000) + (0.1 \times \$50,000) \\ &= \$37,500 + \$20,000 + \$5,000 = \$62,500\end{aligned}$$

Thus, the total project budget estimate is:

$$\textit{total budget} = \$4,000,000 + \$62,500 = \$4,062,500$$

We can apply the $62,500 contingency to specific WBS tasks (e.g., $37,500 to construction, $20,000 to procurement). If the risks don't materialize, the project cost is reduced accordingly.

6.1.4.2.3 Scenario 6.3 Monte Carlo Simulation

In this case, we use some software (e.g., @RISK), task costs in the WBS (e.g., $2M for panels, $1.5M for construction) are modeled with triangular distributions (min, likely, max costs). Simulation shows an 80% confidence level that costs won't exceed $4,500,000. That is:

$$\textit{total budget} = \$4,500,000$$

Therefore, the contingency is:

$$contingency = \$4,500,000 - \$4,000,000 = \$500,000$$

The project manager can spread the $500,000 across high-risk WBS tasks (e.g., construction and procurement).

Technology Readiness Levels are discussed in Chapter 7, but it is important to note here that projects at technology readiness levels 1–3 tend to have contingencies at the 20–30% level.

6.1.4.3 Practical Considerations for Contingencies

To wrap up our discussion of contingencies, let's consider a few practical matters.

First, it is advisable to tune the contingency based on the project phase. Early phases (e.g., feasibility) need higher contingencies (20–30%) due to uncertainty; while later phases (e.g., execution or construction) should have lower contingencies (5–10%) as risks are clarified.

Next, you should monitor and adjust the contingencies as the project unfolds. Track contingency usage via cost reports. If risks decrease, reallocate funds to improve the TCO or IRR, as applicable. Avoid overusing contingencies. Excessive contingencies inflate budgets, reducing competitiveness in bids or skewing cost–benefit trade-offs. Overuse also reduces the credibility of the estimator.

Finally, you should document risks by linking contingencies to the risk register for transparency. Software Tools can help to integrate contingencies into the WBS and track via EVM. Many of these tools also support Monte Carlo simulations and variance analysis.

Contingencies, when calculated rigorously, protect against the financial pitfalls seen in megaprojects, ensuring engineering projects stay on budget and deliver value.

6.1.5 Validate Estimates

The last step in the five-step process of project cost estimation is to validate the estimates, though validation is not a single act—it recurs at all phase gates. In project finance and cost analysis, *estimate validation* ensures that cost estimates for engineering projects are accurate, reliable, and aligned with project objectives. Validation reduces the risk of overruns by confirming that estimates reflect realistic assumptions and data. This process is critical when using cost estimation techniques (analogous, parametric, and bottom-up) and integrates with financial concepts such as NPV, TCO, and contingency planning.

Validation helps ensure that estimates are:

Accurate—reflecting true project costs, minimizing errors.
Credible—supported by data and stakeholder consensus, enhancing trust.
Realistic—accounting for risks, scope, and market conditions, reducing overruns.
Aligned:—capable of supporting financial metrics and project goals.

There are five primary estimate validation techniques:

1. cross-checking with historical data
2. expert judgment review
3. benchmarking against industry standards
4. sensitivity analysis
5. Monte Carlo simulation

Without validation, projects risk underfunding or overfunding (reducing competitiveness in bids).

In the following sections, we consider some key estimate validation techniques, their methodologies, applications, and examples, using the 5 MW Solar Panel Installation Project (base estimate: $4,000,000) from the WBS shown in Figure 6.2 and contingency calculations for context.

6.1.5.1 Cross-Checking with Historical Data

Perhaps the easiest validation technique is to compare the estimate with costs from similar past projects. You can adjust the past project costs for differences in scope, location, or inflation using expert judgment or indices (e.g., Consumer Price Index). This approach is great for early-stage estimates or when historical records are robust, such as in repetitive projects (e.g., solar farms and bridges).

Use this method by following these steps:

1. Gather cost data from past projects (e.g., via project management software or company records).
2. Normalize costs for inflation, scale, or technology differences.
3. Flag discrepancies if the estimate deviates significantly (e.g., >20%) from historical norms.

For example, for the 5 MW solar project ($4M estimate), a similar 4 MW project cost $3.5M two years ago. Adjusting for 5% inflation and 10% efficiency gains yields:

$$\text{Adjusted Historical Cost} = \$3{,}500{,}000 \times (1 + 0.05) \times (1 - 0.10) \times \frac{5}{4} \approx \$4{,}134{,}375$$

The $4M estimate is 4.5% higher, within a reasonable 10% range, validating it. If it were, say, $5M, further scrutiny (e.g., to identify scope creep) would be needed.

Using historical data is a quick approach and it leverages organizational experience and improves analogous estimates. But it requires historical data from similar projects, and outdated data may skew results.

6.1.5.2 Expert Judgment Review

Expert judgment validation relies on experienced engineers, project managers, or consultants to review the estimate, focusing on assumptions, scope, and risks. This approach can be used at all estimation stages, especially for complex or novel projects (e.g., first-of-its-kind infrastructure). The process follows these steps:

1. Assemble a panel of experts (internal or external).
2. Review WBS tasks, cost drivers (e.g., labor and materials), and contingencies.
3. Adjust estimates based on feedback, using tools like the Delphi method for consensus.

As an example, for the solar project, a senior engineer flags that the \$1.5M construction cost (WBS 4.1–4.4) may be low due to recent labor rate increases (20%). Recalculating the cost:

$$\text{Revised Construction Cost} = \$1,500,000 \times 1.2 = \$1,800,000$$

This raises the base estimate to \$4,300,000, prompting a contingency adjustment.

Expert judgment review captures qualitative insights and is very effective for unique projects. It is subjective, however, and depends on expert availability and the ability to identify and mitigate biases.

6.1.5.3 Benchmarking against Industry Standards

Another way to validate a cost estimate is to compare it with industry cost databases or standards for similar project types. This approach is useful with parametric or bottom-up estimates, especially in construction, energy, or manufacturing. The validation procedure is as follows:

1. access databases for unit costs,
2. compare WBS task costs to benchmarks, adjusting for location or complexity,
3. validate if within a standard range (e.g., ±15% of benchmark).

EXAMPLE 6.2 INDUSTRY BENCHMARK VALIDATION OF SOLAR FARM COST ESTIMATES

Industry data suggests per unit cost to build of \$0.8M–\$1.2M/MW for solar farms. For 5 MW:

$$\text{Benchmark Range} = 5 \times \$0.8M \text{ to } 5 \times \$1.2M = \$4,000,000 \text{ to } \$6,000,000$$

The \$4M estimate is at the low end of this range, suggesting our estimate is valid but that there is potential risk if costs trend higher. Therefore, we should adjust the contingency (e.g., 15% → 20%) accordingly to \$800,000.

Benchmarking is a great approach to validation as it is objective, leverages industry-wide data, and improves parametric accuracy. It may not reflect project-specific conditions, however, and industry financial metrics databases can be costly.

6.1.5.4 Sensitivity Analysis

You can use sensitivity analysis to validate an estimate. To do so, see how changes in key cost drivers (e.g., labor rates and material costs) impact the total estimate, identifying sensitive variables. If there is too much sensitivity, then the estimate and/or some of the original assumptions need to be challenged. Such an approach is great for validating bottom-up estimates or projects with high uncertainty (e.g., recent technology deployments).

To conduct a sensitivity analysis, use the following steps:

1. Identify critical WBS cost elements.
2. Vary each by ±10–20% and recalculate the total.
3. Adjust estimates or contingencies if sensitive variables cause significant deviations.

EXAMPLE 6.3 SENSITIVITY ANALYSIS OF PROCUREMENT COST VOLATILITY

In the solar project, we can test the sensitivity of the procurement tasks 3.1–3.3 in he WBS with a cost of $3M. Consider the effects of a 20% increase in material cost:

$$\$3,000,000 \times 1.2 = \$3,600,000.$$

This change yields a new cost estimate of:

$$\$4,000,000 - \$3,000,000 + \$3,600,000 = \$4,600,000.$$

This result suggests adding at least a $600K contingency for material price volatility.

Sensitivity analysis validation is great for a quick "back of the napkin" validation, and it quantifies risk impact and supports overall contingency planning. But sensitivity analysis requires making assumptions about sensitive variable ranges, and this method becomes very complex for projects with a large WBS.

6.1.5.5 Monte Carlo Simulation

Monte Carlo simulation uses statistical modeling to simulate thousands of cost scenarios based on probability distributions for WBS tasks, validating the estimate at a desired confidence level. This kind of validation is especially useful for complex projects with considerable risk.

The process is as follows:

1. Assign risk distributions (e.g., triangular) to task costs (min, likely, max).
2. Run simulations.
3. Validate if the estimate falls within the confidence interval.

Note that the confidence level selection (e.g., P80) should reflect organizational risk appetite.

EXAMPLE 6.4 MONTE CARLO SIMULATION FOR SOLAR PROJECT ESTIMATE VALIDATION

For the solar project, we have a $4M cost estimate. Refer to the WBS and tasks 3.1–3.3 and 4.1–4.4. Let's assume triangular risk distributions for these tasks, which is:

Procurement (WBS 3.1–3.3): min=$2.8M, likely=$3M, max=$3.5M.

Construction (WBS 4.1–4.4): min=$1.3M, likely=$1.5M, max=$1.8M.

We assume all other task costs are fixed. Simulation yields an 80% confidence interval cost of $4,500,000. This suggests that our $4M estimate is too low. There, we can adjust that estimate, say, to $4.2M with a $300,000 contingency to meet 80% confidence level.

Monte Carlo validation is a robust technique for analyzing multiple risks and is well-suited for TCO and EV estimations. However, it does require access to simulation software such as @RISK or Oracle Crystal Ball, which can be expensive, if not using open-source tools. Monte Carlo validation is also data-intensive.

6.1.5.6 Practical Considerations

As with other aspects of cost estimation, with validation, it is always advisable to combine Techniques. For example, use cross-checking and benchmarking for early validation, then sensitivity or Monte Carlo for detailed estimates. Using the right validation techniques strategically can really help avoid costly errors. For example, with the solar project, historical data and expert reviews caught underestimations early.

It's also important to use the right software tools. For example, Microsoft Project integrates WBS costs for cross-checking, @RISK or Crystal Ball runs Monte Carlo simulations, RSMeans or CostX provides benchmarks for relevant projects.

Finally, cost validation should be considered as an iterative process, not a "one and done." You should, for example, validate at each project phase (feasibility, planning, and execution) as scope and risks evolve, which is recommended in the PMBOK Guide.

6.2 BUDGETING

Budgeting is a cornerstone of effective project finance and cost analysis. It ensures that resources are allocated efficiently, financial risks are managed, and projects remain viable despite uncertainties. There are three steps to the budgeting process:

1. allocating costs to project phases or tasks,
2. monitoring actual spending, and
3. adjusting for scope changes or unexpected costs.

These three basic steps provide a robust framework for managing project budgets. Let's discuss each step further.

6.2.1 Allocate Costs to Project Phases or Tasks

Allocating costs involves distributing the total project budget across specific phases or tasks, typically defined in a work breakdown structure, to create a detailed cost baseline. This baseline serves as the foundation for tracking, monitoring, and financial decision-making, ensuring alignment with metrics like NPV and TCO. Monitoring project costs diligently can help to avoid catastrophic overruns. For example, the Channel Tunnel's 80% overrun resulted from poor cost allocation, with inadequate contingencies for geological risks [Flyvbjerg 2014]. A detailed WBS with risk-based allocations could have mitigated this problem.

6.2.1.1 Detailed Process

The detailed process for cost allocation is as follows:

1. Construct the WBS.
2. Estimate Costs: using the techniques discussed in Section 6.1.3.
3. Categorize Costs according to the following categories:
   a. Direct Costs: *labor* (e.g., electricians), materials (e.g., panels), equipment (e.g., cranes).
   b. Indirect Costs: *overhead* (e.g., office utilities), administrative expenses.
   c. Contingencies: *reserves* for risks (e.g., $200,000 from EMV or percentage-based calculations).
4. Allocate these Costs:
   a. Assign costs to WBS elements based on estimates and risk exposure.
   b. Distribute contingencies proportionally or to high-risk tasks (e.g., procurement and construction).
   c. Use cost drivers (e.g., labor hours and material quantities) to ensure accuracy.
5. Validate and Baseline:
   a. Validate costs using techniques such as cross-checking historical data or benchmarking.
   b. Finalize the cost baseline, a fixed budget plan, after stakeholder approval.

EXAMPLE 6.5 ALLOCATING A VALIDATED BUDGET ACROSS WBS ELEMENTS

To illustrate the process, consider our Solar Farm Project. Suppose the board of directors has approved a $4.2M budget. The budget is allocated across the WBS as follows:

WBS 1.: Project Management: $60,000 ($50,000 base + $10,000 contingency; covers salaries, EVM software).

WBS 2.: Design & Engineering: $120,000 ($100,000 base + $20,000; includes CAD tools, permits).
WBS 3.: Procurement: $2,420,000 ($2.3M base for solar panels +$120K for inverters, etc.).
WBS 4.: Construction: $1,545,000 ($1.5M base + $45,000; labor, cranes).
WBS 5.: Testing and Commissioning: $55,000 ($50,000 base + $5,000; testing equipment).

The logic for these allocations is as follows. Procurement costs dominate due to high material costs (e.g., solar panels ~$2M). Contingency is weighted toward procurement and construction (≈ 80% of $200,000) due to risks such as material price hikes and weather delays.

To validate these cost allocations, we can cross-check with industry benchmark (e.g., $0.8M–$1.2M/MW, yielding $4M–$6M for 5 MW) and expert review confirming labor rates (e.g., $40/hour).

Together, these allocations form a validated cost baseline that seamlessly sets the stage for monitoring actual spending against the budget.

6.2.1.2 Cost Allocation Best Practices

It is important to document and communicate your decisions for cost allocations religiously. Record allocations in a budget document and share this with stakeholders (e.g., client, contractors) to align expectations. You should also pay attention to the granularity of your decisions. That is, allocate at the work package level (e.g., WBS 3.1 for panels in the Solar Project) for precise tracking.

Assign larger contingencies to high-risk tasks (e.g., $120,000 to procurement for supply chain volatility). You should maintain stakeholder engagement throughout this process to secure client approval, avoiding disputes like those in the Scottish Parliament Building (980% overrun) [Flyvbjerg 2014].

Finally, software tools are usually required for the cost allocation activity. For example, tools such as Project and Primavera can be used to link costs to WBS tasks, assign resources, and generate cost baselines. Other tools such as CostX or RSMeans provide cost data for materials, labor, and equipment to refine estimates. SAP or Oracle Financials integrate cost allocations with enterprise financial systems. You can even use Microsoft Excel or another spreadsheet program to build spreadsheets for manual cost allocation and contingency tracking.

6.2.2 Monitoring Actual Spending against the Budget

Monitoring compares actual expenditures to the cost baseline to detect deviations, assess performance, and prevent overruns. Earned Value Management (EVM) is a powerful tool that integrates cost, schedule, and scope to provide real-time insights into project health. The Berlin Brandenburg Airport's 257% overrun [Flyvbjerg

2014] lacked robust EVM, missing early detection of technical cost overruns. Regular EVM could have flagged issues like fire safety system costs.

6.2.2.1 Metrics for Monitoring Spending Using EVM

Earned Value Management uses the following metrics:

Planned Value (PV): Budgeted cost of work scheduled up to a given time, based on the WBS baseline.
Earned Value (EV): Budgeted cost of work completed, measured by task completion percentages.
Actual Cost (AC): Actual expenditures incurred, tracked via invoices, payroll, or accounting systems.

I will show you how to use these via an example shortly. Other Key Performance Indicators that are used in EVM include:

Cost Variance (CV):

$$CV = EV - AC \tag{6.3}$$

A positive CV indicates that the project is under budget; a negative CV means the project is over budget.

Schedule Variance (SV):

$$SV = EV - PV \tag{6.4}$$

A positive SV means the project is ahead of schedule; a negative SV means the project is behind schedule.

Cost Performance Index[1] (CPI):

$$CPI = EV/AC \tag{6.5}$$

If the CPI is > 1, then the project is cost-efficient; if the CPI <1 then the project is over budget and adjustments to the project scope, staffing or budget need to be considered.

Another important indicator for EVM is the Schedule Performance Index (SPI):

$$SPI = EV/PV \tag{6.6}$$

If the SPI is >1, then the project is ahead of schedule. If the SPI is <1, then the project is behind schedule, and adjustments to the project scope, schedule or staffing need to be considered.

Finally, the Estimate at Completion (EAC) metric forecasts the final cost of the budget. It is calculated as:

$$EAC = \mathrm{BAC}/CPI \tag{6.7}$$

where BAC is the *budget at completion* and it represents the total approved budget for completing all project work. Of course, if the EAC is significantly over budget, then changes in scope, staffing, schedule, or budget need to be considered.

6.2.2.2 Detailed EVM Process

Let's look at the detailed process for monitoring actual versus budgeted spending using EVM. The steps are:

1. Establish EVM Metrics:
2. Collect data:
   a. Track actual costs via accounting software.
   b. Measure progress using WBS completion metrics.
   c. Update EVM metrics weekly or monthly, depending on project pace.
3. Analyze and Report:
   a. Calculate CV, SV, CPI, and SPI to identify issues.
   b. Generate EVM reports with dashboards or graphs for stakeholders.
   c. Use contingency reserves.
4. Take Corrective Actions: If CPI < 1 or SV < 0, investigate causes.

To see the process in action using the EVM metrics, consider the following example.

EXAMPLE 6.6 APPLYING EARNED VALUE MANAGEMENT TO MONITOR PROJECT PERFORMANCE

For the solar project and a \$4.2M budget, let's assume at 6 weeks (50% timeline of the 12-week project), that the following project data is determined:

PV: \$2,100,000 (50% of \$4,200,000)
EV: \$1,900,000 as determined by the project team based on work completed
AC: \$2,300,000 determined from timesheets, materials receipts, etc.

Now let's calculate the other EVM metrics. That is,

$$CV = EV - AC = \$1,900,000 - \$2,300,000 = -\$400,000.$$

This indicates that the project is over budget by \$400,000:

$$CPI = \frac{EV}{AC} = \frac{\$1,900,000}{\$2,300,000} = 0.83.$$

Since PI is <1, a cost overrun is detected. Next,

$$SV = EV - PV = \$1,900,000 - \$2,100,000 = -\$200,000.$$

This means the project is behind schedule. Next,

$$SPI = EV/PV = \$1,900,000/\$2,100,000 = 0.90.$$

Since the value is <1, it means the project is delayed. Finally, we calculate

$$EAC = \frac{BAC}{CPI} = \frac{\$4,200,000}{0.83} \approx \$5,060,241.$$

This EAC represents a forecasted overrun of $860,241.

In response to this data, the project manager decides to take the following actions:

- Use $150,000 of procurement contingency ($120,000 allocated) to cover material cost increase.
- Investigate delays (e.g., slower panel delivery) and reallocate labor to accelerate construction.
- Report these findings to appropriate stakeholders, showing CPI/SPI trends.

Of course, the calculation of these metrics is not a one-time event. Using appropriate tools, the project manager can monitor these on a daily, weekly or monthly basis, with the time interval dependent on the duration and volatility of the project. That is, long-term and relatively stable projects can be monitored via these metrics less frequently, while short-duration and volatile projects need to be monitored more frequently, even daily.

6.2.2.3 Best Practices and Tools for EVM

To wrap up our discussion on project monitoring, let's look at some best practices and tools that can really help make EVM a practical, everyday part of project management practice.

First, you should conduct granular tracking, that is, monitor the project at the WBS work package level (e.g., WBS 3.1 for panels) to isolate issues. Focusing on the lowest levels of the WBS leads to more tactical decisions and solutions, while higher levels may be too abstract for direct project interventions.

You should provide regular updates to key stakeholders based on the metrics tracks. For example, give weekly EVM updates for high-risk phases (e.g., construction) and less frequent, say, monthly updates for more stable phases (e.g., testing). Use these updates to align with teaming decisions, ensuring partners (e.g., suppliers) address variances.

Be sure to carefully manage the contingency budget so that you don't spend it all in panic early in the project. For example, draw from a $200,000 reserve only for validated risks (e.g., material costs), preserving funds for later phases.

Tools such as Microsoft Project or Oracle Primavera are recommended to automate EVM calculations, track PV/EV/AC, and visualize variances. For larger projects, you may wish to consider tools that are specially designed for EVM, such as Deltek Cobra. Tools such as Power BI or Tableau can be used to create interactive EVM dashboards for stakeholder reporting. But you can use inexpensive options too, such as Microsoft Excel (or equivalents) for manual EVM tracking—you can create custom formulas and charts as needed.

6.2.3 Adjusting for Scope Changes or Unexpected Costs

The last step in the budgeting process is to adjust the scope as needed. This step has been mentioned already, for example, when EVM calculations indicate that the project is over budget or significantly under budget. Changing the scope of a project once it has begun, for whatever reason, is a significant event, however, and should not be undertaken lightly. Your company should have a formal *change control board* (CCB). Running all scope change decisions through the change control board prevents scope creep, fosters good stakeholder relationships, and can avoid costly problems and lawsuits.

A CCB is an organized group of stakeholders responsible for reviewing, evaluating, and approving or rejecting proposed changes to a project. The National Programme for IT's £10B overrun resulted from unadjusted budgets for scope creep [Flyvbjerg 2014]. A formal change control process with EVM could have mitigated losses.

Typical members of a CCB include the internal product owner, senior engineering leaders, including quality control and configuration management staff, key financial personnel, and customer representatives. In many cases, government or regulatory representatives may be involved. But the actual composition of the CCB varies greatly across industries and companies.

Budget adjustments are necessary when scope changes (e.g., client-driven additions) or unexpected costs (e.g., risk events) arise. This step ensures the budget reflects current project realities, maintaining financial viability and supporting decisions like lease-versus-buy or computing TCO (to be discussed later).

6.2.3.1 Scope Change Process

Scope changes rarely arrive with much warning, and when they do, they often bring along a host of financial and scheduling implications. The key is to approach them methodically using the following steps:

The steps for adjusting for scope changes are as follows:

1. Identify Triggers
2. Assess Impact
3. Adjust Budget
4. Implement Change Control
5. Communicate and Approve

When a new requirement surfaces or an unexpected cost emerges, the first task is simply recognizing that a change has occurred. This may be because of a client-driven feature addition or a risk event that wasn't fully anticipated.

Once identified, the project team must assess the ripple effects. That is, how the change alters cost, how it affects the schedule, and what adjustments are required within the work breakdown structure. This analysis also includes revisiting financial metrics such as NPV, IRR, and break-even thresholds to ensure the project still makes sense economically.

With impacts understood, the next step is to determine how to fund the adjustment. Minor issues may be absorbed through contingency reserves, but larger scope changes typically require tapping into management reserves or seeking client approval for additional funding.

Only after the budget is updated and redistributed across the affected WBS elements should the formal change-control machinery begin. The adjustment is documented, submitted to the Change Control Board for review, and, upon approval, integrated into the project plans.

Finally, the revised scope, budget, and schedule are communicated to stakeholders to maintain alignment and prevent misunderstandings down the road. This disciplined process keeps scope changes from derailing project objectives and preserves financial and operational integrity.

EXAMPLE 6.7 EVALUATING THE FINANCIAL AND SCHEDULE IMPACTS OF A MAJOR SCOPE CHANGE

Let's see the scope change process in action using the Solar Project. Suppose at the halfway point in the project, the customer requests the addition of a $600,000 battery storage system that costs $450,000 and has an installation cost of $150,000. The scope change also extends the project timeline by three weeks. During the project, weather delays contributed to unexpected costs of $100,000 due to labor and equipment overrun. What is the impact of these changes?

The battery adds $600,000 to the budget, requires a new WBS task, and delays testing. The weather delay consumes $100,000 of the $165,000 construction contingency (from $200,000 total). These impacts require a recalculation of the EVM metrics. That is:

EVM update (at 50% completion):

New PV: $2,000,000 + $600,000 = $2,700,000.
EV: $2,000,000 (unchanged, battery task not started).
AC: $2,400,000 (includes $100,000 weather overrun).

$$CV = \$2,000,000 - \$2,400,000 = -\$400,000$$

$$EAC = (\$4,200,000 + \$600,000) / 0.83 \approx \$5,783,133$$

The data suggest that a budget adjustment is needed:
New Budget:

$$\$4,200,000 + \$600,000 = \$4,800,000$$

The following additional actions need to be taken:

Use \$100,000 contingency for weather delay, leaving \$100,000.
Fund battery addition via client approval (management reserve).
Update WBS:

- WBS 4.5: Battery Installation (\$600,000).
- Redistribute remaining \$100,000 contingency (\$75,000 to WBS 4.0, \$25,000 to WBS 4.5).

Next, we can calculate the change in the NPV of the project based on this scope change, assuming a 5% discount rate, \$1M annual cash flows for 5 years, and the battery adds \$100,000/year in planned value.

Original NPV:

$$\sum_{t=1}^{5} \frac{1,000,000}{(1.05)^t} - 4,200,000 \approx 121,000$$

New NPV:

$$\sum_{t=1}^{5} \frac{1,100,000}{(1.05)^t} - 4,800,000 \approx 33,000$$

Thus, the project remains viable but less attractive (\$33K NPV vs. \$121K). Therefore, it might be appropriate to negotiate cost-sharing with the customer.

With this information at hand, we submit a change request for the battery addition to the Change Control Board. Upon approval of the CCB and client (for the revised NPV), the scope change is finalized.

Note: for simplicity in this scope-change vignette, we assumed linear PV with time (i.e., 50% elapsed ⇒ 50% of BAC), and then we added the new \$600k scope to PV. In practice, we would keep the same PV basis (time-phased plan vs. WBS-specific scheduled completions) across reporting periods.

By grounding every scope adjustment in a disciplined, transparent process, we ensure the project remains financially sound and ready to move into the detailed example that follows.

6.2.3.2 Scope Change Best Practices and Tools

Scope changes are a serious matter as they can have a significant impact on cost, schedule and even the company's reputation. Therefore, it is important to adhere to the following principles.

Do not use the contingency budget for scope changes. Contingency budgets should be used to deal with risks (e.g., weather), not scope changes, to maintain TCO estimate accuracy. In fact, for any scope changes, recalculate all relevant metrics, such as TCO, NPV, and IRR, to ensure alignment with cost–benefit trade-offs.

Finally, be sure to reconcile and document scope changes with all stakeholders to avoid disputes later.

Various commercial and open-source tools can help with scope management. For example, use Jira or Primavera to track scope changes and update the WBS. Microsoft Excel to recalculate budgets, EVM metrics, and NPV. Microsoft Project or Primavera can be used for WBS allocation, EVM tracking, and change management. For larger, enterprise-level budgets use SAP or Oracle. Collaboration tools like SharePoint are useful to manage stakeholder approvals and change logs.

6.3 BREAK-EVEN ANALYSIS AND COST–BENEFIT TRADE-OFFS

A *break-even analysis* determines the point at which project or investment costs equal the associated benefits, including revenue and reduction of other costs, helping decision makers to assess the viability and risk of some endeavor. The break-even analysis formula is:

$$\text{Break-Even Point (units)} = \frac{\text{Fixed Costs}}{\text{Price per Unit} - \text{Variable Cost per Unit}} \tag{6.8}$$

where

Fixed Costs: are the costs independent of output (e.g., equipment purchase).

Variable Costs: are the costs that vary with output (e.g., materials).

Price per unit is the revenue per unit sold.

EXAMPLE 6.8 BREAK-EVEN ANALYSIS FOR CAPITAL EQUIPMENT INVESTMENT

The senior managers of a manufacturing plant invest $200,000 in a machine to produce a custom stainless-steel flange. Production costs are $50/unit, and the flange sells for $100/unit. Break-even occurs at:

$$\textit{break-even} = \frac{\$200,000}{\$100 - \$50} = 4,000 \textit{ units}$$

Therefore, the company must sell 4,000 units to recover the investment in the machine. Does this mean the company should buy the machine? We cannot know for sure, as there are other factors to consider. For example, the company may feel it cannot sell 4,000 units in some period, or at all. There may be tax implications (both negative and positive) to purchasing the machine. The company may wish to invest the money in a different opportunity with a higher IRR. Whatever the case, the purchase decision should not be based solely on the break-even analysis. A more sophisticated cost–benefit analysis is required.

A *cost–benefit trade-off* analysis is used to evaluate whether benefits justify costs. That is, one compares the tangible benefits of some decisions (e.g., revenue and cost savings) and intangible benefits (e.g., improved safety) against costs.

For example, with the flange machine, the break-even point was 4,000 units. But what if some of those 4,000 units were sold instead of another of the company's products (reducing the sales of the latter)? What if the investment could have been used to pay down some company debt? Is there space in the factory for the machine? There are so many variables here that making the decision simply based on the break-even formula is very naïve.

The form of the cost–benefit trade-off analysis varies differently for every situation. In addition, you would need to use NPV or payback period calculations to quantify long-term value in the cost–benefit analysis (see Chapter 1).

6.4 TOTAL COST OF OWNERSHIP

Another important consideration for purchasing major equipment or software is the *total cost of ownership (TCO)*. The TCO includes all costs over an asset's lifecycle such as purchase price, maintenance, repairs, supplies, insurance, and disposal. But there can be many other contributors to the TCO depending on the type of asset. Engineers and project managers must consider the TCO for informed decision-making.

To compute the TCO, the general formula is:

$$TCO = purchase\ cost + \sum_{t=1}^{n} \frac{operating\ costs_t + maintenance\ costs_t}{(1+r)^t} \frac{+insurance\ costs_t}{(1+r)^t} + disposal\ costs \quad (6.9)$$

where n is the number of years the machine or software is in operation before disposal and r is the discount rate. Note that the disposal cost could be negative, if, for example, the used asset has a positive value and can be sold.

EXAMPLE 6.9 TOTAL COST OF OWNERSHIP CALCULATION FOR A CNC MACHINE

Here is an example of TCO calculation. Suppose a manufacturing firm wishes to buy a CNC machine that costs \$150,000, and has an operating cost of \$50,000 per year, with \$10,000/year maintenance, \$2,000/year insurance, and \$5,000 disposal cost after 10 years. Assuming the discount rate is 3%, then the TCO is:

$$TCO = \$150{,}000 + \sum_{t=1}^{10} \frac{\$50{,}000 + \$10{,}000 + \$2{,}000}{(1+3\%)^t} + \$5{,}000 = \$683{,}873$$

Note that the \$5,000 is assumed to have already been discounted through year 10.

EXAMPLE 6.10 TOTAL COST OF OWNERSHIP ANALYSIS FOR CAD SOFTWARE LICENSING

Here is another example. A CAD software license costs $20,000 upfront, then $5,000/year for updates and renewals, and $1,000/year for training. There is no disposal cost or residual value of the software at the end of its useful life, and no insurance cost. We recast the updates, renewals and annual training costs as operating expenses. Suppose the discount rate is 2.5% and we expect to use this software for only five years. Then the TCO is

$$\text{TCO} = \$20{,}000 + \sum_{t=1}^{5} \frac{\$6{,}000}{(1+2.5\%)^t} \approx \$47{,}875$$

A TCO analysis is useful in any decision-making situation, such as break-even analysis or cost–benefit analysis, as a rigorous quantification of one of several alternatives. Then an IRR analysis can be done to choose between these alternatives.

6.5 CAPITALIZATION VERSUS EXPENSING

Now, let us turn our attention to a frequent problem in engineering projects, that is, an early financial budgeting decision on whether to capitalize or expense certain costs. Capitalization applies to assets with long-term benefits and lifespans (e.g., equipment and high-value software). Costs are amortized or depreciated over time. Expensing applies to short-term costs (e.g., repairs and supplies) and is fully deducted in the year incurred.

For example, a $100,000 machine with a 10-year life could be capitalized and depreciated at $10,000/year. A $50,000 circuit board layout optimization software package might be purchased and capitalized over 5 years at $10,000 per year, while a $2,000 software license that requires annual renewal would be treated as an expense.

Treating any equipment or software purchase as an expense or a capital acquisition affects financial statements and has important tax implications. Capitalization spreads costs over an asset's useful life, and the value of the asset is entered on the company's balance sheet. Expensing records costs immediately on the income statement. But beware, in many cases, tax law dictates what can be capitalized and what can be expensed (and we are not going to get into tax policy).

6.6 MAKE-VERSUS-BUY DECISIONS

Make-versus-buy decisions determine whether to produce a component in-house or purchase it externally. Factors include cost, quality, capacity, and alignment.

There are several factors on which to make this decision. First is based on cost. You should compare in-house production costs (labor, materials, and overhead) to purchase costs (price, shipping, and taxes). The second is based on qualitative factors. That is, consider control over quality, intellectual property, and supply chain reliability.

Finally, you can compute the TCO for the make and buy decisions by including such factors as maintenance, training, and scalability. That is:

$$\text{total cost}_{\text{make}} = \text{fixed costs} + \left(\text{variable cost per unit} \cdot \text{quantity}\right) \quad (6.10)$$

$$\text{total cost}_{\text{buy}} = (\text{purchase price per unit} + \text{additional costs} \cdot \text{quantity} \quad (6.11)$$

EXAMPLE 6.11 MAKE-VERSUS-BUY COST COMPARISON FOR MANUFACTURING COMPONENTS

Suppose a machine manufacturing company needs 1,000 units of a specialized fastener. It can either manufacture the part in-house or purchase it from another company that specializes in fasteners. Manufacturing the 1,000 units in-house costs: \$50,000 (fixed) + \$20/unit (variable). Therefore:

$$\textit{total cost}_{make} = \$50{,}000 + \left(\$20 \cdot 1{,}000\right) = \$70{,}000$$

Conversely, the buying cost is \$30/unit plus \$5,000 for bulk shipping. Adapting equation (6.11) slightly because the shipping costs were given in bulk, not per unit, yields:

$$\textit{total cost}_{buy} = \$30 \cdot 1{,}000 + \$5{,}000 = \textit{\$35,000.}$$

In this example, buying is cheaper, but other factors such as intellectual property rights exposure, quality or delivery risks must also be evaluated in making the final decision.

With a clear understanding of how make-versus-buy decisions shape both cost and capability, we can now turn to the equally strategic question of whether major assets should be leased or purchased.

6.7 LEASE VERSUS BUY[2]

Many engineering and technical projects require expensive equipment such as large tower cranes, high voltage test sets, industrial 3D printers, mass spectrometry systems, and high-performance computing clusters. These devices range from \$100s of thousands to a million or more. For example, supercomputers are very expensive, and few companies own them, choosing to lease them (or the computational time) as project needs demand. But other companies make strategic decisions to avoid leasing altogether and may own one or more supercomputers (even if they sit idle some of the time).

Purchasing an expensive piece of equipment for a single (even a high-value) project might not make sense, but leasing the equipment can carry significant costs too. Therefore, a cost–benefit analysis is required to determine whether to buy or lease any major piece of equipment if it is not already owned (or owned, but unavailable because it is assigned to another project).

For lease versus buy decisions, the questions that must be considered are project, opportunity, and financially based. Some of the questions to be considered include those related to leasing versus buying decisions that apply to capital equipment or software used infrequently or for specific projects. Leasing avoids large upfront costs but may have higher TCO.

We can recast the standard TCO formula (6.9) for leasing as follows

$$\text{TCO}_{\text{lease}} = \sum_{t=1}^{n} \frac{\text{lease payment}_t}{(1+r)^t} + \textit{additional costs} \tag{6.12}$$

Additional costs include maintenance, insurance, and disposal for buying include lease terms and residual value for leasing. We can use the TCO for a lease comparison of lease payments (discounted to present value) versus the purchase TCO. Other considerations include usage: infrequent use (e.g., cranes and specialized scientific equipment) favors leasing and flexibility—leasing allows upgrades or termination, while buying locks in ownership.

EXAMPLE 6.12 EVALUATING LEASE VERSUS BUY FOR A TOWER CRANE

A construction firm needs a tower crane for a two-year project. Buying costs \$600,000, with \$25,000/year maintenance, \$5,000/year insurance, and \$15,000 disposal cost. Leasing costs \$80,000/year. The firm uses a 6% discount rate.

Calculating the TCO for buying the equipment yields:

$$\text{TCO}_{\text{buy}} = \$600{,}000 + \frac{\$25{,}000 + \$5{,}000}{1.06} + \frac{\$25{,}000 + \$5{,}000}{1.06^2} + \frac{\$15{,}000}{1.06^2} \approx \$668{,}352$$

The TCO for leasing the equipment is:

$$\text{TCO}_{\text{lease}} = \frac{\$80{,}000}{1.06} + \frac{\$80{,}000}{1.06^2} \approx \$146{,}671$$

The results of these calculations show that leasing is cheaper for the two-year project, especially given infrequent use. However, if the firm anticipates frequent crane use, buying may be justified.

There are many real-world factors to consider in the lease versus buy decision. The following questions can help make such a decision:

If purchasing, how will it be financed?

- What is the lead time for delivering the equipment? Will it be a cash purchase or paid via an equipment loan?

- What are the terms for the loan (are they favorable, given your company's credit situation)?
- Could the cash for purchase be used for other, more important needs?
- Can the company lease the equipment to others, even competitors, when not needed?
- What is the total cost of ownership (energy, maintenance, repair, insurance, and costs)?
- Is there a favorable resale value for the used equipment?

If leasing,

- What will be the availability of the equipment (certain leased equipment can be scarce at key times)?
- Are the leasing terms favorable, given your company's credit profile?
- What kind of support does the leasing firm provide (e.g., training and maintenance)?
- General questions:
  - What is the required skill set to use the equipment?
  - Are these skill sets already available in-house, or can someone be hired or trained, or will temporary employees be brought in for the duration of the project?
  - Will this piece of equipment be for one-time use versus ongoing use going forward?
  - What kind of downtime will this equipment experience?
  - What are the tax advantages of depreciating the equipment, versus recording the leasing costs as an expense?

The buy versus lease decision is a very complicated one that depends on the situation. Be very thorough in making such a decision because omitting an important cost consideration can be disastrous. Also, in the absence of all the facts, don't be quick to judge those who choose to buy versus lease (or vice versa).

6.8 EXERCISES

6.1 Create a context diagram for a 3-MW wind turbine installation project. Identify at least six external entities, the system boundary, and briefly explain how the diagram can help prevent scope creep.

6.2 Develop a three-level WBS for an industrial water-treatment plant upgrade. Ensure the decomposition follows the rules given in the chapter (no mixing phases/components, appropriate task size, etc.).

6.3 Select one task from your WBS in Exercise 6.2 and create a complete work package, including duration, resources, risks, predecessor/successor tasks, and objective acceptance criteria.

6.4 A 10-mile rural pipeline project completed three years ago cost $7M. For a new 12-mile pipeline:
   a. Increase for inflation by 4%/year,
   b. Increase cost by 15% due to rockier soil conditions.

   Compute the analogous estimate and briefly discuss the method's limitations.

6.5 A data-center cooling contractor uses a parametric model of \$150 per sq. ft. of cooled floor area. For a 22,000-sq-ft facility:
  a. Estimate base cost,
  b. Apply a 12% complexity factor,
  c. Then state one benefit and one limitation of parametric estimation.

6.6 Using a small WBS you design (four to six leaf nodes), assign realistic cost estimates to each leaf. Roll up the total cost from the bottom up and state two advantages of bottom-up estimation.

6.7 Choose an engineering project of your own (e.g., microgrid, drone platform, and software tool). Explain how you would sequence analogous → parametric → bottom-up estimation across the project lifecycle and why each stage is appropriate.

6.8 A complex robotics deployment project has a base cost of \$8M. Industry benchmarks suggest a 20% contingency. Compute the contingency, total budget, and describe when a percentage-based approach is most appropriate.

6.9 A project has the following risks:
  a. 40% chance of a \$300,000 supply chain delay
  b. 15% chance of a \$120,000 design rework
  c. 10% chance of a \$70,000 site hazard remediation

  Compute the EMV contingency and relate it to when risk-based methods are preferred.

6.10 Given a project with a \$9M base estimate and triangular cost distributions for the major cost drivers, suppose the simulation shows an 85% confidence cost of \$10.4M. Compute the required contingency and explain why Monte Carlo methods are suited for high-uncertainty projects.

6.11 A new substation upgrade project is estimated at \$6M. A similar project two years ago cost \$5.2M. Adjust the historical cost using 5% annual inflation and evaluate whether the new estimate is within a reasonable range.

6.12 Create a scenario where an expert challenges one of your cost drivers (e.g., labor, electronics, and permitting). Revise the estimate accordingly and explain how expert review compensates for model limitations.

6.13 Use an industry benchmark (choose any reasonable published cost per unit, e.g., \$/MW, \$/sq ft, and \$/mile). Compare your estimated cost for a project to benchmark ranges and determine whether contingencies should rise or fall.

6.14 Given a procurement package of \$2.5M, analyze the effects of a ±20% price fluctuation.

  Recompute the total project estimate and discuss whether additional contingency is warranted.

6.15 For a project with a total validated budget of \$5.5M, allocate costs to five WBS elements and justify the distribution using cost drivers (labor hours, material intensity, risk exposure).

6.16 At week 8 of a 16-week project, the following metrics are available:
  a. PV = \$1.8M
  b. EV = \$1.5M

c. AC = $1.9M

Compute CV, SV, CPI, SPI, and EAC and describe what actions a project manager should take based on these.

6.17 Create a scenario where a new customer requirement adds $400,000 of equipment and extends the project schedule by 2 weeks. Determine how PV, EV, AC, and EAC must be updated, referencing a change control board.

6.18 Create a WBS, estimate costs, and apply a 15% contingency for the following:
   a. $100,000 pedestrian river bridge
   b. $1M custom agentic AI system
   c. $15M robotic forklift fleet

6.19 For the solar farm WBS, compute a risk-based contingency using:
   a. 30% × $200,000
   b. 15% × $80,000
   c. 10% × $50,000

   Compare it to a 15% percentage-based contingency and discuss NPV impact.

6.20 Validate the $4M estimate using:
   a. Benchmarking at $1M/MW
   b. Sensitivity with 15% procurement cost increase

   Recalculate contingency and NPV (5% discount rate, $1M/year for 5 years).

NOTES

1 Not to be confused with Consumer Price Index (CPI) which is a key economic indicator that measures the average change over time in the prices paid by urban consumers for a market basket of goods and services.

2 Some of this section has been adapted from Laplante and Kassab (2022) with permission.

REFERENCES

Flyvbjerg, B. (2014). What you should know about megaprojects and why: An overview. *Project Management Journal*, *45*(2), 6–19. https://doi.org/10.1002/pmj.21409

Laplante, P. A., & Kassab, M. (2022). *Requirements engineering for software and systems* (4th ed.). CRC Press/Auerbach Publications (Taylor & Francis).

Project Management Institute. (2021). *A guide to the project management body of knowledge (PMBOK® Guide)* (7th ed.). Project Management Institute.

7 Corporate Finance and Decision-Making

Corporate finance provides a structure that enables firms to plan, fund, and scale their operations. It links strategic goals to measurable financial plans through budgeting and forecasting, and helps organizations

- allocate resources,
- anticipate cash needs, and
- assess the feasibility of major technical initiatives.

These practices form the backbone of effective project execution and long-term planning, which is one theme of this chapter.

Building on this foundation, firms must make disciplined capital-allocation decisions; choosing when and where to invest in technology, facilities, intellectual property, or acquisitions. Therefore, this chapter also introduces corporate capital gains taxation, explaining how gains are taxed. It further explores how mergers, acquisitions, and partnerships shape the growth trajectories of engineering-focused companies, including unconventional moves in open-source ecosystems.

The chapter concludes with financing pathways for both established firms and startups, ranging from commercial paper and private equity to crowdfunding and early-stage personal-network funding. It highlights the critical role of nondilutive grant programs, such as SBIR/STTR and international equivalents, in helping technology-oriented companies bridge the gap between research and commercialization.

This collection of topics offers a comprehensive framework for understanding financial decision-making across budgeting, taxation, capital raising, and innovation funding. It is a highly complex and technical chapter, however, and you may find yourself referring to previous chapters and the glossary. So take it slow.

7.1 BUDGETING AND FORECASTING IN ENGINEERING FIRMS

Budgeting and forecasting form the foundation of financial discipline in EST organizations. Because these firms operate within long development cycles, high technical uncertainty, and capital-intensive workflows, they require financial systems that can translate strategic intent into executable plans. Effective budgeting aligns resources with engineering priorities, while forecasting provides forward visibility into feasibility, cash needs, and milestone readiness. Together, they ensure that innovation efforts are supported by the financial structure necessary to advance prototypes, manage risk, and prepare for commercialization [Berk and DeMarzo 2020].

DOI: 10.1201/9781003733850-7

7.1.1 Strategic Planning to Financial Plans

Translating strategic goals into measurable operating and financial plans is an important aspect of C-level oversight. Planning horizons (near-term execution vs. long-term roadmaps) and engineering program alignment should be conducted on a regular basis (e.g., annually and semi-annually) by senior management in conjunction with corporate directors.

These strategic plans provide the basis for corporate operational plans and then engineering programs. These programs require budgets, which are then used to forecast revenue and growth (Figure 7.1).

One widely used technique for providing input to this planning process is a SWOT analysis.

A *Strengths, Weaknesses, Opportunities, and Threats (SWOT) analysis* is a study of internal and external factors and positive and negative forces and is a standard tool used in strategic planning. The SWOT analysis framework can be illustrated by a two-dimensional array (Table 7.1).

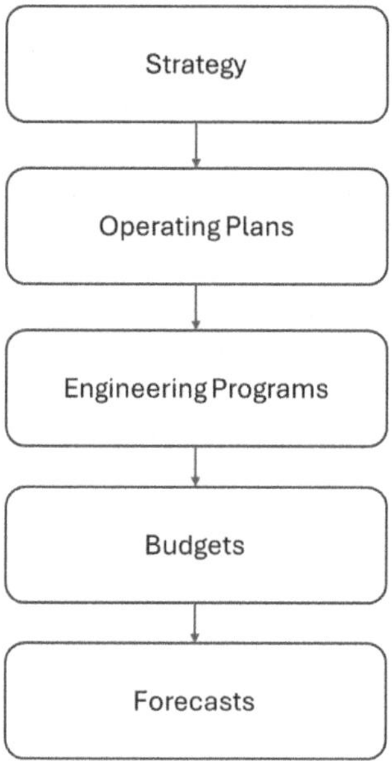

FIGURE 7.1 Strategic planning to financial.

TABLE 7.1
SWOT Analysis Forces and Factors

	Positive Force	Negative Force
External Factor	Opportunities	Threats
Internal Factor	Strengths	Weaknesses

Opportunities are based on positive external forces and threats upon negative ones. Strengths are based on positive internal forces and weaknesses on negative ones.

While a traditional SWOT looks broadly at internal and external forces, an EST-focused financial SWOT emphasizes factors that shape a firm's ability to fund innovation, manage cash-intensive development cycles, and scale technology to market. A financial SWOT analysis for EST companies applies the classic SWOT analysis to the unique financial structure and operating environment of technical organizations.

From a **strengths** perspective, many EST companies benefit from high-value intellectual property, skilled technical teams, and differentiated R&D capabilities that translate into competitive financial advantages. Strong patent portfolios, proprietary algorithms, specialized facilities, or established industry partnerships can enhance a company's valuation and help secure capital on favorable terms. Internally, these strengths often improve access to grants, venture financing, or strategic investment.

Weaknesses, however, frequently arise from the financial realities of technology development. Engineering programs can be expensive, highly uncertain, and slow to commercialize, creating long cash-conversion cycles. Weaknesses in this quadrant might include under-diversified revenue streams, dependence on a single flagship technology, insufficient cash runway to complete critical milestones, or limited financial controls around R&D spending. These factors can constrain the ability to attract investors or maintain liquidity during long development phases.

On the **opportunities** side, EST firms operate in rapidly evolving markets where new funding models, public-sector programs and emerging technologies can expand financial pathways. *Nondilutive grant* programs, like the US SBIR/STTR or international equivalents, offer early-stage capital without equity loss, improving financial resilience. These will be discussed shortly. Likewise, industry trends such as digital transformation, cleantech investment, or AI-driven automation can open new markets for technology firms, enabling improved revenue projections or stronger investor interest.

Finally, **threats** encompass external financial pressures that can undermine an EST company's trajectory. These include volatile capital markets, tightening venture funding cycles, rapid competitor innovation, and regulatory or geopolitical shifts that affect supply chains or commercialization pathways. High-tech industries are particularly vulnerable to economic downturns, which may reduce R&D budgets, slow procurement, and shrink investor appetite for long-horizon bets. Threats may also arise from the cost structure of high-tech operations. For example, inflation in materials, specialized labor shortages, or the expense of compliance in regulated industries.

Conducting a financial SWOT in this context involves both an environmental scan (industry trends, regulatory shifts, and capital market conditions) and internal financial analysis (cash flow health, funding mix, IP value, and cost structure). By systematically identifying internal financial drivers alongside external market forces, EST firms can prioritize investments, mitigate risks, and align financial strategy with their technical roadmap [Wheelwright and Clark 1992].

7.1.2 Budgeting Frameworks

Budgeting in EST organizations requires systematic frameworks that account for the unique financial dynamics of technical development. These dynamics include:

- long R&D horizons,
- uncertainty in milestone timing,
- specialized labor,
- equipment-intensive workflows, and
- coordination of multiple, parallel projects.

Unlike traditional service or manufacturing firms, many EST organizations must manage both *operating budgets* for day-to-day technical work and *capital budgets* that fund major infrastructure, facilities, equipment, or intellectual property (Figure 7.2).

Effective budgeting frameworks create discipline around resource allocation, align spending with engineering priorities, and provide early visibility into feasibility constraints. The timing of budget preparation also needs to be considered as a trade-off between effort and accuracy and external reporting mandates

7.1.2.1 Operating Budgets

In EST companies, an *operating budget* typically includes labor (often the largest cost driver), materials and consumables, prototype-build expenses, testing resources, and cloud computing or data-processing costs for software-heavy engineering. These budgets often integrate program-management structures, such as work packages or engineering change orders, to track expenditures at the subsystem or workstream level.

Because engineering programs are iterative, operating budgets usually allocate contingency for redesigns, re-testing, or updates to system requirements [Berk and DeMarzo 2020]. Here are some typical operating budget structure examples:

- **A semiconductor startup** budgets operating costs around design cycles (circuit development, simulation, verification), prototype builds, and

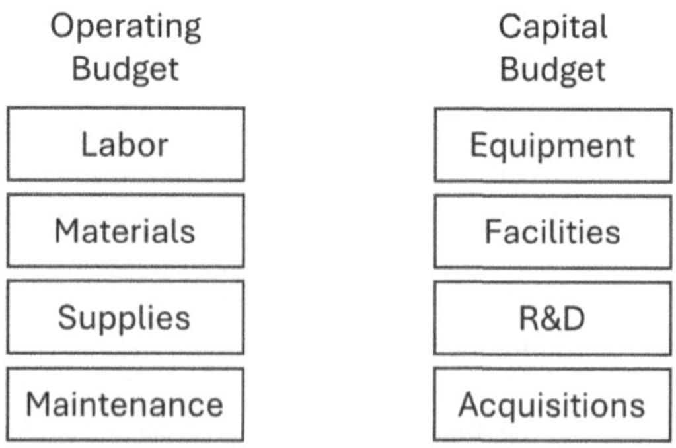

FIGURE 7.2 Operating versus capital budget structure.

external testing. A 10–15% cost overrun buffer is common due to unpredictable validation outcomes.

- **A biotech therapeutics company** allocates operating budgets across assay development, preclinical testing, lab consumables, and compliance. Because assays often require repeated optimization, budgeting frameworks include rolling-review mechanisms to recast expenses monthly or quarterly.
- **A robotics firm** budgets technical labor across hardware engineering, controls development, firmware integration, and field-testing pilots. Operating budgets may include on-site deployment teams, a cost category that grows as robots enter real-world environments.

Operational budgeting also interacts closely with financing strategy, since large programs are often funded through grants, venture dollars, or government/industry partnerships.

7.1.2.2 Capital Budgets

Capital budgets cover long-lived assets such as

- laboratory renovations,
- pilot-scale manufacturing lines,
- specialized equipment (e.g., 3D-metal printers, cryogenic freezers, and lithography tools), or
- IP acquisition.

EST companies require capital budgeting that can accommodate multi-year engineering timelines, regulatory dependencies, and lead times that may stretch six to 18 months [Damodaran 2015].

EXAMPLE 7.1 SOME CAPITAL BUDGET OUTLINES

A battery-technology company may require $10M–$30M to build a pilot manufacturing line. The capital budget includes equipment procurement, installation, facility retrofits, and yield-ramp support. Delays in equipment availability can shift the entire technology roadmap.

An aerospace firm building a propulsion test facility must budget for wind tunnels, safety systems, and custom fixtures. Because test facilities require permitting and compliance, capital budgets include regulatory timelines as embedded constraints.

A quantum-computing startup budgets for dilution refrigerators, custom cryogenic wiring, and microwave test equipment. Since many components have multi-month lead times and require specialized installation, capital budgets include phased spend plans aligned with system-bring-up milestones.

In technology-intensive organizations, capital budgeting is less about approving spending than about carefully sequencing irreversible investments to manage technical risk, lead times, and cash runway.

7.1.2.3 Budget Timelines

There are several approaches to the timing for developing budgets, both capital and operational (Table 7.2).

Traditional budgeting is conducted on an annual basis with limited revisions throughout the year. Another approach, called *rolling budgeting*, involves continuously updating the budget on a regular basis, for example, quarterly. Milestone budgeting involves preparing the budgets because of some significant event or engineering evidence (e.g., proof-of-concept demonstration and Phase 1 completion). There are also hybrid approaches that combine rolling and gate-based approaches [Berk and DeMarzo 2020].

7.1.2.4 Gatekeeping Processes

Because engineering programs evolve with new data, budgeting frameworks must also incorporate structured review and *gatekeeping processes* for the budget. Common approaches include:

- monthly or quarterly budget checkpoints,
- technical-milestone gates (i.e., a *stage-gate process*), and
- integrated portfolio reviews led by finance and engineering.

These reviews ensure resource allocations are aligned with real-time technical progress. Regular reviews also allow teams to reallocate funds from underperforming to high-priority initiatives and prevent runaway spending on programs that fail to meet risk-adjusted performance expectations.

7.1.2.5 Budget Preparation Approaches

The underlying approach to budget preparation is an important consideration as it can have a significant impact on company success. Organizations typically choose among several alternative budgeting approaches, or a combination thereof, depending on stability, uncertainty, and strategic objectives.

TABLE 7.2
Comparison of Budgeting Approaches

Approach	Description
Traditional	Fixed annual budgets with limited revision
Rolling	Continuous quarterly updates
Milestone-Based	Gates tied to engineering evidence
Hybrid	Mix of rolling + gate-based

EXAMPLE 7.2 SOME GATEKEEPING PROCESSES

A deep-tech materials company runs stage-gate reviews where each prototype generation must meet target performance metrics (e.g., tensile strength and thermal stability). Failure to meet targets can freeze additional spending until redesign plans are approved.

A medical-device firm uses FDA regulatory milestones (e.g., pre-submission, investigational Device Exemption approval, and clinical trial enrollment) as budgeting gates. Each milestone releases the next round of operating and capital budgets.

An AI-infrastructure startup conducts monthly cloud-expenditure reviews to prevent budget overruns from unexpected training workloads. Budgets are re-baselined based on model-training efficiency and inference-cost projections.

Incremental budgeting starts with the prior period's budget or actual spending and applies percentage increases or decreases for inflation, growth, or known changes. Only new or incremental expenses are justified. This approach is simple and has low administrative burden, but it can perpetuate inefficiencies as managers promote a "use it or lose it" culture.

Activity-based budgeting allocates resources based on the cost of activities required to produce expected outputs. Engineering tasks, test cycles, or production runs are identified first, and budgets are built from the activity drivers upward. It provides a strong linkage between technical work and cost drivers but requires detailed process modeling and reliable forecasting of activity volumes. It is best suited for manufacturing, operations, and engineering-heavy workflows.

Value-based budgeting evaluates each expenditure based on whether it delivers sufficient value to customers, stakeholders, or strategic objectives relative to its cost. It emphasizes strategic contribution rather than historical precedent. While it encourages alignment with long-term value creation, value judgments can be subjective and hard to quantify.

Finally, in *zero-based budgeting,* every expense must be re-justified for each budgeting period, rather than being carried forward or adjusted from prior-year spending. Each function, project, or cost center begins with a budget of "zero," and funding is allocated only to activities that can be. This is a powerful approach for cost discipline, strategic realignment, and elimination of legacy inefficiencies as it forces explicit linkage between spending and current strategy or technical goals. It also helps eliminate redundant or low-value activities and improves transparency and accountability, especially during restructuring or cash-constrained periods.

Zero-based budgeting is time- and resource-intensive, however, and can disadvantage long-horizon activities (e.g., exploratory R&D) whose value is harder to quantify annually. Because it is often impractical to apply across all functions every year, many companies use it selectively, periodically or to targeted cost areas, rather than as a universal annual process.

In practice, EST organizations rarely rely on a single budgeting method. A common hybrid approach is to:

- Use incremental budgeting for stable support functions.
- Apply zero-based budgeting selectively during cost-realignment or funding stress.
- Employ activity-based budgeting for major programs and platforms.
- Maintain rolling forecasts to stay aligned with technical milestones and uncertainty.

This blended approach balances discipline, flexibility, and strategic alignment, which is essential in environments characterized by long development cycles and capital-intensive innovation [Berk and Demarzo 2020].

7.1.3 Forecasting Methods and Assumptions

Forecasting in EST organizations requires integrating financial modeling with the realities of technical development. In service or retail firms, historical data often predicts the future with reasonable accuracy. Effective forecasting must therefore combine traditional financial tools with engineering-specific assumptions about:

- yield,
- reliability,
- schedule risk,
- regulatory timelines, and
- technology readiness [Dixit and Pindyck 1994].

These are essential considerations.

7.1.3.1 Revenue Projection

Revenue projections for EST companies must reflect adoption curves that depend on successful technical validation, regulatory approvals, and customer qualification.

EXAMPLE 7.3 SOME REVENUE FORECASTING SCENARIOS

Semiconductor startup: A fabless chip company forecasting revenue for a new AI accelerator must assume wafer yields, tape-out success, and customer design-win cycles. A one-quarter delay in achieving a functional prototype may shift revenue by 12–18 months due to long qualification windows with manufacturers.

Biotech firm: A synthetic biology company cannot forecast revenue without modeling clinical or regulatory milestones. Phase I and II trial outcomes dramatically reshape revenue projections, requiring probability-weighted forecasts.

Robotics firm: A robotics manufacturer may use pilot deployment results to estimate unit economics and customer adoption curves, building separate scenarios for success or failure of early field trials.

Because commercialization often arrives years after initial R&D investments, forecasts typically use scenario-based adoption models rather than linear growth projections [Dixit and Pindyck 1994].

7.1.3.2 Cost Forecasting

Cost forecasts for EST companies hinge on engineering program structure: materials costs, specialized labor, prototype fabrication, equipment depreciation, and testing infrastructure.

EXAMPLE 7.4 SOME TECHNICAL ASSUMPTIONS

Advanced manufacturing: A battery-technology firm must model unit costs assuming cell yield improvements over time. Initial pilot-line yields (e.g., 30–40%) are not representative of production-scale economics. Cost forecasts must incorporate expected yield ramps as processes mature.

Aerospace/defense engineering: Flight-testing costs scale with the number of test hours needed to validate performance envelopes. Forecasts must include contingency budgets for redesign cycles if early tests fail or reveal integration issues.

Software-heavy engineering companies: An AI or cybersecurity firm must forecast cloud compute expenses, which scale with model-training cycles or customer usage patterns, often becoming a major line item before revenue materializes.

Technical assumptions (e.g., expected yields, cycle time, or learning curve improvements) directly affect both R&D and future margin expectations [Wheelwright and Clark 1992].

7.1.3.3 Cash-Flow Forecasting

Cash-flow forecasting is especially critical for EST firms because they often face long periods of negative cash flow before reaching commercialization or a major funding milestone.

EXAMPLE 7.5 SOME FUNDING GATES

Deep-tech startups: A quantum computing startup may plan cash flow around discrete milestones, since each is tied to investor money releases. For example, successfully demonstrating qubit coherence times.

Government-funded innovation companies: Firms depending on SBIR/STTR grants must model reimbursement timeframes, as delays in federal disbursement can materially affect cash runway despite awards being secured.

Clean-energy hardware firms: A hydrogen-electrolyzer company may require large capital purchases (e.g., precision catalysts and industrial reactors). Cash forecasts incorporate lead times for equipment, partial prepayment requirements, and potential supply-chain disruptions.

Forecasting models typically incorporate milestone-based funding gates, grant funding timing, prototype-build schedules, and expected delays [Damodaran 2015].

7.1.3.4 Milestone-Linked Forecasting

Because technical outcomes fundamentally shape cost and revenue timing, EST firms often rely on *milestone-linked forecasting*, where financial forecasts explicitly depend on engineering progress checkpoints. These milestones often include:

- Completion of a prototype, that is, Rev A and Rev B.
- Passing regulatory or safety testing.
- Achieving target performance specifications (e.g., energy density, throughput, and accuracy).
- Securing customer pilots or letters of intent.
- Scaling production from lab → pilot → full manufacturing.

These milestones can have significant financial impact, for example, as shown in Table 7.3.

TABLE 7.3
Example Engineering Milestones and Financial Impacts

Milestone	Financial Impact
Rev A	Prototype verification, limited budget release
Rev B	Yield targets achieved, investment cash unlock
Regulatory Pre-Submission	Compliance budgets released
Pilot Readiness	Deployment funds activated

Therefore, a change in any milestone triggers an automatic reforecast of revenue timing, cash burn, and resource allocation [Cooper 1990].

7.1.4 Cash Planning and Liquidity Readiness

Cash planning and liquidity readiness are foundational to the financial stability of EST organizations, which often endure long development cycles, high upfront costs, and uncertain milestone timing. Effective cash-runway management requires building forward-looking visibility into burn rates, milestone-linked cash inflows (such as the release of a *tranche*—a portion of a total investment) and the timing of major engineering expenditures.

EST firms must maintain clear views of operating cash needs for:

- labor,
- prototypes,
- testing,
- cloud compute, and
- regulatory work,

while also modeling large, large, lump-sum cash requirements for

- equipment,
- facilities, or
- key long-lead materials.

A well-structured cash-runway view helps leadership anticipate when liquidity will tighten, supports decisions on deferring or accelerating spend, and ensures that mission-critical programs remain funded through validation, pilot deployment, and early commercialization.

Cash-runway forecasts must align with upcoming decisions on whether to commit capital to new technology initiatives, defer capital expenditures until technical evidence improves, or sequence investments to avoid liquidity compression during critical R&D phases. For example, advancing a major engineering build too early may shorten the runway and force premature fundraising, while deferring investment too long may slow competitive positioning.

Aligning liquidity planning with capital allocation requires integrating engineering milestone forecasts, cash release schedules, and sensitivity analyses that reflect technical risk. By building disciplined cash-planning systems that link real-time technical progress with financial readiness, EST organizations not only prevent unexpected shortfalls but also strengthen their ability to execute long-term innovation strategies [Damodaran 2015].

7.1.5 Scenario, Sensitivity, and Risk

Scenario planning in EST organizations provides a structured way to prepare for uncertainty in technical outcomes, market conditions, and resource availability. Because

EST firms operate with long development cycles, breakthrough-dependent milestones, and high exposure to regulatory or supply-chain variability, financial plans must incorporate multiple potential futures rather than rely on a single deterministic forecast.

Effective scenario planning develops a *base case* tied to expected engineering performance, an *upside case* for accelerated technical or commercial validation, and a *downside case* that reflects delays, failures, or external shocks to the system. These scenarios help executives evaluate how engineering risks translate into budget requirements, cash-runway implications, and shifts in capital allocation long before those risks materialize.

Sensitivity analysis complements scenario planning by examining how changes in specific technical or financial drivers affect feasibility and long-term margins. Typical drivers include unit cost, manufacturing yield, cycle time, cloud compute intensity, or schedule slip. In EST firms, even small variations in these metrics can propagate across programs. For instance, a minor reduction in forecast yield may raise prototype costs, delay revenue recognition, and compress cash runway. Meanwhile, schedule shifts can influence tranche-based financing or trigger procurement reprioritization.

By using sensitivity models to stress-test key assumptions, organizations can identify which engineering parameters have outsized financial impact and implement targeted mitigation strategies. For example, a photonics-chip company may model how a ±5% variation in coupling efficiency affects system-level performance and, ultimately, market competitiveness. Even small technical changes can cascade into major financial impact.

Together, scenario planning and sensitivity analysis form a risk-aware forecasting discipline that helps EST organizations maintain readiness, support evidence-based decision-making, and ensure that capital is deployed where it creates the greatest strategic advantage.

7.1.6 Metrics and Review Cadence

Effective budgeting and forecasting in EST organizations depend on well-chosen metrics and a disciplined review cadence that keeps financial plans synchronized with technical progress. Because EST programs evolve through iterative builds, validation cycles, and readiness gates, leadership must track a blend of technical, financial, and operational indicators that collectively signal feasibility, execution health, and emerging risks.

Technical metrics often include

- prototype pass rates,
- yield trends, reliability test results,
- cloud-compute efficiency, and milestone readiness levels.

Financial indicators such as

- run-rate accuracy,
- variance to plan,
- burn-rate trends, and
- capital-spend utilization

provide an early warning when development cycles deviate from assumptions. Operational metrics, such as

- cycle time,
- procurement lead-time stability, and
- resource-loading balance

help teams understand whether programs remain executable within schedule, staffing, and equipment constraints. Together, these metrics offer a real-time view of both engineering maturity and financial feasibility.

A structured review cadence ensures that these metrics drive timely decision-making rather than retrospective reporting. EST organizations commonly implement

- monthly financial reviews,
- engineering sprint[1] or phase-exit reviews, and
- quarterly portfolio reviews.

These reviews are conducted with staff from Finance, Engineering, Product, and Operations.

Monthly reviews prioritize short-term burn-rate tracking, contingency usage, cloud-compute guardrails and immediate procurement risks. Milestone-based reviews, exits, regulatory pre-submission readiness checks, or yield-validation gates trigger a deeper assessment of whether technical progress justifies releasing the next budget tranche or adjusting the forecast. Quarterly portfolio reviews examine multi-program resource allocation, reprioritize funding toward high-leverage initiatives, and align capital spending with organizational strategy.

By maintaining predictable review rhythms supported by objective, milestone-aligned metrics, EST organizations can course-correct early, protect runway, and ensure resources flow to programs with the strongest evidence of technical readiness and commercial potential [Damodaran 2015].

7.1.7 An Extended Case Study in Budgeting and Forecasting

Here is a hypothetical case study to tie together the discussions to this point. The details are given in Exhibits 7.1(A), (B), and (C). These are *pro forma* documents and provided to illustrate some of the elements of a budget, but all the information contained therein is simulated.

Helios Dynamics, a mid-stage engineering company, is evaluating Project Atlas, a next-generation module designed to boost system throughput by 30% through a new control algorithm and custom sensor board. Engineering estimates two prototype cycles (Rev A and Rev B) and a 12-month validation plan, while Finance requires the investment to stay within the current $5.2M allocation without reducing the runway below 12 months.

Quarterly burn (planned): Q1 $0.95M; Q2 $1.10M; Q3 $1.15M; Q4 $1.06M (front-loaded for equipment; mid-year peak during Rev B builds).

Gate releases: 30% of prototype and contingency budgets released at Rev A exit; remaining 70% at Rev B entry pending test results.

EXHIBIT 7.1(A)
Sample Engineering Program Budget (Summary View)

Category	Description	FY Budget (USD)	% of Total	Notes
Labor (Fully Loaded Cost)	8 FTE engineers (EE/ME/FW/QA), 1 PM	1,760,000	44	Avg FLC $200k/FTE
Materials & Consumables	PCBs, sensors, fasteners, adhesives, lab supplies	280,000	7	10% contingency embedded
Prototype Builds	Rev A (2 builds), Rev B (2 builds)	520,000	13	Includes non-recurring engineering for custom fixtures
Test & Validation	Environmental, EMC, HALT/HASS, field pilots	350,000	9	External lab fees + fixtures
Cloud/Compute	Training/inference, simulations, data pipelines	240,000	6	Spot/reserved mix; budget guardrails applied
Equipment & Tools (CapEx)	3D metal printer access, HIL rigs, test jigs	600,000	15	Capitalized; see CapEx schedule
Facilities & Utilities	Lab expansions, safety compliance	120,000	3	Includes calibration services
Regulatory & Compliance	Pre-submissions, audits, and documentation	90,000	2	External consultants
Travel & Deployment	Pilot site installs, customer trials	70,000	2	Heavier in Q3–Q4
Program Management	Tools, governance, training	50,000	1	Includes product lifecycle management software
Contingency (Program)	Unplanned redesigns, supply shocks	180,000	5	Released by gate approvals
Total		**4,260,000**	**100**	

EXHIBIT 7.1(B)
Cash Flow Forecast and Runway Chart for a Technical Initiative

Context: Same "Project Atlas." Opening cash: $5.2M restricted to this program. Nondilutive grant reimbursement lag: 60 days. VC tranche of $2.5M contingent on Rev B entry.

Quarterly Cash Flow (Summary)

Quarter	Cash In	Cash Out	Net Cash Flow	Ending Cash	Key Drivers
Q1	$0.30M (grant reimbursement)	$0.95M	**–$0.65M**	**$4.55M**	Equipment deposits; lab setup
Q2	$0.35M (grant)	$1.10M	**–$0.75M**	**$3.80M**	Rev A builds + external testing
Q3	$2.85M (VC tranche + grant)	$1.15M	**+$1.70M**	**$5.50M**	Tranche released at Rev B entry
Q4	$0.30M (grant)	$1.06M	**–$0.76M**	**$4.74M**	Field pilots & validation

EXHIBIT 7.1(C)

Scenario Matrix Linking Milestones to Funding Gates

Objective: Make funding releases contingent on technical evidence. Milestones are "go/no-go" gates for budget tranches, not just calendar dates.

Scenario	Milestone (Gate)	Evidence/Key Performance Indicator	Funding Released	Budget Impact	Notes/Risks
Base Case	**Rev A exit (G1)**	Functional prototype; pass 80% test cases; energy efficiency within—5% of target	Release 30% of Prototype & Contingency	Proceed to limited field pilots	Minor redesigns acceptable
	Regulatory Pre-Sub (G2)	Pre-submission accepted; no major findings	Release 50% of Compliance budget	Lock documentation & labeling	If major findings, pause release
	Rev B entry (G3)	Yield ≥ 70% on pilot build; EMC pre-check passed	VC tranche \$2.5M + 40% of CapEx	Scale test rigs; expand pilots	Slippage delays tranche
	Customer LOIs (G4)	≥ 3 LOIs covering 2 market segments	Release 50% of Deployment & PM	Prepare for limited production	Letters of intent must be non-binding but specific
Upside	**Performance Over-achieve**	+10% beyond spec; field MTBF > 2,000 hours	Pull forward 20% of Q4 spend to Q3	Accelerate market trials	Consider early purchase commitments
Downside	**Rev A Fail**	< 60% test pass; thermal runaway incidents	Freeze Contingency; redesign sprint	Slip Rev B by 1–2 sprints	Invoke vendor re-qualification
	Supply Shock	Lead time +10 weeks on key component	Reallocate to validation tasks	Defer CapEx by one quarter	Evaluate redesign to alternative BOM
	Regulatory Delay	Additional testing requested	Hold Compliance release	Extend cash runway; add buffer	Update commercialization roadmap

Planned runway: ~18 months at the current burn. If Rev B entry slips by one quarter (VC tranche delayed), runway compresses to ~13–14 months; mitigation includes deferring non-critical CapEx and tightening cloud spend guardrails.

Runway guardrails:

- Trigger a spending freeze on new prototype variants if the test failure rate > 20% in two consecutive sprints.
- Auto-scale compute to keep monthly cloud ops < 8% of total burn; force retraining windows to off-peak rates.
- Convert vendor payment terms from Net 30 → Net 60 during tranche gaps where feasible.

Tensions emerge as Engineering pushes for early access to a high-precision human-in-loop rig to accelerate the discovery of failure modes, while Finance proposes renting shared equipment to defer capital expenditure until Rev A exit. Product teams push for pilots across two verticals, whereas Operations warns this would strain deployment and cloud budgets.

By Rev A exit, Atlas-X achieves 82% test pass rate and is within—3% of the energy target, allowing Finance to release additional prototype funding for a focused redesign and a single-vertical pilot. Despite a vendor delay requiring a BOM adjustment, runway is preserved through deferred CapEx.

At Rev B entry, improved yields unlock the VC tranche tied to the milestone, enabling controlled pilot expansion while keeping a 15-month runway intact.

This extended case study illustrates how budgeting and forecasting serve as the structural backbone for engineering-driven financial decision-making. It shows how technical uncertainty, milestone-linked progress, and cross-functional negotiation shape both the timing and scale of resource allocation. Finally, the interplay between Engineering, Finance, Product, and Operations demonstrates why EST organizations require disciplined operating and capital budgets, iterative reviews, and structured gatekeeping to prevent overspend, preserve runway, and ensure that financial plans remain synchronized with evolving technical data.

7.2 CAPITAL ALLOCATION AND RESOURCE MANAGEMENT

Engineering, science, and organizations need to make evidence-based decisions about deploying limited capital across competing initiatives while balancing short-term liquidity and long-term strategic value. In this section, we look at capital allocation frameworks to manage uncertainty, sequence CapEx, and prioritize portfolios under runway constraints and milestone-linked funding.

7.2.1 Capital Gains

Capital gains are profits realized when a company sells an asset for more than its original cost. These play a subtle but important role in corporate decision-making, particularly for high-tech organizations. Because these firms often hold specialized equipment, technical infrastructure, or legacy intellectual property with fluctuating strategic value, leadership must consider not only whether an asset has outlived its operational usefulness, but also whether selling it at the right moment can strengthen the company's financial position.

The timing of realizing capital gains or losses directly affects cash availability, net taxable income, and the organization's ability to fund new initiatives without resorting to external financing. For EST firms operating under runway constraints or preparing for milestone-linked investment decisions, the thoughtful management of capital gains becomes a practical tool for optimizing liquidity while keeping long-term roadmaps intact.

In technology-intensive environments, capital gains decisions frequently intersect with broader capital allocation strategy because divesting legacy tools or non-core assets can free resources for higher-priority engineering programs. But the treatment differs by corporate structure, for example, for US C-corporations, capital gains are taxed at the ordinary corporate rate; capital losses offset only capital gains with limited carrybacks/carryforwards.

Practically, selling older equipment to support a new technical initiative can help firms reallocate capital toward areas with stronger strategic value, accelerating progress toward critical milestones such as prototype validation, yield improvement, or pilot deployment. This also allows companies to manage the financial burden of long-lived assets more effectively by reinvesting gains into capabilities that enhance competitiveness or align with updated technical readiness.

EXAMPLE 7.6 LIQUIDITY-AWARE BUDGET ADJUSTMENT AFTER A MILESTONE SLIP

An engineering firm is preparing to launch a custom precision control module and identifies several legacy test fixtures and an older lithography station that no longer align with its updated technology roadmap. Rather than continuing to maintain or store these under-utilized assets, the company chooses to sell them on the secondary equipment market, generating a modest but meaningful capital gain.

The proceeds are redirected into a new technical initiative, for example, funding early prototype builds, acquiring specialized sensors, and securing additional lab capacity needed to validate the updated design. This divest-and-reinvest approach not only frees financial resources for higher-value engineering work but also streamlines the firm's equipment base, ensuring capital is concentrated on programs with stronger strategic and technical potential.

Capital gains are not simply a tax or accounting consideration. Rather, they become an integrated part of evidence-based resource management, enabling organizations to sequence investments more efficiently and support programs that deliver the greatest technical and financial impact [Jensen 1986].

7.2.2 Decision Frameworks for Allocating Capital in EST Organizations

EST organizations allocate capital under conditions of significant technical and market uncertainty. To navigate this complexity, firms rely on structured decision frameworks that ensure investments are aligned with both strategic objectives and measurable technical evidence.

As shown in Table 7.6, approaches such as stage-gate, net present value, real options, and technology readiness level (TRL)-based methods provide different lenses for evaluating engineering programs and sequencing capital commitments. These frameworks help organizations compare initiatives that vary widely in readiness, risk, and potential impact.

The stage-gate model, operationally, is suited to hardware-centric development cycles where prototypes advance through defined engineering checkpoints. This framework ties capital release to evidence, thereby controlling downside risk while supporting steady technical progress.

By contrast, NPV analysis is most effective when revenue models are predictable, enabling leadership to quantify long-term financial returns. *Real options* complements these methods by treating early investments in uncertain, high-impact technologies as low-commitment "options" that can be expanded, paused, or abandoned as engineering evidence emerges. This approach is particularly valuable in deep-tech environments where the timing and likelihood of technical breakthroughs are difficult to forecast.

Finally, TRL-based frameworks help EST firms align capital sequencing with technical maturity by mapping investment decisions to demonstrated readiness levels. In early TRL phases, capital is typically limited to feasibility studies and prototype exploration, while higher TRL levels justify larger commitments such as pilot deployments, manufacturing investments, or commercialization efforts.

In practice, organizations often blend multiple frameworks such as Stage-Gate for engineering discipline, Real Options for uncertainty management, and TRL mapping to ensure that capital intensity increases only when technical readiness supports it. This integrated approach allows firms to balance liquidity constraints, competitive timelines, and long-term innovation strategy across their portfolios.

7.2.3 Capital Budgeting for Technology-Intensive Initiatives

Capital budgeting in technology-intensive organizations requires disciplined decisions about when to commit capital, when to defer it, and when to phase it over time.

TABLE 7.6
Capital Allocation Frameworks

Framework	Sample Use Case
Stage-Gate	Hardware prototyping cycles
NPV	Predictable revenue models
Real Options	High technical uncertainty
Technology Readiness Level-Based	Deep-tech readiness gating

Because EST firms operate with long development cycles, uncertain milestone timing, and irreversible equipment purchases, early CapEx decisions must be anchored in technical evidence rather than enthusiasm or schedule pressure. Premature investment in facilities, specialized tools, or production infrastructure can compress runway and lock the organization into a path before feasibility is demonstrated, making sequencing as important as the spend itself.

A second dimension involves managing long-lead assets that often require six to 18 months of procurement and integration. Long-lead assets include tooling, pilot-line equipment, and regulated test facilities. The timelines created introduce natural coupling between engineering plans, supply-chain conditions, and capital-allocation decisions. If pulled forward too aggressively, the organization absorbs unnecessary burn or risks stranded assets if engineering slips. If delayed too far, programs can stall, causing milestone bottlenecks and cascading roadmap impacts. Effective capital budgeting, therefore, aligns asset timing to milestone-based technical evidence rather than calendar-driven targets.

Finally, capital budgeting must assess return not just in traditional financial terms but in strategic and technical readiness terms. Investments in infrastructure can accelerate iteration cycles, increase yield learning, and materially shorten time-to-market. Typical infrastructure investment include pilot manufacturing lines or custom test platform.

But these benefits must be weighed against alternatives like renting shared equipment, outsourcing early builds, or deferring CapEx until uncertainty narrows. By evaluating strategic contribution, readiness, and liquidity impact together, EST firms can sequence capital in a way that preserves runway, mitigates risk, and strengthens the overall ability to advance the technical roadmap [Dixit and Pindyck 1994].

7.2.4 Portfolio Management and Resource Prioritization

Managing multiple engineering programs under a unified capital plan requires treating the portfolio as an integrated system rather than a collection of independent projects. Leadership must balance near-term liquidity constraints with long-term strategic priorities, ensuring that platforms, incremental improvements, and exploratory initiatives compete for resources on equal footing. Prioritization hinges on distinguishing platform investments, which shape the company's long-horizon technical trajectory, from incremental product enhancements that deliver faster but narrower returns.

To support this process, EST organizations often use heatmap or scoring frameworks that position each program along two axes: technical readiness and strategic value (Figure 7.3). This evidence-driven view enables capital to flow toward initiatives with the strongest combination of feasibility, impact, and alignment with the corporate roadmap, while gating or deferring programs that lack technical maturity or strategic leverage.

Programs with high readiness and high strategic value merit accelerated funding and execution, while breakthrough but low-readiness initiatives are treated as milestone-gated options to manage risk. The framework supports evidence-based capital allocation, portfolio rebalancing, and runway-aware decision-making under technical and financial uncertainty [Wheelwright and Clark 1992].

EXAMPLE 7.7 APPLYING PORTFOLIO HEATMAPS TO STRATEGIC CAPITAL ALLOCATION AT ALPHABET

Consider the example of Alphabet, the holding company for Google, and one of the largest technology companies by revenue. Figure 7.3 depicts a hypothetical heatmap for the company.

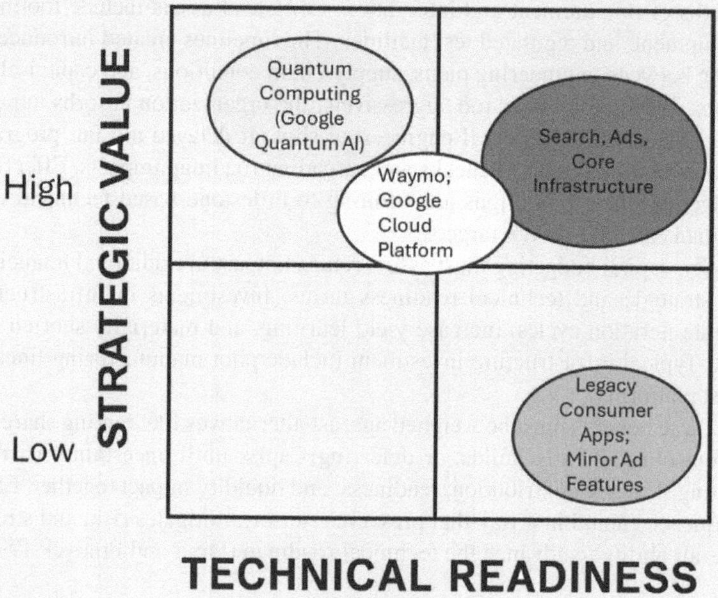

FIGURE 7.3 Portfolio Heatmap for Alphabet.

Let's discuss each of the four quadrants.

High Readiness/High Strategic Value: Search, Ads, Core Infrastructure

These programs represent Alphabet's economic engine. It is technically mature, highly scalable, and deeply embedded in the firm's long-term strategy. Capital allocation here emphasizes optimization, reliability, and incremental innovation, with large but predictable operating and capital budgets. These initiatives justify continued investment because they fund the rest of the portfolio while sustaining competitive dominance.

Low Readiness/High Strategic Value: Quantum Computing (Google Quantum AI)

Quantum computing remains technically immature but strategically transformative. Alphabet treats this category as a real option, allocating constrained capital tied to research milestones rather than full commercialization assumptions

- **Medium Readiness/Medium–High Strategic Value:** Waymo; Google Cloud Platform

These initiatives sit in the middle of the heat map, where uncertainty is highest and resource debates are most intense. Programs like Waymo required years of capital before proving technical feasibility, prompting Alphabet to restructure, spin out, or re-scope investment as evidence evolved.

Google Cloud illustrates the opposite trajectory. In this case the technology matured from medium to high readiness as technical scale, margins, and customer adoption improved, justifying accelerated investment.

- **Low Strategic Value Quadrants:** Legacy Consumer Apps; Minor Ad Features

Low-value programs, especially those with weak differentiation, are candidates for sunsetting, consolidation, or maintenance-only funding. Even if technically mature, they do not justify significant capital under runway-aware or return-focused allocation frameworks. This is where companies actively reclaim resources for higher-leverage initiatives.

Reallocating funds from underperforming programs toward higher-leverage initiatives ensures that capital consistently flows to the areas with the strongest technical evidence and strategic impact. Likewise, cross-functional governance creates the accountability and transparency needed to keep capital decisions aligned with both near-term feasibility and long-term roadmap priorities.

7.2.5 Liquidity Constraints and Runway Aware Capital Allocation

Runway-aware allocation starts with cash-flow visibility, not capital-framework theory. Every major spend request is evaluated against the forward cash projection so that approvals do not shorten the runway below pre-set guardrails. Major spend requests include grant reimbursement lags, tranche timing, and lump expenditures. In practice, finance and engineering review the same weekly cash view, highlight near-term inflections (e.g., equipment deposits and field-test bursts), and adjust release timing before liquidity tightens.

The operating discipline is simple: sequence spend to cash, not the calendar. When milestone dates move, cumulative burn rises and runway compresses; the countermeasure is to defer non-critical purchases, tighten variable-cost levers (e.g., cloud usage and external testing cadence), and re-phase procurement so cash outlays align with the updated path. This keeps the organization from locking into fixed costs while technical timing is still shifting.

Between investor tranches, companies are most exposed. Because tranche release often depends on evidence produced by the very spend under review, teams plan "bridge-light" executions. These could include smaller batches, shared equipment rentals, and contract-based flex capacity that maintains progress without committing to irreversible CapEx. If tranche risk rises, managers automatically trigger spend throttles. For example, they can freeze new system versions, convert payment terms to Net-60 where feasible, and re-prioritize toward activities that unlock the gate.

Finally, decision guardrails convert cash awareness into action:

- a minimum runway threshold (e.g., 12–18 months) under Base Case assumptions;
- an auto-reforecast when any critical milestone date slips beyond a defined tolerance; and
- pre-approved contingency levers (defer CapEx, cap variable spend, and re-scope pilots) that can be pulled in a single review cycle.

These mechanisms keep capital flowing to high-leverage work while preserving liquidity through inevitable schedule volatility [Brealey et al 2020].

7.2.6 Build, Buy, or Partner: Strategic Pathways for Capability Expansion

Choosing whether to build, buy, or partner begins with understanding what each pathway enables in terms of capability, timing, and strategic fit. EST firms must decide whether organic development provides the required technical depth, whether an acquisition accelerates the roadmap, or whether a partnership reduces risk while still delivering access to critical technology. For example, a robotics company evaluating a new perception module may determine that in-house development preserves IP control but delays time-to-market, whereas acquiring a small vision-systems startup could compress development time by a full year.

These decisions also require weighing valuation, IP control, and integration risk, as each factor materially shapes long-term financial and technical outcomes. Acquiring a team with strong IP may offer strategic leverage, but the integration of their processes, codebase, or lab infrastructure can introduce hidden complexity. Likewise, partnerships often deliver faster access to capability but may dilute architectural control or lock the firm into external roadmaps. For instance, a biotech company might license an assay platform to accelerate a diagnostic product, only to discover integration barriers that constrain future iterations.

Capital budgeting must also account for the investment profile and cost structure associated with each pathway. Internal development typically spreads cost over time but consumes engineering bandwidth; acquisitions require upfront capital but can eliminate multi-year R&D cycles; partnerships shift costs toward shared development or revenue-based structures. A cleantech hardware firm, for example, may choose to partner with a manufacturing specialist to avoid the CapEx of building a pilot line, reducing risk while preserving runway for later-stage commercialization.

Finally, the decision between building, buying, or partnering must reflect the company's strategic horizon and competitive position. EST firms with strong core competencies often choose to build to maintain differentiation, while firms under schedule pressure or facing capability gaps may lean toward acquisitions or ecosystem partnerships. A quantum-computing startup, for instance, may partner with a cryogenics vendor to secure critical subsystems quickly while focusing internal resources on qubit architecture. This approach ensures capital and talent are deployed where they create the highest leverage [Wheelwright and Clark 1992].

7.2.7 Governance, Accountability, and Decision Rights

Capital-allocation decisions in EST organizations must be grounded in well-defined decision rights that clarify who has authority to approve, sequence, or defer major investments. Boards, CFOs, CTOs, and cross-functional committees each play distinct roles, and the structure must ensure that technical evidence and financial feasibility shape decisions rather than organizational hierarchy or legacy expectations. For example, a materials-science firm may require that all investments above a certain CapEx threshold receive joint approval from Engineering and Finance to prevent either group from over-optimizing for speed or cost alone. In smaller companies, it is very common that any expenditure exceeding a certain amount, say, $50K, even if budgeted, receives Board of Directors' approval.

Clear documentation and evidence requirements are equally important. Capital requests must articulate the specific technical milestone being funded, the expected readiness impact, and the cash-flow implications under different scenarios. This creates a transparent, repeatable process that forces teams to justify spending with objective data rather than intuition. A robotics startup, for instance, may require a formal test-readiness package before approving the next tranche of tooling or pilot-deployment costs. The package might include yield data, pass-rate trends, and reliability results.

Governance structures also help prevent "pet project" funding, which can quietly consume resources without contributing to strategic objectives. By maintaining cross-functional reviews and explicit gating criteria, organizations can detect when enthusiasm for an internal initiative begins to outweigh the technical evidence supporting it. A common example appears in software-heavy engineering teams, where a favored experimental model architecture can absorb cloud resources disproportionately unless guardrails and periodic reevaluation are enforced.

Finally, effective governance ensures that technical evidence, not calendar pressure, drives capital release [Berk and DeMarzo 2020].

7.2.8 Case Study: Capital Allocation Under Technical and Financial Uncertainty

Here is a brief story of a fictitious company's struggles to allocate capital under technical and financial uncertainty. It ties together many of the concepts in this chapter.

NovaFlux Systems, a mid-stage energy-storage company, is evaluating whether to advance its next-generation thermal-management module, Astra-2, while still completing validation on its current design. Engineering pushes to initiate Astra-2 early to meet customer interest, but Finance warns that the company has only 14 months of runway and a major VC tranche tied to Astra-1's next milestone. Moving forward would pull $1.2M of CapEx forward for a specialized thermal-cycling rig with uncertain technical payoff.

During a cross-functional review, early data suggests Astra-1 needs several more weeks to reach its yield target, while procurement reports a six-month lead time for the required rig. Product argues for immediate action to preserve pilot opportunities, but Operations cautions that splitting focus could jeopardize Astra-1's milestone and delay the tranche.

Leadership opts for an "option-sized" commitment: $150k for feasibility-scale fixtures that allow Astra-2 exploration without triggering full CapEx. The larger investment is deferred until Astra-1 passes its validation gate.

Six weeks later, improved reliability unlocks the tranche, enabling NovaFlux to proceed with Astra-2 from a stronger financial position. The staged approach preserves runway, protects optionality, and ties capital release to technical evidence [Copeland and Antikarov 2003].

7.3 EVALUATING MERGERS, ACQUISITIONS, AND PARTNERSHIPS

Mergers, acquisitions, and partnerships become pivotal inflection points for EST companies, reshaping technical capability and financial trajectory in ways that internal development alone cannot achieve. These decisions must be grounded in evidence and strategic fit, ensuring that any external collaboration aligns with the firm's long-term technical and financial objectives.

Technical due diligence is as important as financial due diligence. EST firms must verify technology maturity, assess integration feasibility, and identify risks that may offset the expected strategic benefit, as integration challenges can easily outweigh the value of the acquired capability. Deal structures, therefore, need to reflect technical uncertainty and roadmap maturity, often incorporating milestone-based payouts, staged commitments, or option-like mechanisms that protect the firm if engineering results diverge from expectations.

7.3.1 Strategic Rationale for Mergers, Acquisitions, and Partnerships

Engineering, science, and technology companies pursue mergers, acquisitions, and partnerships when these moves strengthen technical capability, accelerate roadmap execution, or meaningfully reduce development risk. Because these firms operate in capital-intensive, uncertainty-heavy environments, leadership looks for transactions that provide clear, evidence-based leverage rather than incremental benefit. For example, an aerospace controls company might acquire a small sensor-calibration startup to eliminate a critical dependency and shorten a validation cycle by several months.

These strategic decisions often hinge on whether external collaboration can deliver capabilities more efficiently than internal development. In some cases, acquiring a specialized team accelerates progress toward a key milestone; in others, partnering with a domain expert reduces risk without requiring full ownership. A robotics startup, for instance, may forge a co-development partnership with a perception-software supplier to avoid delaying field-trial readiness.

Market access also plays a central role. EST firms frequently operate in niches where early customer traction or regulatory positioning can determine long-term competitiveness. Acquiring a company with established customer relationships can significantly compress time-to-market. The same is true when partnering with a company that has a proven compliance pathway. For instance, a clean-energy hardware company might enter a joint venture with a utilities integrator to speed adoption of its pilot-ready modules.

The rationale for M&A or partnership must align with the company's broader technical and financial strategy. EST organizations should prioritize moves that create measurable synergy while avoiding transactions that introduce integration complexity without clear readiness or roadmap benefits. Inadvertent complexity can be added through unique IP, complementary infrastructure, or accelerated milestones. Effective deals strengthen both near-term execution and long-term strategic position, ensuring capital and talent flow to the areas with the highest leverage [Koller et al 2020].

7.3.2 Types of Transactions Relevant to EST Companies

There is a range of transaction structures that can be used for M&A depending on the capability and level of control required. Full acquisitions are the most comprehensive, giving the acquiring company ownership of technology, talent, and infrastructure. These moves are often reserved for situations where proprietary IP or tight architectural integration is essential. For example, when a semiconductor company acquires a niche photonics team to secure a critical component in its stack.

Partial acquisitions and asset purchases offer a more targeted alternative. Instead of absorbing an entire company, an EST firm may buy specific IP blocks, datasets, or specialized tooling that directly advance a technical initiative. For example, a battery-materials startup might purchase a proven electrode design rather than acquire the entire lab that developed it. This allows firms to capture high-value assets while avoiding the integration burden of a full acquisition.

Joint ventures and co-development agreements come into play when both parties bring meaningful, complementary capabilities, but neither wants full ownership. These structures let EST firms share risk, pool expertise, and accelerate development timelines. A clean-energy hardware company, for instance, might form a joint venture with a manufacturing specialist to build a pilot-scale production line without taking on the full cost or operational load alone.

Finally, ecosystem and open-source-driven partnerships represent a growing category, especially in software, robotics, and embedded systems. These collaborations give EST companies access to large developer communities, rapid prototyping tools, or niche capabilities that evolve faster than internal teams can match. The challenge, however, lies in integrating open-source components into proprietary architectures without compromising reliability, security, or long-term maintainability.

7.3.3 Financial and Valuation Considerations

Valuing technical assets with uncertain readiness levels requires assessing both their engineering maturity and their financial implications (Figure 7.4).

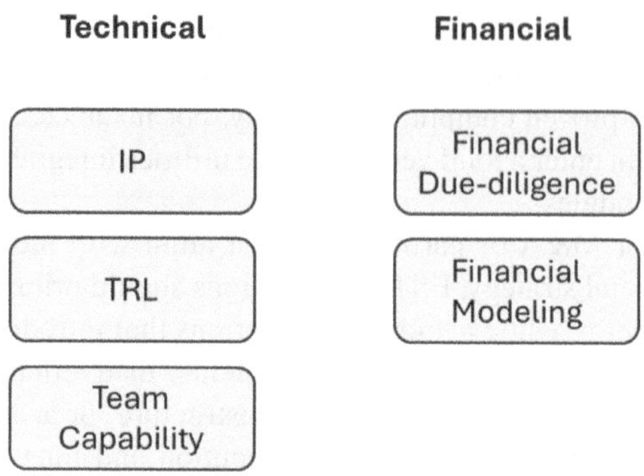

FIGURE 7.4 Technical versus financial due diligence layers.

Technical due-diligence at each layer provides the foundation for understanding whether an asset can realistically progress toward commercialization. Layers include IP strength, TRL, and team capability. Due diligence at each of these layers determines feasibility and the timing and scale of investment needed to advance the asset.

Financial due diligence complements this view by linking those technical realities to valuation models, risk profiles, and capital requirements. In practice, R&D pipeline strength, prototype maturity, and expected development pathways directly shape valuation because they influence revenue timing, cost structure, and the probability of achieving meaningful milestones [Mankins 2009].

For example, a photonics-chip startup evaluating its newest coupling module may have strong early IP and promising TRL indicators, but inconsistent prototype yields introduce valuation uncertainty. Technical diligence would flag yield instability as a material readiness risk, while financial diligence would incorporate that risk into scenario-based modeling. This dual approach lowers the projected value until reliability improves.

The Venn diagram of Figure 7.5 highlights that high-value M&A and partnership outcomes in EST organizations emerge when cost synergies, capability synergies, and roadmap acceleration reinforce one another.

Rather than focusing solely on cost reduction, the figure underscores that the most impactful deals sit at the intersection of all three dimensions, where shared infrastructure, specialized technical expertise, and reduced development timelines compound into meaningful strategic leverage.

For example, an engineering firm acquiring a small sensor-calibration startup may capture cost benefits through shared facilities while simultaneously gaining unique technical know-how that enhances internal capability. If that expertise also shortens the validation cycle for an upcoming product line, the deal also unlocks roadmap acceleration.

7.3.4 Technical and Operational Due Diligence

In practice, technical due diligence precedes everything else. First we assess technology maturity through TRLs, yield stability, and reliability metrics. Then we evaluate

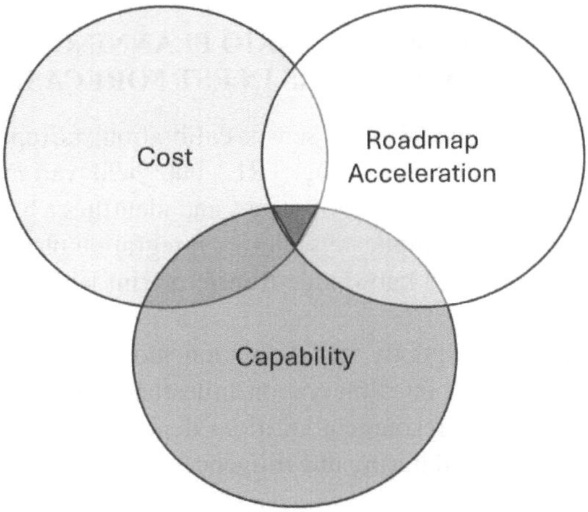

FIGURE 7.5 Cost, capability, roadmap, acceleration synergy map.

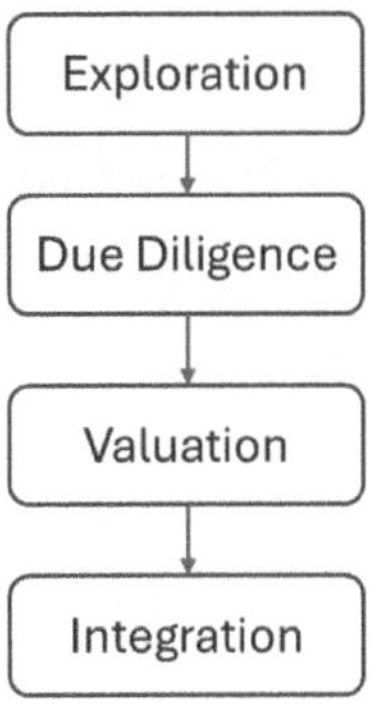

FIGURE 7.6 M&A pipeline.

team capability, knowledge-transfer plans, and cultural fit to determine whether the acquired know-how will transfer.

Integration risk is evaluated just as rigorously while IP position and freedom-to-operate guard against hidden constraints that can surface after close. For example, data pipelines, production workflows, and regulatory interfaces must be mapped early. This evidence-first posture keeps the process anchored to feasibility rather than narrative. Figure 7.6 frames this as a disciplined pipeline: Exploration → Due Diligence → Valuation → Integration.

During the *exploration phase*, we screen opportunities against strategic and technical fit. In *due diligence,* we pressure-test the engineering claims and operating realities; only then do we translate findings into *valuation* assumptions. Finally, *integration* planning incorporates risk controls into post-deal execution. In this way, the same evidence that supported the deal also governs how capital, people, and systems are combined.

EXAMPLE 7.8 SCENARIO PLANNING AND SENSITIVITY ANALYSIS IN EST FORECASTING

An EST firm considers acquiring a sensor-calibration startup. Early Explore work flags strong IP and promising TRL, but yield variance is high. In Diligence, the team confirms reliability gaps and identifies a brittle data-ingestion pathway that would complicate scale-up; integration planning proposes a staged pipeline refactor and knowledge-transfer sprint with the target's senior engineers.

Those findings flow directly into Valuation (down-weighting projections until yield stabilizes) and into Integration milestones tied to closing. In this way, tranche releases and resource allocations depend on hitting the yield and data-pipeline gates defined during due diligence.

By grounding valuation and integration planning in rigorous, evidence-driven technical diligence, EST organizations ensure that acquisition decisions remain aligned with feasibility, strategic fit, and long-term execution success.

7.3.5 Risk Assessment: When M&A or Partnerships Fail

Failures in mergers, acquisitions, or partnerships often stem from misjudging the true complexity of integration. Organizations may overestimate expected synergies or underestimate the engineering, operational, or cultural work required to combine two teams. Even when the technical fit appears strong, mismatches in development processes, communication norms, or decision-making styles can quietly erode the anticipated benefits. Regulatory issues, export-control constraints, or incompatible compliance expectations can also surface late in the process, slowing integration and reducing strategic value.

External dependencies can introduce additional fragility. Supplier disruptions, manufacturing incompatibilities, or fragmented ecosystem relationships may undermine what initially looked like a straightforward combination.

For instance, an EST firm acquiring a niche sensor-calibration startup might expect immediate capability gains, only to discover that the startup's data-pipeline tools are incompatible with the acquirer's production workflow. As teams attempt to resolve integration gaps, early delays dilute the projected synergies and force the company to reassess timelines, investment needs, and overall deal value.

7.3.6 Structuring Deals

Structuring transactions in EST organizations requires aligning financial incentives, ownership considerations, and collaborative mechanisms with the underlying technical realities. Earnouts and milestone-linked payouts help ensure that compensation reflects actual technical readiness rather than optimistic projections. Equity versus cash decisions also matter, particularly in early-stage deals where liquidity is limited and alignment with long-term roadmap execution is essential. In many cases, equity provides stronger strategic alignment, while cash offers simplicity and reduces future dilution pressures [Koller et al 2020].

Partnership structures, for example, licensing, technology-sharing agreements, or co-manufacturing, must be designed so that capabilities can be integrated without slowing engineering progress. Governance structures then anchor these arrangements by defining decision rights, steering-committee roles, and IP ownership rules that prevent ambiguity as programs scale.

For example, a robotics firm licensing a perception module may tie milestone payments to demonstrated field reliability, while a clean-energy hardware company entering a co-manufacturing partnership might retain IP ownership but give the partner limited rights to operate shared tooling. These structures keep incentives aligned, reduce integration risk, and ensure that each party contributes meaningfully to readiness and execution.

7.3.7 Unconventional Partnerships and Ecosystem Plays

Companies sometimes acquire or partner with open-source or community-driven platforms to accelerate capability development. These ecosystems provide immediate access to large developer communities, faster prototyping cycles, and broader adoption pathways. These benefits can meaningfully extend a firm's technical reach and shorten early discovery and iteration timelines.

However, integrating open-source assets into proprietary architectures introduces real challenges. Firms must reconcile differing governance expectations, align licensing and contribution models, and ensure that community-driven components fit within internal reliability, security, and lifecycle standards. These tensions make open-source partnerships strategically powerful but operationally non-trivial.

EXAMPLE 7.9 AN UNUSUAL ACQUISITION OF AN OPEN-SOURCE ECOSYSTEM

Qualcomm's unusual acquisition of Arduino, merging a high-performance giant with one of the most influential open-source hardware ecosystems, caught analysts by surprise. The deal was particularly surprising because Arduino was not rumored to be for sale. Qualcomm sought to expand its reach into low-cost, special-purpose electronics, positioning the acquisition to pair its Dragonwing system-on-a-chip platform with Arduino's massive developer base, exemplified by the launch of the new Arduino Uno Q at the time of the announcement.

While Qualcomm framed the acquisition as an ecosystem-expanding strategy that would give developers new access to advanced processing capabilities, the move prompted concern about whether a large corporate owner could preserve Arduino's long-standing open-hardware ethos [Smith 2025].

With many more high-quality open-source ecosystems in play, it is likely that this kind of acquisition will become more common.

7.4 RAISING CAPITAL: COMMERCIAL PAPER, VENTURE CAPITAL, LOANS, AND CROWDFUNDING

Financing decisions sit at the center of how EST companies advance their technical roadmaps, manage liquidity, and navigate long development cycles. Because these organizations often face extended periods of negative cash flow, capital-raising strategies must be tightly aligned with technical readiness and milestone timing rather than generic growth assumptions.

This section examines the major financing pathways available to EST firms and shows how each instrument shapes runway, risk posture, and long-term strategic flexibility.

7.4.1 Commercial Paper and Corporate Debt Instruments

Commercial paper (CP) is short-term, unsecured corporate debt used to cover near-term liquidity needs without drawing down longer-tenor facilities. Because it relies on reputation and credit strength rather than collateral, CP is typically available only to established firms with strong balance sheets, predictable cash flows, and solid market credentials. That combination keeps issuance costs low and placement fast, but it also narrows eligibility. That is, if investors doubt a firm's ability to refinance or generate cash on schedule, the paper won't be approved, or it will price unfavorably.

In practice, companies use CP to fund inventory builds, bridge timing gaps in receivables, and smooth working-capital swings. These are short-duration use cases where speed and cost matter most.

The trade-offs are straightforward. Investors expect high credit quality, program documentation, and continuous access to money markets. Credit issuers accept rollover risk and the discipline of daily pricing. If market conditions tighten or ratings slip, rollover can fail and liquidity compresses quickly, forcing an immediate pivot to backup lines or internal cash.

Relative to *bank credit lines* and *asset-backed lending*, CP is less costly and more straightforward, but it lacks the protections that banks provide. Committed *revolving credit*[2] facilities function as safety nets for failed rollovers and are often required as formal liquidity backstops. Asset-backed lending, by contrast, ties borrowing capacity to pledged assets (e.g., receivables/inventory) and stays available even when credit rating is unstable. However, this form of financing comes at the cost of collateral monitoring and tighter covenants [Brealey et al 2020].

EXAMPLE 7.10 USING COMMERCIAL PAPER TO ADVANCE PARTS IN MANUFACTURING

A top-tier manufacturer of consumer drone technology issues 60-day commercial paper to pre-buy components ahead of a seasonal ramp, then repays from collections. If spreads widen unexpectedly, it draws on its committed revolving line of credit to retire maturing notes and waits for markets to normalize.

Commercial paper provides established firms a fast, low-cost way to bridge short-term liquidity needs. However, the company must maintain credit strength.

7.4.2 Venture Capital and Equity Financing

Venture capital funds the step-function investments technical programs require when revenue is distant and uncertainty is high. Equity rounds trade dilution for runway, governance support, and access to networks that help recruit talent, secure pilots, and navigate later financing.

The core dynamic is evidence for capital. Investors release funding at milestones that de-risk technical and commercial assumptions, and valuation tracks progress rather than calendar time. Terms such as board representation, protective provisions, and information rights formalize that exchange of risk and control.

Practically, rounds are tranched to engineering gates, so cash arrives when the data shows readiness to scale. Typical gates include prototype validation, yield targets, and field reliability. Term sheets often balance price with structure. That is, liquidation preferences, anti-dilution, and milestone-linked closings align incentives but can also constrain flexibility if progress slips. The objective is to scope a round that funds the next proof point with enough buffer for iteration, while keeping dilution low so the team remains motivated and future rounds stay viable [Gompers and Lerner 2004].

EXAMPLE 7.11 STAGED VC FINANCE WITH GATES

A solar farm controls startup closes a $6M Series A equity round with $3M wired at close and $3M contingent on achieving MTBF and pilot-deployment key performance indicators. Hitting the gate lifts the post-money, triggers the second tranche, and green-lights manufacturing fixtures. Missing the gate forces a re-plan on scope, timing, or terms.

In practice, venture capital supplies the staged, evidence-driven fuel that lets EST companies convert technical proof into runway, so long as funding remains tightly synchronized with milestone readiness rather than the calendar.

7.4.3 Loans, Venture Debt, and Structured Credit

Traditional bank loans and *venture debt* serve different roles in EST companies. Bank loans rely on collateral, cash-flow coverage, and covenant discipline, making them better suited to firms with predictable revenues and assets to pledge. Venture debt complements equity by extending runway between rounds; it's underwritten primarily on investor backing and progress against engineering milestones, not full profitability, and often sits alongside a smaller warrant package that prices the lender's risk.

In practice, teams use venture debt to bridge to the next technical or commercial proof point, finance equipment, or add a buffer after an equity round without immediately taking more dilution. Lenders protect themselves with *covenants* and may tie draw schedules to milestone evidence, mirroring the tranche logic used by VCs.

Typical covenants include:

- minimum cash on hand,
- performance revenue key performance indicators (such as Z-score), or
- material adverse change clauses.

The trade-off is straightforward: provide runway at a lower immediate dilution cost, offset by fixed repayments and operational constraints if progress slips [Brealey et al 2020].

EXAMPLE 7.12 VENTURE DEBT AS A MILESTONE-GATED BRIDGE FINANCING TOOL

A telemedicine startup closes a $12M Series A and adds a $3M venture-debt facility to fund prototype systems and field support. Draws are unlocked when the team hits defined reliability key performance indicators; covenants require 9 months of minimum cash and limit new CapEx without lender consent.

Delivering pilots on schedule lets the company refinance or retire the facility at the next round; missing the gate tightens covenant headroom and can force scope reductions to stay compliant.

These debt instruments extend runway while preserving flexibility, reinforcing the principle that financing must stay synchronized with technical readiness and milestone risk rather than arbitrary timelines.

7.4.4 Crowdfunding and Community-Driven Financing

Crowdfunding gives EST companies a way to validate demand and bring in early cash without immediately accessing institutional capital. Two models dominate: equity crowdfunding and pre-order (rewards/commerce) crowdfunding:

- *Equity crowdfunding*—sells securities to a broad base of investors under regulated exemptions.
- *Pre-order (rewards/commerce) crowdfunding*—exchanges early product commitments for working capital and market signal.

Done well, crowdfunding can reduce risk assumptions about willingness to pay, generate community momentum, and extend runway ahead of formal rounds [Brealey et al 2020].

EXAMPLE 7.13 CROWDFUNDING CAMPAIGN FOR A STARTUP

A small telematics startup launches a pre-order crowdfunding campaign to finance a limited pilot build of a new sensing module. Early traction validates pricing and use cases, but the team gates spending to engineering evidence. This locks supplier terms only after reliability tests pass, so cash burn stays aligned with readiness.

Crowdfunding has downsides, however. For example, there are operational visibility and delivery obligations. Public campaigns create a timeline that the team must meet, and missed milestones are highly visible. Further, equity campaigns add cap-table complexity and ongoing communications overhead; pre-order campaigns pull forward logistics, support, and manufacturing risk.

For technical hardware in particular, the gap between prototype demos and production readiness can stress budgets if yield, reliability, or certification work runs long. It can also turn a demand win into a fulfillment liability as customers may demand a product that is appealing but not ready for market.

7.4.5 Choosing the Right Financing Mix

Choosing the right financing mix starts with the risk profile and maturity of the program. Early, uncertainty-heavy phases benefit from nondilutive or option-sized capital; later phases that have clearer revenue paths can accommodate larger equity or structured debt. The governing principle is simple: align the instrument to the technical and market risk you bear, not the calendar you wish you had.

Liquidity planning is a first-order constraint, not an afterthought. It is best to sequence rounds and draw schedules to avoid runway cliffs, model reimbursement lags and tranche timing, and ensure minimum cash thresholds are preserved through validation gates. In practice, this means mapping every major spend to the forward cash view and reforecasting automatically when milestones move.

Senior leaders often prefer hybrid financing models. For example, low venture debt alongside equity. This blend helps to extend the runway with less immediate dilution, while retaining the ability to pause, resize, or defer CapEx as evidence evolves. This approach balances cost of capital with control and preserves headroom if gate timing slips, consistent with standard corporate-finance guidance on capital structure trade-offs [Brealey et al 2020].

Financing should support, not dictate, the technical roadmap. A good strategy is to tie cash release to engineering evidence and milestone readiness, mirroring how sophisticated investors tranche capital to proof points rather than dates.

For example, iRobot advanced early robotics work with US SBIR awards (see Section 7.6.2) before scaling with private capital, which illustrates how staged, evidence-aligned funding can bridge from feasibility to commercialization [SBIR 2017].

7.4.6 Case Study: Financing Strategy Under Uncertainty

NeuralFlight Systems, an AI-hardware startup, is developing *AeroEdge*, an edge-AI coprocessor for autonomous drones. Early lab tests show strong inference-latency performance, but yield variability and flight-test uncertainty press heavily on the company's nine-month runway. Leadership must choose a financing path that aligns with technical milestones rather than optimistic timelines.

The next phase centers on three gates: G1 (flight-ready integration), G2 (customer-pilot readiness), and G3 (field reliability). Each requires specialized testing and support costs that could accelerate burn if schedules slip.

Venture capital firms offer a Series A with milestone-linked tranches, that is, $5M upfront and $3M released after G2. While this would extend runway, the dependency loop is risky: progress toward G2 requires spending the funds tied to achieving it.

Venture debt provides a $2M facility with draws contingent on G1 and moderate covenants. It adds runway with limited dilution but introduces repayment pressure and cash-minimum requirements that could constrain engineering flexibility. A preorder crowdfunding proposal could validate demand and generate early cash, but it risks committing the company to delivery timelines before technical readiness is proven.

After review, NeuralFlight adopts a hybrid strategy: draw only the first $750k of the venture-debt facility once G1 evidence is secured, rent instead of buying test equipment to limit irreversible spend, and defer crowdfunding until after G2 results. This preserves optionality, controls burn, and ties capital release directly to readiness.

When AeroEdge clears G1 six weeks later, NeuralFlight moves forward with strengthened runway and reduced financing risk.

7.4.7 Financing Pathways

EST firms require external financing because their long development cycles, heavy R&D intensity, and delayed revenue models make it difficult to self-fund progress through early technical milestones. These organizations must secure capital well before commercial viability is proven, creating a dependence on staged funding that aligns with evidence rather than calendar-driven expectations.

Debt, equity, and alternative financing instruments each play distinct roles in this process, with trade-offs that reflect liquidity needs, risk tolerance, and the level of technical uncertainty inherent in engineering-centric development.

7.4.7.1 Financing Option Evolution

Early proof-of-concept work is typically supported by pre-seed capital, while prototype and validation stages transition toward prescale financing that enables more robust testing and early deployment (Figure 7.7).

Growth-stage capital becomes viable only once the technology has demonstrated reliability, market fit, and a credible pathway to scale.

EXAMPLE 7.14 FINANCING PATHWAYS ALIGNED TO TECHNICAL MATURITY

SkyMind Robotics begins with a $150k pre-seed grant to prove the basic feasibility of its autonomous inspection drone. After demonstrating a working proof-of-concept, the company raised a $1.2M seed round to build full prototypes and conduct early flight tests. As technical readiness improves and customer pilots validate performance, SkyMind secures a $6M Series A to fund pilot production and regulatory certification.

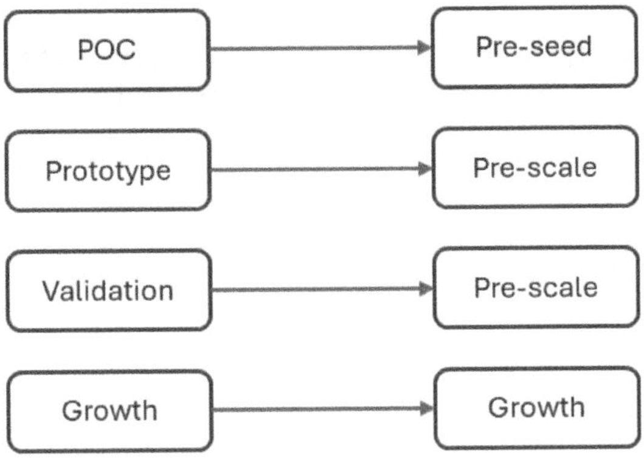

FIGURE 7.7 Financing options by maturity.

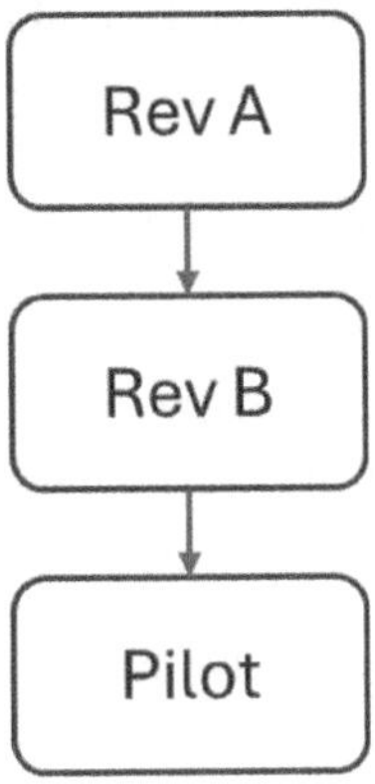

FIGURE 7.8 Tranche-based VC funding.

As has been emphasized repeatedly, maturity-linked progression ensures that resources are allocated in step with demonstrated feasibility, enabling EST firms to advance through POC → prototype → pilot → growth in a financially disciplined manner.

7.4.7.2 Tranche-Based Funding versus Milestone Gates

Tranche-based venture financing works when cash releases are explicitly tied to engineering evidence, not the calendar. In practice, investors wire an initial amount at close and stage subsequent tranches at objective gates, so capital arrives only when technical risk is reduced. This structure keeps dilution and spend synchronized with proof, and it forces automatic reforecasting when milestones slip.

Figure 7.8 illustrates this cadence briefly.

Here, funding is triggered sequentially as the program advances Rev A → Rev B → Pilot. Consider this quick example.

EXAMPLE 7.15 TWO-STAGE VC INVESTMENT IN A HARDWARE AI STARTUP

Suppose a hardware AI startup closes an $8M Series A investment from a VC. The payments are $5M at close and $3M upon achieving Rev B yield and pre-compliance checks. The first tranche is paid and development proceeds.

However, when Rev B fails to meet its schedule, the second tranche is delayed, forcing the team to defer noncritical CapEx and throttle cloud spend to protect runway.

The implication is straightforward: if outcomes move, cash timing moves with them [Gompers and Lerner 2004].

7.4.7.3 Burn Rate Sensitivity to Technical Slip

Delays magnify cost. Even modest schedule slips raise cumulative burn and compress runway because fixed costs continue while variable spend rarely scales down perfectly with time. The practical takeaway is simple: when milestones move, cash needs move faster. This forces earlier fundraising or scope trade-offs if spending is not reduced.

Figure 7.9 highlights this effect visually.

Here, the burn-curve steepens as slip accumulates, demonstrating how a "one-sprint" delay can ripple into higher cumulative cash outlays and reduced runway headroom. Therefore, it is imperative to plan countermeasures in advance so that schedule volatility doesn't automatically become a liquidity problem. For example, pausing noncritical variants, deferring CapEx, and tightening cloud or external-test cadence [Koller et al 2020].

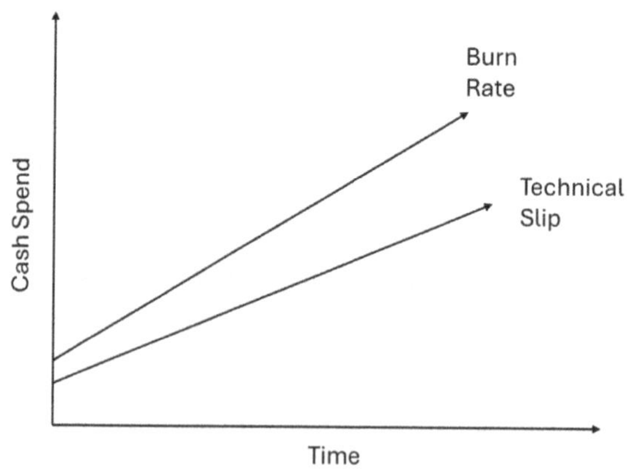

FIGURE 7.9 Burn rate sensitivity.

EXAMPLE 7.16 VC STAGED FUNDING FOR A BIOTECH FIRM

A biotech therapeutics startup closes a $10M Series A, structured so that $6M is wired at close and the remaining $4M is released only after the company achieves Phase I clinical safety results. When an unexpected assay-validation issue pushes the Phase I timeline back by two months, the second tranche automatically slides.

To preserve runway, the team pauses non-critical CRO spending and defers a planned equipment purchase. This is exactly the type of transaction-driven adjustment described in Figure 7.8. This ensures spending stays aligned with evidence rather than optimistic schedules, keeping capital sequencing tied to technical readiness rather than calendar targets.

Even modest schedule slips can steepen the burn curve and compress the runway, underscoring why EST firms must plan countermeasures in advance so that timing volatility doesn't automatically turn into a liquidity problem.

7.4.7.4 Cashflow Impact by Financing Type

Financing instruments shape cashflow very differently, so it is essential to design the mix to match readiness and runway goals. Equity financing typically provides the largest immediate inflow but at the highest long-term cost through dilution. Debt financing adds moderate inflow with fixed repayments and covenant constraints. Grants arrive in smaller tranches but are nondilutive and cash-efficient.

Figure 7.10 provides a visualization of how each financial instrument affects both liquidity and long-term value.

The operating rule is to map each dollar of financing to near-term proof points and reforecast when milestones move.

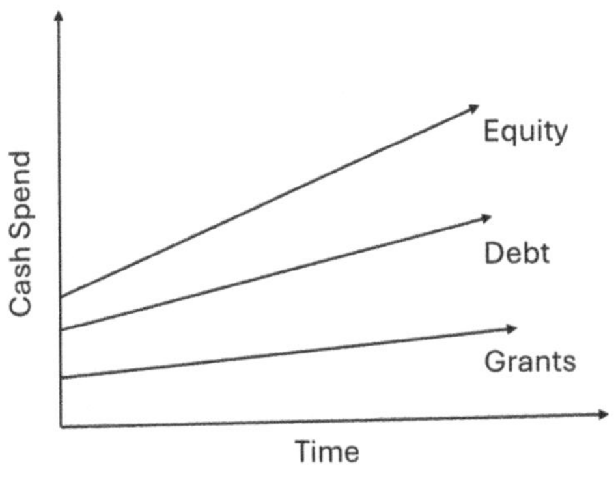

FIGURE 7.10 Cashflow impact: equity versus debt versus grants.

EXAMPLE 7.17 COMPARING CASHFLOW PROFILES ACROSS EQUITY, DEBT, AND GRANT FINANCING

A digital health company pursuing FDA approval uses a $750k state innovation grant to fund verification testing. It also uses a small venture-debt line of credit to bridge submission-to-clearance, and a Series A wired after clearance to finance commercialization.

The effect mirrors Figure 7.10's cashflow profiles while tying releases to objective gates (V&V complete → submission accepted → clearance). The mix minimizes dilution before proof, protects runway during review, and deploys equity only when risk is materially reduced.

In practice, it is appropriate to take equity draws when validation gates unlock scale decisions, lean on modest debt for bridge liquidity between gates, and use grants to offset early technical risk without adding dilution pressure. This keeps spending synchronized with evidence and preserves optionality if schedules slip.

7.5 FINANCIAL PLANNING FOR STARTUPS

This section outlines the foundational tools, milestone-linked budgeting practices, and liquidity disciplines that enable young engineering-driven companies to advance from feasibility to prototype while preserving runway and strategic optionality.

7.5.1 The Financial Reality of EST Startups

Early-stage EST startups face long development cycles, uncertain milestone timing, and front-loaded costs with little early revenue. Therefore, financial planning must anchor to technical readiness rather than calendar targets. In practice, this means treating every major spend as an option on future evidence, sequencing cash commitments to objective gates, and preserving runway as a first-order constraint, consistent with standard corporate-finance guidance on capital budgeting under uncertainty.

EXAMPLE 7.18 PRESERVING CASH IN A BIOTECH STARTUP

A biotech assay-development startup is evaluating whether to purchase an automated high-throughput screening system. Instead of committing $450k upfront, the team contracts weekly access at a shared facility and releases internal budget only after early assays meet predefined reproducibility thresholds. This is a conservative approach that preserves runway and ties irreversible spending directly to technical evidence.

Because cash is scarce and irreversibility is high, leaders should bias toward milestone-gated budgets, rolling reforecasts, and liquidity guardrails that trigger automatic throttles when timelines slip [Dixit and Pindyck 1994].

7.5.2 Establishing the Early-Stage Financial Foundation

Establishing a rigorous early-stage financial foundation begins with defining the minimum runway, baseline burn, and evidence requirements that anchor all subsequent budgeting and investment decisions. We can set the foundation by defining an explicit runway target (e.g., 6–18 months), building a baseline burn that isolates the most impactful cost drivers, and mapping these to the minimum evidence needed for the next gate.

Next, we convert that foundation into an operating discipline. This includes rolling reforecasts, clear liquidity guardrails (minimum cash thresholds), and a short evidence checklist per gate, so spend releases are pre-approved but only triggered when proof arrives.

Finally, make the foundation auditable using dashboards reviewed on a predictable cadence (monthly financials; milestone reviews tied to gates). Produce one-page capital requests that spell out the milestone being funded, and the cash-flow impact if timing slips.

EXAMPLE 7.19 MILESTONE-LINKED BUDGETING AND EVIDENCE-DRIVEN FUNDING IN PRACTICE

A cybersecurity startup training a new anomaly-detection model targets a 12-month runway and caps monthly cloud spend within guardrails. It funds only a $150k "Gate 1" tranche to reach defined inference-accuracy and cost-per-scan thresholds. If those targets aren't met, the team pauses external testing and defers a planned data-pipeline tool purchase.

This kind of oversight keeps capital synchronized with technical readiness and protects runway when schedule risk materializes.

7.5.3 Early Funding Pathways for Technical Startups

Early-stage EST startups depend on funding sources that match their maturity, uncertainty profile, and evidence needs, making early pathways fundamentally different from traditional growth financing. Nondilutive programs such as SBIR/STTR are particularly valuable at this stage because they fund feasibility and prototype-development work without forcing premature dilution, while also imposing a milestone-oriented structure that keeps spending synchronized with technical readiness. These pathways help teams convert early scientific or engineering intuition into objective proof points that later unlock equity or debt under more favorable terms.

As companies advance toward prototype and validation gates, pre-seed capital, *angel investors*,[3] and strategic micro-VCs often complement nondilutive funding by providing flexible risk capital that can be released in tranche form. Because technical variance remains high in these phases, the blend of nondilutive and small equity checks gives startups room to iterate without collapsing runway, while keeping governance proportional to the firm's actual readiness.

Later, as evidence hardens startups can prepare for institutional capital by building credible forecasts, tightening financial controls, and documenting progress in a way that external investors can evaluate. The sequencing of these pathways matters: when used correctly, nondilutive awards bridge feasibility, early equity funds validation, and structured debt or larger rounds follow only when technical and commercial viability are demonstrated [NRC 2015].

EXAMPLE 7.20 STRATEGIC FUNDING USING SBIR GRANTS

A clean-energy materials startup developing a novel catalyst begins with a Phase I government grant program award to establish basic feasibility, then secures a small pre-seed round to fund initial prototype batches. Once early testing shows promising durability metrics, the team applies for a Phase II grant to support extended validation before pursuing a priced seed round. This approach delays dilution until evidence meaningfully reduces technical risk.

In practice, early funding pathways matter most when they are sequenced to evidence so scarce capital advances readiness without forcing premature dilution.

7.5.4 Managing Burn Rate and Capital Efficiency

Managing burn rate in early-stage EST startups requires disciplined control of the few cost drivers that materially influence runway. These are typically engineering labor, prototype and test spending, and cloud or compute intensity. Because development cycles are long and milestone timing is uncertain, even modest schedule slip can propagate into higher cumulative burn, compressing runway and forcing premature fundraising unless teams actively throttle variable costs and defer irreversible commitments.

Classic corporate-finance guidance reinforces this posture. Organizations should commit capital gradually, preserve flexibility under uncertainty, and ensure that spending follows demonstrated technical evidence rather than planned timelines.

Capital efficiency also depends on making deliberate build-versus-buy choices, using contract engineering selectively, and avoiding premature scale or over-engineering before feasibility is proven. Liquidity guardrails, rolling reforecasts, and milestone-linked spend approvals help teams maintain alignment between technical readiness and financial capacity, preventing runway erosion as programs evolve. For hardware- or lab-intensive startups, this often means sequencing CapEx behind early validation signals and relying on shared infrastructure or short-term contracts until uncertainty narrows [Dixit and Pindyck 1994].

EXAMPLE 7.21 KEEPING CASH BURN LOW

A synthetic-biology startup developing a new enzymatic assay keeps burn low by outsourcing early wet-lab validation to a contract research organization rather than hiring full-time staff or purchasing specialized equipment.

Only after the assay meets predefined reproducibility and sensitivity thresholds does the team authorize internal hiring and lab build-out. This approach maintains capital efficiency while ensuring investment follows evidence.

Disciplined burn-rate management ensures that capital is deployed only when warranted by technical evidence, preserving runway and protecting early-stage EST teams from premature scale or irreversible commitments.

7.5.5 Preparing for Institutional Capital

Institutional investors evaluate early-stage EST companies through a tight lens of technical readiness, validated demand, cost/schedule discipline, documentation quality, and credible forecasting. In this way, the prep work is about turning evidence into investor-grade artifacts rather than polishing narratives. In practice, this means packaging milestone data, reliability/yield trends, and burn/runway visibility into a clean data room and operating cadence that investors can diligence quickly.

Scenario-based forecasts, milestone-linked budgets, and liquidity guardrails are essential to show how you'll protect runway if timing slips. Pair those with an explicit explanation of how the next tranche of capital converts into readiness and anchor valuation logic to those gates rather than to calendar aspirations. This is an approach aligned with mainstream corporate-finance practice and investor expectations around gated risk reduction.

Finally, signal institutional discipline in governance and reporting using the following:

- predictable monthly financials,
- milestone exit reviews,
- one-page capital requests that spell out the gate being funded,
- cash-flow impact if dates move, and
- predefined contingency levers.

This evidence-first posture maps directly to how professional investors underwrite staged financing and covenant-style safeguards, making your raise more legible and your plan more credible [NRC 2015].

7.5.6 Case Study: A Startup Navigates Early Financial Planning

NovaSense Analytics entered its first year with a promising but unproven environmental-monitoring algorithm, pushing the team to secure early validation before committing meaningful capital. Initial lab tests showed encouraging signal-detection accuracy, but variability across field samples created uncertainty about the feasibility of hitting commercial-grade reliability targets. With only eight months of runway, leadership recognized that technical proof had to drive every financial decision, not simply enthusiasm.

As burn-rate pressure mounted, the team adopted milestone-linked budgeting, allocating a small tranche solely to achieve defined evidence thresholds. The thresholds are:

- reproducibility across three field environments; and
- stable calibration under temperature drift.

Any budget for scaling the data-pipeline infrastructure or onboarding new engineers was withheld until those metrics were met.

When validation took longer than planned, NovaSense activated predefined throttles to keep the runway above its guardrail. In this case, deferring a nonessential cloud-compute upgrade and renegotiating vendor terms.

Facing a choice between raising an early equity round at a steep discount or extending runway through nondilutive means, the company opted to pursue an SBIR Phase I grant before approaching venture capital. The grant proposal centered on the very validation work the team was already performing, allowing NovaSense to fund critical experiments without surrendering ownership prematurely.

Three months later, with grant funding secured and validation data improving, the company re-entered fundraising conversations from a position of strength. That is, armed with evidence, disciplined financial controls, and a clear pathway to its next technical gate.

7.6 GRANT FUNDING TO COMMERCIALIZATION

Nondilutive grant funding plays a critical role in helping engineering- and science-driven startups bridge the gap between early feasibility work and investable prototypes, particularly when equity is too early or expensive to pursue. Programs such as the US SBIR/STTR system and their international counterparts provide a structured, phased pathway that supports feasibility, prototype development, and commercialization readiness, giving EST companies a disciplined route to advance technical maturity while preserving founder equity.

7.6.1 US SBIR/STTR Programs

The US Small Business Innovation Research (SBIR) and Small Business Technology Transfer (STTR) programs remain the primary nondilutive funding mechanisms for engineering- and science-driven startups.[4] Federal agencies, like the DoD, NSF, NIH, DOE, and NASA, allocate a mandated portion of their R&D budgets to these programs. This creates a predictable and mission-aligned pathway for early technical validation. These programs also create credible third-party validation that investors and partners recognize [National Research Council 2015].

For EST companies operating in long development cycles with little early revenue, this structure offers a disciplined entry point to begin building prototypes, generating evidence, and maturing intellectual property without immediate equity dilution. Both programs follow a phased progression designed to align funding with technical readiness (Figure 7.11).

Phase I supports feasibility work allowing companies to demonstrate proof-of-concept and establishes the technical basis for further investment. Phase I typically provides a ~$50k–$275k award over six to 12 months.

Successful Phase I outcomes position firms for Phase II. This is a larger ~$750k–$1.8M award that funds prototype development, testing, and early validation over roughly two years.

Phase III derives from prior SBIR work, receives no SBIR dollars, and may be awarded non-competitively with a statutory preference for the SBIR firm. Although Phase III offers no direct funding, it enables commercialization by prioritizing awardees for government procurement or facilitating transition to private capital.

For engineering-intensive companies, SBIR/STTR funding serves as a bridge across the so-called "valley of death." That is the gap between promising research and investable prototypes. These awards help derisk technology through structured milestones, while external review provides third-party validation that is often persuasive for venture investors and strategic partners. Historical examples such as Qualcomm's early Code Division Multiplex Access (CDMA) development and

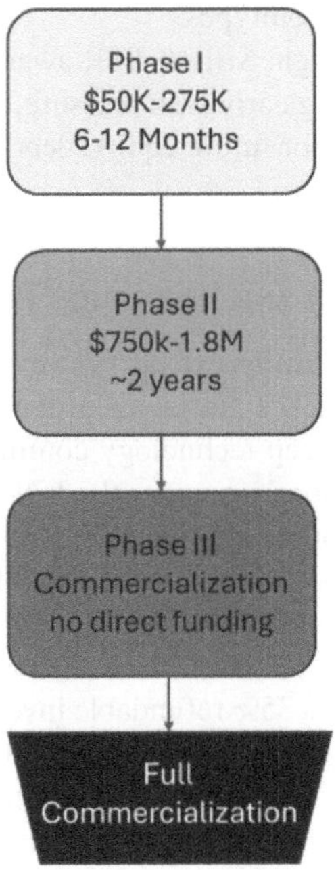

FIGURE 7.11 SBIR/STTR funnel.

iRobot's initial robotics work demonstrate how nondilutive awards can underpin breakthrough technologies long before commercial markets or private investors are ready to fully engage [SBIR 2017].

EXAMPLE 7.22 USING PHASED SBIR FUNDING TO BRIDGE FEASIBILITY AND PROTOTYPE VALIDATION

A therapeutics startup pursuing a novel assay can use an NIH SBIR Phase I to show technical feasibility, then a Phase II to fund prototype validation and early testing. The Phase I/II cadence finances proof, establishes external review credibility, and delays dilution until clinical or regulatory evidence is in hand. This is precisely the feasibility-to-prototype bridge that these programs are designed to provide.

However, reliance on SBIR/STTR funding introduces operational burdens that EST companies must plan for. Reporting requirements are stringent, reimbursement timing can be slow, and milestone expectations may not always perfectly match engineering realities. Companies must also navigate IP considerations, including government rights clauses and expectations around publication.

Nondilutive funding enables EST organizations to advance early-stage technical readiness while preserving runway and ownership, making it a foundational bridge from feasibility to investable prototypes.

When used effectively, though, SBIR/STTR awards complement internal capital allocation decisions by reducing early cash pressure, extending runway, and creating a clearer technical foundation for future equity, debt, or strategic financing [Brealey et al 2020].

7.6.2 International Grant and Innovation Programs

Global programs can meaningfully extend runway for EST firms when equity is premature or expensive. The EU's Horizon Europe/EIC Accelerator offers sizable nondilutive awards aimed at deep technology commercialization (commonly up to €2.5M), while multilateral initiatives under the UN/UNDP/UNIDO umbrella fund cleantech and infrastructure pilots aligned to the SDGs. These are often structured as staged pilots that de-risk field deployment. The EIC Accelerator is a grant of up to €2.5M plus an equity component of ~€1–€10M (higher under STEP ScaleUP [European Commission n.d.].

Canada's SR&ED provides a 35% refundable investment tax credit for Canadian-Controlled Private Corporations on qualified expenditures (15% base rate above it). 2024–2025 enhancements increase limits and extend refundability to some public corporations. Industrial Research Assistance Program commercialization grants create a blended nondilutive path that can substitute for early equity [NRCC 2023].

Strategically, international funding should be mapped to milestone gates rather than calendar expectations. EU and UK grants typically emphasize TRL advancement, consortium strength, and impact, which pair naturally with milestone-linked budgeting and portfolio gating in EST organizations.

Canadian SR&ED works differently. Benefits accrue retroactively through the tax system, so teams forecast claims alongside IRAP cash inflows to smooth liquidity between engineering builds and partner pilots. Designed well, the mix complements tranche-based equity by pulling forward specific proof points (prototype → pilot) without locking into irreversible CapEx.

EXAMPLE 7.23 USING INTERNATIONAL GRANT PROGRAMS TO ADVANCE TRL AND DE-RISK COMMERCIALIZATION

1) A robotics sensing startup secures an EIC Accelerator grant to fund environmental and EMI validation, then pairs it with a strategic manufacturing partner for pilot builds. This advances TRL while delaying equity until pilot reliability data is in hand.
2) A Canadian AI-hardware venture sequences SR&ED credits to offset R&D payroll and wins an IRAP grant for field pilots, using the predictable nondilutive inflows to gate CapEx and time a later equity round around proven reliability metrics.

Operationally, these grant programs carry real obligations, such as regular reporting, specific deliverables, and local-content or partnership requirements. Horizon Europe and Innovate UK awards often require multi-party governance and periodic reviews. UN-aligned pilots can introduce host-country procurement and compliance steps. SR&ED/IRAP demand disciplined cost tracking to substantiate claims and milestones.

Therefore, teams should plan for reimbursement timing and documentation overhead up front, building buffers into runway models and aligning consortium roles to internal decision rights so funding releases remain synchronized with engineering evidence for these reviews.

7.6.3 From Grants to Commercialization: The Development Pathway

Grant milestones should be treated as explicit evidence gates on the road from feasibility to commercialization. In practice, Phase I establishes technical feasibility; Phase II funds prototype development and validation; and the post-award Phase III period is where commercialization and procurement pathways come into focus. Mapping these stages to TRLs keeps capital release synchronized with demonstrated readiness rather than calendar targets. This type of discipline is well supported by TRL guidance and commercialization studies.

Well-timed Phase II and European pilot programs can attract strategic partners or investors by converting technical proof into credible market signals. The external review and staged nature of SBIR/STTR create third-party validation that lowers perceived risk and improves financing terms. International analogs like Horizon Europe's EIC are similar [NRC 2015].

EXAMPLE 7.24 LEVERAGING EIC FUNDING AND STRATEGIC PARTNERSHIPS FOR EVIDENCE-ALIGNED SCALING

A robotics sensing venture wins an EIC Accelerator grant for environmental and EMI validation, pairs it with a strategic manufacturing partner for pilot builds, and delays equity until pilot reliability data is in hand.

International nondilutive programs, such as the EU's EIC Accelerator and Canada's SR&ED/IRAP, also reinforce a milestone-first posture by phasing awards to TRL advancement and predictable reimbursement. This approach pairs naturally with tranche-based equity decisions.

7.6.4 Risks, Constraints, and Operational Requirements

Grant programs deliver real leverage, but they come with reporting and compliance overhead that must be planned, not discovered. Teams should budget time and tooling for progress reports, audits, and documentation that agencies require, and assume a steady cadence throughout the award. Treat these obligations as part of the work breakdown structure, not as administrative afterthoughts.

Another significant problem is that milestone rigidity can diverge from engineering reality. When grant deliverables or review gates no longer map neatly to what the lab is learning, there needs to be an effort to re-scope early and push for evidence-based amendments rather than forcing the build to match the calendar. The operational rule stays the same as elsewhere in this chapter. Namely, cash release and spending should follow technical readiness, with gates tied to objective proof rather than dates.

Here is another problem—reimbursement lags and cash-timing matter. Even when awards are secured, disbursements can arrive weeks to months after costs are incurred, compressing the runway if we have not modeled the lag. Companies need to incorporate these delays directly into the forward cash view and trigger auto-reforecasts when milestone dates move, so runway thresholds are protected without creating stress.

Finally, IP and program-market alignment deserve explicit scrutiny. Government rights clauses and publication expectations should be reviewed up front, and grant-dependency needs to be avoided by pressure-testing that funded milestones still support commercialization paths. When alignment is weak, managers need to scale back scope or pair grants with partner pilots that keep the roadmap commercially relevant [SBIR 2025].

EXAMPLE 7.25 ADJUSTING GRANT DELIVERABLES AND SPEND TIMING AFTER TECHNICAL FINDINGS

A sensing-hardware startup secures a Phase II award with quarterly reporting and a fixed deliverable sequence. Midway, thermal test results force a redesign that would slip the planned EMI validation. The team petitions to shift the gate by one review cycle and re-phases spending to match the new build order, modeling a 60-day reimbursement lag to preserve its 12-month runway.

The result is that milestones are realigned to evidence, cash timing updated, and publication/IP terms are reviewed before pilot-customer data is shared.

By proactively managing these risks and aligning grant execution to technical evidence, EST organizations/ensure that nondilutive funding strengthens their path from feasibility to commercialization.

7.6.5 Case Studies and Pathway Illustration

The following two concise examples illustrate how nondilutive grants can be sequenced to advance technical readiness and de-risk commercialization, consistent with the pathways outlined for this section.

EXAMPLE 7.26 ADVANCING TRL THROUGH PHASED GRANTS AND STRATEGIC PILOT INTEGRATION

A deep-tech sensing startup begins with an NIH SBIR Phase I to prove the feasibility of a novel biosensing stack, then secures Phase II to fund prototype validation and environmental/EMI testing. The startup treats each grant milestone as an explicit evidence gate tied to TRL advancement.

The team times hiring and external lab work behind these gates, uses the peer-reviewed Phase II outcomes to open a strategic pilot with a hospital network, and defers irreversible CapEx until reliability targets are demonstrated under field conditions.

Operationally, the company incorporates reporting, reimbursement lags, and IP terms into its plan from day one. Quarterly reports are scheduled as sprint reviews, a 60-day disbursement delay is modeled into the runway, and publication/government rights are reconciled with partner data-sharing before pilot launch.

The result is disciplined progress that preserves optionality, aligns cash release with evidence, and positions the firm for a priced round off validated performance.

EXAMPLE 7.27 PAIRING INTERNATIONAL GRANTS WITH PARTNER PILOTS TO ENABLE EVIDENCE-FIRST SCALE-UP

A robotics sensing venture wins an accelerator award to complete environmental validation and pairs it with a strategic manufacturing partner to execute a constrained pilot build. The grant funds de-risk the validation work while the partner shortens lead times and spreads integration risk, allowing the team to postpone equity until field reliability and yield data are in hand. This pairing demonstrates how international programs can extend runway and move a system from prototype toward commercialization without forcing premature dilution.

Governance ensures that milestone-driven grant deliverables are mapped to TRL gates; partner commitments are gated to reliability KPIs; and internal CapEx is sequenced behind proof (e.g., tooling only after pilot MTBF is met).

The company maintains a blended cash view that includes grant reimbursement timing and partner payment terms, ensuring runway thresholds hold through validation and pilot deployment. This illustrates the evidence-first, liquidity-aware posture described in this chapter.

Together, these examples show how firms can convert phased grants into investable proof points by coupling evidence-gated execution with runway-aware cash planning. Across budgeting, capital allocation, financing, and grants, the throughline is constant: sequence spend to engineering evidence, not the calendar, and re-forecast automatically when milestones move.

7.7 EXERCISES

7.1 Define the primary purpose of budgeting and forecasting in engineering-, science-, and technology-driven (EST) organizations. Explain why these functions differ from budgeting in traditional service firms.

7.2 Create a short financial SWOT analysis (one item per quadrant) for a hypothetical early-stage robotics company.

7.3 Explain the difference between operating budgets and capital budgets. Provide two examples of cost items that belong in each.

7.4 Given an EST company with long prototype cycles, describe which budgeting timeline (traditional, rolling, milestone-based, and hybrid) is best suited and justify your answer.

7.5 Describe a gatekeeping process (e.g., Stage-Gate) and explain how evidence-based gates help prevent overspending in EST firms.

7.6 Select one budgeting approach (incremental, activity-based, value-based, zero-based). List two advantages and one limitation when applied to an EST program.

7.7 Identify three technical assumptions that must be incorporated into revenue or cost forecasting for deep-tech companies (e.g., yields, regulatory timing, and cycle times). Explain why excluding one of them could distort forecasts.

7.8 Using a hypothetical example, describe how milestone-linked forecasting reshapes cash flow planning when a critical milestone slips by one quarter.

7.9 Explain the importance of liquidity readiness. List two actions an EST company can take when the runway drops below a guardrail threshold.

7.10 Write a short scenario demonstrating how a ±5% change in a key technical parameter (e.g., yield, training efficiency, and cycle time) affects a company's financial feasibility.

7.11 Given a quarterly burn plan (Q1–Q4), identify two early warning metrics that signal the need for a mid-year reforecast.

7.12 Using the portfolio heatmap logic described in Section 7.2.4 and the example of Alphabet's portfolio, categorize and prioritize the following four initiatives.
- Platform A: High Readiness/High Strategic Value (e.g., Core Infrastructure).
- Platform B: Low Readiness/High Strategic Value (e.g., Quantum Computing).
- Feature C: High Readiness/Low Strategic Value (e.g., Legacy App features).
- Research D: Low Readiness/Medium Strategic Value.

Then answer the following questions:

a. Which initiative should receive the next marginal dollar of capital to sustain the "economic engine" of the firm?

b. Explain why Platform B should be treated as a "Real Option" with milestone-gated funding rather than receiving full commercialization assumptions.

c. Based on the source's guidance for "Low Strategic Value" quadrants, what action should leadership take regarding Feature C to improve overall capital efficiency?

7.13 Explain how TRL-based frameworks influence the sequencing of CapEx in long-lead engineering programs.

7.14 Describe the trade-offs among build, buy, and partner decisions for an EST company developing a novel sensing subsystem.

7.15 Write a brief due-diligence checklist (five items) for evaluating a potential acquisition in the EST industry.

7.16 Explain one common failure mode in M&A or partnerships involving technical startups and outline how better technical diligence could have mitigated it.

7.17 Compare commercial paper, venture capital, and venture debt along three dimensions: cost of capital, risk, and timing fit for EST milestones.

7.18 Provide a short example of tranche-based financing. Identify (a) the triggering milestone and (b) the mitigation step if the milestone slips.

7.19 Explain how burn-rate sensitivity to technical delay can create a "runway cliff." Produce a 2–3 sentence scenario to illustrate this effect.

7.20 A startup's verification milestone slips by two sprints, triggering an automatic reforecast that shows the runway falling below the minimum threshold in nine months.

Describe two immediate liquidity-preservation actions the company could take and explain how each action protects the runway without halting technical progress.

7.21 Drawing on the NovaSense Analytics case study (Section 7.5.8) and the SBIR/STTR funnel, analyze the financing strategy for a deep-tech materials startup. Answer the following questions:
   a. Explain why the company should pursue an SBIR Phase I grant before closing a dilutive pre-seed round. How does this sequence anchor the valuation to technical evidence rather than "calendar aspirations"?
   b. Identify the primary risk of pursuing these in the wrong order (e.g., premature dilution). Explain how taking equity too early impacts the founders' ownership and long-term strategic flexibility.
   c. How does securing a phased grant improve the terms of the eventual dilutive round when the company finally approaches institutional investors?

7.22 Describe how nondilutive grants (such as SBIR/STTR or international equivalents) can be integrated with equity and debt to create a milestone-first financing strategy. Give one brief example.

NOTES

1 A brief period (e.g., 1–4 weeks) during which an engineering team focuses on completing a clearly defined set of technical work.

2 A credit mechanism that allows a borrower to draw, repay, and re-borrow funds repeatedly up to a limit.

3 An individual who invests personal funds in very early-stage companies.

4 Congressional authority for SBIR/STTR was still pending as of February, 2026. But due to its longevity and importance, expect this program to be reauthorized or at least replaced.

REFERENCES

Berk, J., & DeMarzo, P. (2020). *Corporate finance* (5th ed.). Pearson.

Brealey, R. A., Myers, S. C., & Allen, F. (2020). *Principles of corporate finance* (13th ed.). McGraw–Hill Education.

Cooper, R. G. (1990). Stage-gate systems: A new tool for managing new products. *Business Horizons, 33*(3), 44–54. https://doi.org/10.1016/0007-6813(90)90040-I

Copeland, T., & Antikarov, V. (2003). *Real options: A practitioner's guide.* Texere.

Damodaran, A. (2015). *Applied corporate finance* (4th ed.). Wiley.

Dixit, A. K., & Pindyck, R. S. (1994). *Investment under uncertainty.* Princeton University Press.

European Commission. (n.d.). *EIC accelerator.* European Innovation Council. https://eic.ec.europa.eu/eic-funding-opportunities/eic-accelerator_en, accessed 2/16/2026.

Gompers, P., & Lerner, J. (2004). *The venture capital cycle* (2nd ed.). MIT Press.

Jensen, M. C. (1986). Agency costs of free cash flow, corporate finance, and takeovers. *American Economic Review, 76*(2), 323–329.

Koller, T., Goedhart, M., & Wessels, D. (2020). *Valuation: Measuring and managing the value of companies* (7th ed.). Wiley.

Mankins, J. C. (2009). Technology readiness assessments: A retrospective. *Acta Astronautica, 65*(9–10), 1216–1223. https://doi.org/10.1016/j.actaastro.2009.03.058

National Research Council (NRC). (2015). *SBIR/STTR at the department of energy.* The National Academies Press.

National Research Council Canada (NRCC). (2023, December 20). *About the NRC industrial research assistance program.* Government of Canada. https://nrc.canada.ca/en/support-technology-innovation/about-nrc-industrial-research-assistance-program

Small Business Innovation Research Program (SBIR). (2017). *iRobot: SBIR/STTR success story.* U.S. Small Business Administration. https://www.sbir.gov/sites/default/files/documents/SBA_SuccessStories_iRobot.pdf

Smith, M. S. (2025, October 18). Qualcomm's Arduino acquisition stirs open–source debate. *IEEE Spectrum.* https://spectrum.ieee.org/qualcomm-arduino-acquisition-open-source

U.S. Small Business Administration. (2025). *SBIR/STTR program overview.* https://www.sbir.gov/about

Wheelwright, S. C., & Clark, K. B. (1992). *Revolutionizing product development: Quantum leaps in speed, efficiency, and quality.* Free Press.

8 Financial Modeling and Data Analysis

Financial models are decision tools, not spreadsheets with pretty dashboards. This chapter provides a disciplined approach that starts with a clear separation of inputs, calculations, and outputs, supported by modular structure, transparent logic, and version control so changes are traceable and audit-ready. It then grounds the financial models rigorously using time structures, revenue/cost/working-capital blocks, and three-statement integration.

Next, we will see how to create scenarios, run sensitivities, and use probability models and Monte Carlo simulation so risk is communicated as clearly as returns. Along the way, quality control and data rigor are emphasized. This rigor is achieved through validation checks, scenario toggles, and exploratory analysis that uncover outliers, schema drift, and feature signals before they contaminate forecasts.

The chapter closes with the governance you need to trust the numbers. To do this, we will review documentation standards, independent review, and ethical guardrails that keep sophisticated models aligned with purpose and compliant with oversight. The main theme is simple: build transparently, analyze uncertainty honestly, and operate under controls. In this way your model becomes a reusable, auditable asset, not a source of surprises.

8.1 BUILDING FINANCIAL MODELS WITH SPREADSHEETS

Financial spreadsheets serve as structured analytical tools that transform assumptions and inputs into clear, decision-ready outputs. A disciplined modeling approach improves auditability and reduces errors. This section introduces the principles and architecture that support reliable spreadsheet-based financial models, setting the foundation for effective forecasting and analysis.

8.1.1 Modeling Philosophy and Best Practices

A sound financial model begins with a clear distinction between assumptions, inputs, and outputs. Assumptions capture the economic logic or strategic beliefs driving the model, while inputs reflect measurable, user-controlled variables such as prices, volumes, or cost drivers. Outputs then summarize the model's conclusions and must remain mechanically linked to, but visually separated from, all inputs. Maintaining this separation both improves auditability and reduces the likelihood of hidden logic or embedded "hard-codes" that undermine the integrity of the model.

Equally important is the principle of *separation of concerns*, which dictates that inputs, calculations, and reporting each occupy distinct and clearly labeled areas of

DOI: 10.1201/9781003733850-8

the workbook. This structure mirrors good engineering design and makes the model more modular, maintainable, and scalable.

EXAMPLE 8.1 ENGINEERING DISCIPLINE IN FINANCIAL MODEL DESIGN

Siemens routinely embeds process discipline in their engineering and financial planning models, ensuring that design specifications, computational layers, and presentation dashboards remain logically decoupled. This approach reduces error propagation and allows teams to modify assumptions or extend features without reworking foundational components [Siemens 2024].

Finally, effective financial modeling requires a commitment to version control and transparent change tracking. Even in spreadsheet environments, maintaining a clear record of updates, parameter changes, and methodological revisions is essential for both internal governance and external review. A concise change log, which is updated alongside each iteration, ensures that collaborators understand why results shift over time. More importantly, this practice reflects the broader engineering and analytics culture of reproducibility, where the credibility of insights depends on the traceability of the underlying process [Bailey and López de Prado 2021].

8.1.2 Spreadsheet Architecture

A well-structured financial model begins with a workbook architecture that promotes clarity, navigability, and error-resilience. Core elements such as workbook structure, model maps, and navigation aids establish a logical flow that users can follow without ambiguity. Clearly defining where inputs reside, how calculations are sequenced, and where outputs are consolidated ensures that the model behaves predictably and reduces the risk of users modifying formulas unintentionally. These foundational practices align with the need for a transparent layout, as indicated in the original outline of structure and navigation considerations.

Robust spreadsheet architecture also depends on tools that enforce consistency and prevent silent errors. Techniques such as named ranges, data validation, and formalized error checks help maintain internal discipline and prevent common modeling pitfalls. By standardizing how data flows through formulas and ensuring that every referenced value is intentional and traceable, modelers can avoid subtle logical breaks that accumulate over time. This approach aligns with the emphasis on validation and structured referencing included in the section framework.

It is also very important to maintain change tracking discipline with any financial spreadsheet. For example, every financial spreadsheet should have a simple change log table at the front or end (e.g., see Table 8.1).

Scalable models rely heavily on dynamic ranges and structured references, which allow spreadsheets to adjust automatically as datasets grow or analytical requirements evolve. This dynamic design mirrors the approach used by firms like

TABLE 8.1
Sample Change Log Table

Date	Version	Author	Change Description	Impacted Cells/Sheets
2025-12-01	1.0	Phil	Initial build—base case assumptions	All
2025-12-10	2.1	Jane	Updated tariff rate to 15%; added sensitivity	Assumptions!B12:B15

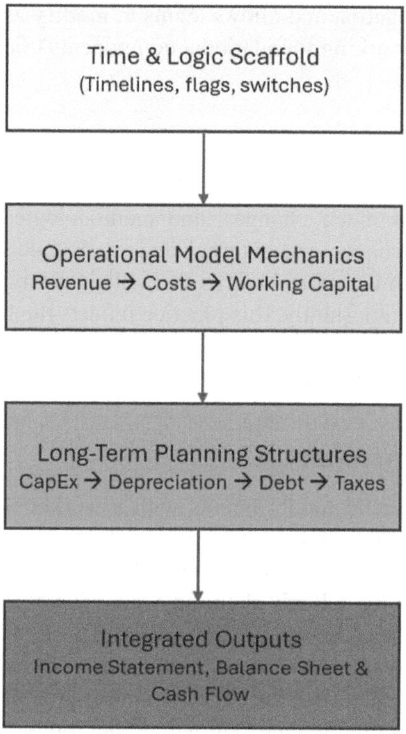

FIGURE 8.1 Financial planning core building blocks.

NVIDIA, whose data-driven engineering workflows require models that adapt to rapidly changing inputs and large-scale data environments [NVIDIA 2024].

By treating the financial model as a living engineering artifact that is documented, modular, and versioned, you will reduce errors, build stakeholder confidence, and make it reusable across projects. A disciplined model becomes a strategic asset, not a source of confusion [Berkowitz 2019].

8.1.3 Core Building Blocks

Effective financial models rely on a set of foundational building blocks that structure how economic activity is translated into forecasted outcomes (Figure 8.1).

Time-based elements, such as monthly, quarterly, or annual timelines, along with clearly defined flags and switches, create the scaffolding that governs model behavior. These features ensure that calculations activate only under the appropriate conditions and that the temporal progression of the model remains consistent and auditable.

Revenue, cost, and working-capital mechanics form the next layer of structure. Revenue models typically decompose performance into price × volume, while more advanced settings incorporate cohorts, churn, or retention dynamics. Cost modeling then distinguishes fixed and variable components, accommodates step-cost behavior, and incorporates driver-based logic that ties expenses directly to business activity.

Working capital schedule ensures the model captures the timing of cash movements with sufficient realism. This means accurately maintaining accounts receivable, accounts payable, and inventory ledgers. Together, these components form the operational core from which liquidity, profitability, and efficiency metrics emerge [Bailey and López de Prado 2021].

Long-term planning elements further strengthen the model by introducing capital structure and tax considerations. CapEx schedules, paired with appropriate depreciation or amortization conventions, govern how future investment influences capacity and expense recognition, while debt and interest modeling capture revolvers, term loans, and covenant mechanics.

Incorporating taxes and loss *carryforwards* provides additional realism by aligning projected earnings with regulatory requirements. Companies operating in capital-intensive industries, like Lockheed Martin, routinely model multi-year production cycles for aerospace programs. They then depend on these building blocks to maintain accurate, scenario-ready forecasts across long time horizons [Lockheed 2024].

8.1.4 Integrating the Core Financial Statements

Integrating the income statement, balance sheet, and cash flow statement is a core requirement of any rigorous financial model. The objective is to ensure that all three statements are mechanically linked so that operational decisions flow consistently through revenue, profitability, cash generation, and financial position. Establishing these connections requires disciplined structuring of supporting schedules. For example, working capital, depreciation, and financing artifacts should tie each line item back to a logical economic driver. This integration aligns with the section's emphasis on linking the three primary statements within a unified modeling framework.

Because these statements are interdependent, *circularity management* becomes an essential part of the modeling process. For example, interest expense, depending on ending debt, which is itself influenced by cash flow, is a circular relationship. Circular relationships require controlled iteration to avoid unstable or misleading results.

High-quality integration enhances the model's usefulness for forecasting and decision-making. Organizations operating in complex engineering or technology environments, such as General Electric, rely on robust three-statement models to evaluate capital-intensive projects, assess liquidity under multiple scenarios, and stress-test financing strategies [General Electric Company 2024].

By aligning operational forecasts with balance sheet capacity and cash flow constraints, these firms ensure that strategic initiatives are grounded in a complete financial perspective. Well-integrated financial statements serve not only as an analytical foundation but also as a governance tool that helps stakeholders understand the implications of major decisions [Bailey and López de Prado 2021].

8.1.5 Quality Control and Audit

Quality control in financial modeling begins with a disciplined set of *balance checks* and *reasonableness tests* designed to flag mechanical inconsistencies before they propagate through the model. These checks ensure that the model's core relationships—such as the balance sheet balancing, cash reconciling across statements, and ratios remaining within plausible ranges—hold under all modeled conditions. The original section highlights the importance of these structural tests as foundational to maintaining model integrity.

Beyond structural accuracy, robust models incorporate *scenario toggles* and *sensitivity panels* to validate how outputs respond to changes in key assumptions. These tools help identify nonlinear behaviors, assumption sensitivities, and potential error amplification, especially in complex or long-horizon forecasts. Incorporating these diagnostic mechanisms not only strengthens the model's analytical rigor but also provides decision-makers with transparent visibility into which drivers matter most [Bailey and López de Prado 2021].

Effective audit practices extend beyond formula checks to include peer review, documentation clarity, and traceable revisions. Engineering-driven organizations such as Texas Instruments, known for their rigorous design validation processes, apply similarly structured review cycles to their financial and operational planning models [Texas Instruments 2024].

By combining transparent logic flows with periodic audits and version-controlled updates, organizations create models that are both technically sound and dependable for strategic decision-making. This level of governance reinforces trust, reduces operational risk, and ensures that the model remains robust as assumptions evolve.

8.2 USING DATA ANALYTICS FOR INVESTMENT DECISIONS

Investment decisions in EST finance depend on analytical workflows that begin with high-quality, well-structured data. After organizing this data, the next step is to conduct an exploratory data. Effective data analytics ensures that valuation, forecasting, and portfolio models rest on reliable market, fundamental, and alternative datasets—each with its own ingestion, validation, and transformation challenges. This section outlines the data sources, exploratory methods, and analytical tools that convert raw information into credible financial insight.

8.2.1 Data Sources and Ingestion

High-quality investment analytics begin with high-quality data. In EST finance contexts, data ingestion is not merely a preprocessing step but a foundational design decision. The reliability of downstream valuation, forecasting, and portfolio analytics depends on properly selected, validated, and structured data from diverse sources. Investment models consume four primary data classes: market data, fundamental data, alternative data, and metadata/reference data. Each class introduces specific ingestion challenges such as timeliness, completeness, survivorship bias, and schema consistency.

8.2.1.1 Market Data

Market data includes prices, volumes, order-book information, and index levels. These datasets often arrive at high frequency and require pipeline designs that handle timestamp alignment, missing intervals, and outliers. *Fundamental data* like financial statements, guidance, segment disclosures, and regulatory filings provide the structural backbone for most valuation models.

These data originate from mandated corporate reporting and must be checked for restatement history, reporting lags, and accounting-standard changes. *Alternative data*, now widely used in institutional investing, expands analytical reach by capturing behavioral, transactional, or environmental signals not found in traditional financial statements. Examples include satellite imagery, credit-card panel data, job-posting trends, supply-chain telemetry, and web-scraped sentiment.

As noted in contemporary financial modeling literature, locating and integrating appropriate data sources is critical, as these sources often materially influence forecasting accuracy and scenario realism.

8.2.1.2 Data Quality

A robust ingestion workflow must address three engineering concerns. First, we should improve data quality, such as handling missing fields, inconsistent units, schema drift, and transformation errors, through validation checks, summary statistics, and automated alerts.

Second, data lineage and documentation must be clarified through clear tagging of sources, timestamps, and transformations so that analysts can audit and replicate decisions.

Finally, scalability and modularity must be enforced. For example, the pipeline should allow drop-in replacement of data sources (e.g., switching providers or adding new alternative datasets) without rewriting downstream models. EST-focused teams often maintain a "data assumptions" document akin to a financial model's assumptions sheet, capturing not only data sources but also caveats, update frequencies, and known limitations [CFI 2025].

EXAMPLE 8.2 MULTI-SOURCE INGESTION FOR A CLEAN-ENERGY INVESTMENT SCREEN

A clean-energy fund builds a screening model for early-stage battery manufacturers. Its ingestion pipeline pulls:

- **Market data**: Daily equity prices for peers and suppliers, used to calibrate comparable-company volatility.

- **Fundamental data**: SEC filings for R&D intensity, CapEx trends, and segment disclosures.
- **Alternative data**: Satellite data on factory construction progress and job-posting trends for engineering roles to proxy hiring momentum.

During ingestion, analysts detect a schema change in the supplier-price dataset. That is, volumes are now reported in MWh instead of kWh. Because the data lineage system logs all transformations, the discrepancy is flagged automatically, preventing a 1,000× scaling error in cost-curve projections.

The final integrated dataset feeds into the model's scenario layer, improving forecast credibility and producing materially different investment conclusions.

Managing data quality ensures that every dataset feeding the model is reliable, traceable, and resilient enough to support credible analytics and trustworthy investment decisions.

8.2.2 Exploratory Data Analysis

Exploratory data analysis (EDA) is the critical first pass over any dataset. EDA is an analytical "shake-down" that reveals structure, irregularities, and directional signals before more formal modeling begins. In EST-finance contexts, where datasets may mix technical variables (e.g., energy yields, throughput, and degradation rates) with financial indicators (e.g., margins, volatility, and unit economics), EDA establishes whether downstream results will rest on solid ground or misleading inputs.

EDA focuses on three foundational activities: summary statistics, distribution and outlier analysis, and feature engineering for model-ready signals. These are described next.

8.2.2.1 Summary Statistics and Data Quality Signals

The first task of EDA is to characterize each variable using basic descriptive statistics like mean, median, variance, skewness, kurtosis, and frequency counts for categorical fields. These statistics provide a rapid sense of central tendencies, dispersion, and irregularities. As described in foundational EDA literature, summary statistics highlight missing values, inconsistent entries, unexpected ranges, and anomalies that could bias later modeling steps.

For example, if a variable representing operational throughput in a clean-energy deployment exhibits an impossible negative minimum, summary statistics immediately flag the issue. These early checks prevent compounding errors during forecasting, especially when data feed valuation models or risk analytics.

8.2.2.2 Distributions, Outliers, and Structural Patterns

Visualizing distributions using histograms, density plots, boxplots, and violin plots reveals shape (normal, skewed, multimodal), tail risk, and boundary anomalies. Outlier detection methods, such as interquartile range rules or z-score thresholds, help distinguish genuine structural shocks (e.g., extreme price spikes or operational downtime events) from data-entry errors. EDA aims to uncover relationships and patterns not immediately visible through summary statistics alone; for instance, scatterplots or time-based visuals can surface correlations, nonlinear trends, or seasonal effects.

In financial-technical datasets, these patterns are especially useful. For example, correlating battery cycle-counts with degradation rates or linking market volatility to utilization dips.

8.2.2.3 Feature Engineering: Lags, Moving Averages, and Factor Creation

Once distributions and patterns are understood, analysts begin constructing derived variables such as lagged features (e.g., t-1 demand), moving averages (e.g., 7-day smoothed energy output), and engineered factors (e.g., utilization-adjusted cost index). Feature engineering is a core objective of EDA because it transforms raw data into meaningful predictors that encode domain relationships.

As noted in practical EDA guidance, feature engineering improves downstream learnability by capturing structure such as shocks, trends, seasonality, or cross-factor interactions. These engineered features form the basis of scenario modeling, forecasting models, and portfolio analytics that rely on stability and predictive strength.

EXAMPLE 8.3 EDA FOR A BATTERY-STORAGE INVESTMENT DATASET

A fund evaluating early-stage battery-storage projects begins its EDA on a dataset that includes hourly energy-output logs, ambient temperature readings, degradation rates, CapEx line items, and market price series [DASCA 2024].

1. **Summary Statistics:**
   Descriptive metrics reveal that degradation rate entries contain several missing values and one outlier reading at 75%, which is far above expected engineering ranges, prompting investigation.
2. **Distribution Analysis:**
   Histograms show that hourly energy output is bimodal, reflecting day-night cycles. Boxplots reveal several zero-output intervals that correspond to scheduled maintenance windows rather than sensor failure.
3. **Feature Engineering:**
   Analysts engineer moving-average smoothed output curves, temperature-adjusted efficiency factors, and lagged degradation variables. These derived features later significantly improve variance-reduction in the financial forecast model.

Through this workflow, the team ensures the dataset is not only clean and internally consistent but also enriched with explanatory features essential for accurate scenario modeling and investment screening.

These diagnostics de-risk downstream modeling by ensuring the signals we model are real, stable, and interpretable.

8.2.3 Valuation Analytics

Valuation analytics convert financial and operational insights into actionable estimates of enterprise value. In EST-finance contexts, analysts rely on two complementary valuation frameworks: Discounted cash flow (DCF) analysis and Relative (comparable-company) valuation. Each method captures a different dimension of value.

DCF emphasizes intrinsic economics rooted in cash-flow fundamentals. Relative valuation reflects prevailing market sentiment.

8.2.3.1 DCF Inputs: Growth, Cash Flows, Reinvestment, and WACC

DCF valuation models forecast future cash flows and discount them at a rate consistent with the underlying risk. As noted in foundational corporate-finance guidance, DCF models are highly sensitive to long-term growth assumptions, discount rates such as *weighted average cost of capital (WACC)*, and terminal-value structure—where even small assumption changes can create large valuation swings.

Core DCF inputs include:

- **Growth drivers:** revenue expansion, margin improvement, and scaling of operational efficiency.
- **Reinvestment needs:** CapEx cycles, working-capital behavior, and maintenance costs.
- **Discount rates:** WACC or cost of equity consistent with the risk profile, capital structure, and currency of cash flows.

Because of this sensitivity, analysts routinely perform scenario analysis and use multi-variable sensitivity tables to quantify how uncertainty propagates through the valuation range. This practice is essential in EST domains where technical milestones, manufacturing yields, and regulatory timelines can materially alter forward cash-flow trajectories.

8.2.3.2 Relative Valuation: Comps Selection and Multiples Normalization

Relative valuation benchmarks the target against a peer group using market-derived multiples, like P/E, EV/EBITDA, *price-to-to-sales (P/S)*, or *price-to-book (P/B)*. This method is fast, market-reflective, and often more stable than DCF when cash-flow forecasts are volatile or uncertain. However, its accuracy depends on strict comparability, as mismatched peers introduce noise stemming from differing growth rates, risk profiles, and accounting treatments.

Multiples (market-based valuation ratios) must also be normalized across firms to avoid distorted conclusions. For example, adjusting them for differences in margin structure, capital intensity, or accounting policy. When applied correctly, comps offer a powerful cross-check on the reasonableness of intrinsic (DCF-derived) estimates.

8.2.3.3 Reconciling DCF and Relative Valuation

Both approaches illuminate different aspects of value. DCF provides a structured, economically grounded view, while comps capture real-time market pricing. Divergence between the two often signals assumption mismatches, peer-group misalignment, or shifts in market risk appetite. Best practice is to present valuation ranges rather than point estimates, transparently documenting key assumptions and sensitivities.

EXAMPLE 8.4 VALUATION ANALYTICS FOR A ROBOTICS-AUTOMATION FIRM

A mid-stage robotics-automation firm seeks funding to scale production. Analysts perform both intrinsic and market-based valuation:

1. **DCF Analysis:**
   - Revenue projections incorporate robotics unit-economics and expected adoption curves.
   - CapEx ramps reflect tooling, automation cells, and factory expansion.
   - WACC reflects both technology risk and cyclicality in industrial-equipment demand.
   - Sensitivity testing shows that a ±0.5% change in WACC shifts valuation by nearly 10%, demonstrating the need for careful discount-rate calibration.
2. **Relative Valuation:**
   - Peer group includes precision-automation and industrial-controls firms.
   - Analysts compute EV/EBITDA and P/S multiples and normalize for differing margin structures and capital intensity.
   - Normalized comps yield a valuation modestly higher than the DCF midpoint, suggesting strong market sentiment in the automation sector.
3. **Integrated Output:**
   - Final valuation presented as a range, with DCF providing the fundamental floor and comps reflecting market-pricing context.
   - The board uses this dual-method range to negotiate investor terms and evaluate dilution [DashboardsExcel 2025].

A well-structured valuation process benefits from using both DCF and relative valuation frameworks since each highlights different dimensions of a firm's economics and market positioning. When these approaches are reconciled into a transparent valuation range, decision-makers gain a more robust foundation for investment judgments and sensitivity testing.

8.2.4 Portfolio Analytics

Portfolio analytics provide the quantitative foundation for evaluating risk, return, and structural exposures in an investment portfolio. In EST-finance contexts, analytics serve as the lens through which decision-makers understand how complex drivers interact. The three essential pillars of portfolio analytics are

- mean-variance intuition,
- risk decomposition,
- and factor-exposure diagnostics.

Let's look at each of these in turn.

8.2.4.1 Mean-Variance Intuition and Portfolio Structure

Mean-variance analysis frames the fundamental trade-off between expected return and portfolio volatility. While simple in concept, it becomes mission-critical when modeling portfolios containing illiquid or technologically uncertain assets. Modern analytics platforms enhance this framework by providing transparency into historical performance, volatility clustering, tracking error, and driver-level attribution. Tools supporting portfolio managers emphasize the importance of evaluating both absolute and benchmark-relative outcomes, highlighting the role of asset-class, sector, or geographic exposures in shaping long-run efficiency.

In practice, investors use mean-variance insights to determine whether incremental risk is compensated by incremental expected return, particularly when instruments have asymmetric payoff distributions or technology-linked regulatory risks.

8.2.4.2 Risk Decomposition and Return Attribution

Risk is rarely monolithic. Portfolio analytics allow risk to be decomposed into underlying components such as interest-rate sensitivity, credit spreads, equity-market beta, style factors, or idiosyncratic exposures. Decomposition can be performed in isolation (examining the stand-alone risk of a factor group) or as a contribution to overall risk, using a factor covariance matrix and defined factor-group inclusion matrices.

This decomposition enables EST-finance teams to understand how much of the total portfolio variance comes from macroeconomic drivers versus technology- or project-specific exposures. This is critical for sectors such as energy transition, biotech, semiconductors, and advanced manufacturing.

Furthermore, robust analytics systems provide performance attribution tools that evaluate historical outcomes by sector, security selection, or style tilt. This approach bridges operational outcomes with financial performance.

8.2.4.3 Factor-Exposure Diagnostics

Modern portfolio analytics rely heavily on *factor-based analysis*, where portfolios are modeled as bundles of risk premiums rather than collections of individual securities. Factors such as value, momentum, size, or sector-specific dynamics can be measured using linear-regression frameworks that relate portfolio returns to explanatory variables.

Factor-exposure diagnostics support two critical functions:

1. **Identifying hidden risks**—exposures that behave like "stealth bets," such as unintended momentum or interest-rate sensitivity.
2. **Distinguishing skill from compensated risk**—ensuring that observed outperformance is not simply the result of exposure to well-known risk premiums, but true growth.

EXAMPLE 8.5 PORTFOLIO ANALYTICS FOR A CLEAN-TECH/INDUSTRIAL INNOVATION FUND

A multi-strategy investment fund focused on clean-tech and industrial automation evaluates its portfolio using a three-layer analytics framework:

MEAN-VARIANCE REVIEW

Historical return data show that the fund exhibits elevated volatility relative to its benchmark during periods of supply-chain disruption. Scenario analysis reveals that a modest reduction in a high-beta robotics holding significantly improves the portfolio's risk-adjusted return.

RISK DECOMPOSITION

Using factor-group decomposition, analysts identify that 38% of total risk stems from "Growth/Innovation" exposures and 22% from interest-rate sensitivity. The "Rates" factor group is isolated through group-matrix analysis, confirming that long-duration clean-tech holdings amplify rate-shock vulnerability.

FACTOR-EXPOSURE DIAGNOSTICS

Regression analysis indicates moderate unintended exposure to the "momentum" factor, driven by strong performance in next-gen battery manufacturers. Additional diagnostics confirm that many of the gains previously attributed to security selection are explained by size and value-premiums.

Integrating these layers, the fund adjusts weightings to reduce uncompensated risks, enhance factor alignment, and improve resilience under future rate-shock scenarios.

For EST-focused portfolios, factor analysis reveals how technical cycles, regulatory shifts, or sectoral innovation waves influence performance, enabling more nuanced portfolio construction and scenario planning [Bloomberg 2025].

8.2.5 Performance Measurement and Attribution

Performance measurement and attribution provide the analytical backbone for evaluating investment outcomes in EST-finance contexts, where portfolios often blend technology-driven assets, long-duration infrastructure, and innovation-exposed sectors. The goal is not simply to compute returns, but to understand why those returns occurred and whether they resulted from repeatable skill or circumstance. A complete evaluation framework incorporates performance measurement, performance attribution, and performance appraisal, each layering insight onto the last.

8.2.5.1 Performance Measurement: Raw Returns and Baseline Signals

Performance measurement begins with the calculation of portfolio returns over a defined period. These are typically expressed as percentage gains or losses relative to the initial value. This step is essential for grounding all subsequent analytics, providing the "what happened" before exploring "why it happened."

In EST-relevant portfolios, performance measurement captures not just overall gains, but also the volatility and drawdown patterns that emerge from exposures to innovation cycles, regulatory developments, and supply-chain disruptions.

8.2.5.2 Performance Attribution: Breaking Down the Sources of Return

Attribution analysis explains the drivers of active return, that is, the difference between portfolio performance and benchmark performance. The most widely used framework is the *Brinson attribution model*, which decomposes excess return into three components:

Allocation Effect—Did the manager overweight or underweight sectors relative to the benchmark (e.g., renewable energy, robotics, and advanced materials)?

Selection Effect—Did the chosen securities outperform their sector peers?

Interaction Effect—The cross-term that captures simultaneous weighting and selection differences.

This decomposition creates a transparent link between decision-making and outcomes. This allows investors to determine whether outperformance came from being in the right sectors, picking the right names, or benefiting from a favorable combination of both.

8.2.5.3 Performance Appraisal: Distinguishing Skill from Noise

Once attribution highlights how the returns were generated, *performance appraisal* evaluates whether the performance was the result of skill or merely randomness. Statistical techniques, such as significance testing help determine whether manager outperformance is replicable or simply coincidental.

EXAMPLE 8.6 PERFORMANCE ATTRIBUTION FOR A GRID-MODERNIZATION EQUITY PORTFOLIO

A portfolio manager oversees a grid-modernization strategy, which includes investing in smart-meter companies, transmission-automation manufacturers and utility-software vendors. During a certain quarter, he evaluates a 6.2% portfolio return versus a 3.4% benchmark return:

PERFORMANCE MEASUREMENT:

The portfolio outperformed its benchmark by 2.8 percentage points, driven by strong quarter-over-quarter revenue growth in automation hardware suppliers.

ATTRIBUTION ANALYSIS:

Allocation Effect: Overweighting the grid-software subsector (15% vs. 6% benchmark) contributed positively when regulatory approvals accelerated spending.

Selection Effect: Within transmission-automation, chosen firms delivered superior returns (8.2% vs. sector benchmark 4.1%), driven by advanced sensor deployments.

Interaction Effect: The manager both over-weighted and selected top performers in grid-software, amplifying the positive contribution.

PERFORMANCE APPRAISAL:

Statistical review indicates that the strong selection effect aligns with a documented multi-quarter pattern of effective technology picking. This suggests genuine process-driven gains rather than random variance.

Performance appraisal is especially valuable in EST-heavy portfolios where temporary surges (e.g., AI adoption waves and green-energy policy announcements) may create performance spikes that don't reflect sustainable advantage.

8.2.6 Practical Pitfalls

Even the most sophisticated financial-analytics workflows can fail if foundational modeling pitfalls go unaddressed. In EST-finance contexts three issues repeatedly undermine model validity. These are: overfitting and data-snooping, look-ahead bias and leakage, and regime shifts that break historical assumptions. Let's look at all three in some detail.

8.2.6.1 Overfitting and Data-Snooping Bias

Overfitting occurs when a model captures noise rather than true structure, often because analysts over-optimize parameters, test too many variations of a strategy, or unintentionally tune models to historical anomalies. As documented in quantitative-finance research, "optimal" designs often prove to be *false discoveries,* for example, high-performing simulations that collapse in real-world trading.

Data-snooping bias, where repeatedly re-testing strategies on the same historical dataset inflates the likelihood of finding misleadingly "good" results that reflect chance rather than skill, worsens this problem. In competitive financial markets, where the probability of true persistent growth is low, this bias can lead to a structural overestimation of strategy quality.

8.2.6.2 Look-Ahead Bias and Information Leakage

Look-ahead bias occurs when future information like prices, fundamentals, earnings releases, or corporate actions is inadvertently included in a model's training, testing, or signal-generation logic. As explained in backtesting methodology research, look-ahead bias is one of the most pervasive sources of artificially inflated performance because it allows a model to "see the future" during simulation.

Practical causes include:

- Using end-of-day values to make start-of-day decisions.
- Failing to align time stamps across assets.
- Incorporating fundamentals updated after the period being modeled.
- Training machine-learning models on datasets that contain post-event observations.

These leakage pathways can occur subtly. For example, when technical indicators use the full day's trading range for intraday signals or when ML pipelines construct features using non-lagged variables.

8.2.6.3 Regime Shifts and Instability of Relationships

Market regimes are a general term for macro cycles, volatility regimes, policy shocks, and technological adoption curves. These can change the underlying statistical relationships in financial and technical datasets. Methods or models that appear robust in one environment may fail catastrophically when the distribution of returns, correlations, or factor exposures changes.

Recent evaluations of AI-driven and LLM-based trading strategies show that models performing well in narrow, short-horizon tests often deteriorate significantly when evaluated over broader universes or longer horizons due to regime sensitivity.

These strategies tend to be overly conservative in bull markets and overly aggressive in bear markets, demonstrating that stability under regime shift is a core requirement, not an optional refinement [Bailey and López de Prado 2021].

EXAMPLE 8.7 PITFALLS IN A BATTERY-TECHNOLOGY FACTOR STRATEGY

A quantitative team develops a long-short strategy targeting battery-materials suppliers using historical factor returns and supply-chain indicators.

1. **Overfitting:**
   Early models exhibit extremely high Sharpe ratios[1] (>3). Subsequent out-of-sample testing reveals that the model accidentally relied on a narrow window containing temporary lithium-price dislocations—fitting noise rather than structural drivers.
2. **Look-Ahead Bias:**
   Feature engineering pipelines accidentally incorporate end-of-day commodity-cost updates when generating early-morning trading signals. This leaks future information into the backtest, inflating accuracy
3. **Regime Shift:**
   When evaluated across a decade-long horizon, such as periods of oversupply, geopolitical distortions, and industrial-policy cycles, the strategy's profitability collapses. Analysts discover that factors relevant in early "technology-scarcity" years become irrelevant in later "scale-manufacturing" years.

This example highlights how fragile untested strategies can be. They also show how critical it is to incorporate out-of-sample testing, feature-lagging, timestamp validation, and multi-regime evaluation before deployment.

8.3 SCENARIO ANALYSIS AND MONTE CARLO SIMULATIONS

This section introduces the tools analysts use to navigate uncertainty in financial models, particularly in EST environments where technical, regulatory, and market variables interact in nonlinear ways. It outlines how scenario analysis, sensitivity testing, and probabilistic methods like Monte Carlo simulation transform static forecasts into dynamic, risk-aware decision frameworks. Together, these techniques ensure that strategic conclusions rest not just on expected outcomes but on a transparent understanding of the full distribution of possible futures.

8.3.1 Scenarios versus Sensitivities: Definitions and Use Cases

Scenario analysis and sensitivity analysis are the two foundational tools for exploring uncertainty in financial models, especially in EST environments where technical, regulatory, and market variables interact in non-linear ways. Although both methods quantify how assumptions shape outcomes, they differ fundamentally in scope, purpose, and interpretive value.

8.3.1.1 Sensitivity Analysis: Single-Variable Impact

Sensitivity analysis measures how a model's outputs (e.g., NPV, IRR, and free cash flow) change when one assumption is varied at a time while all others remain constant. In other words, it identifies which variables act as "single-point levers" in the model. This approach is ideal for determining which assumptions are most critical, such as unit price, cost inflation, utilization, discount rate, or yield improvement. Sensitivity analysis helps reveal model fragility by showing which inputs drive disproportionate swings in valuation.

Analysts typically perform sensitivity analysis through incremental adjustments. For example, modifying one assumption by ±5%, ±10%, or ±20% and observing how valuation metrics shift. This method is fast, transparent, and well-suited for executive decision-making when isolating specific risk drivers. Sensitivity analysis is also widely used in forecasting, budgeting, and valuation to test robustness under small, controlled perturbations.

8.3.1.2 Scenario Analysis: Multi-Variable, Narrative-Driven Cases

Scenario analysis evaluates how a model behaves under multiple simultaneous changes across key variables. Rather than isolating a single factor, scenarios represent qualitatively distinct futures. Scenario examples include:

- regulatory shifts,
- supply-chain constraints,
- pricing shocks, or
- changes in adoption curves.

Investors use scenario analysis to assess the combined effects of interacting risks.

A typical scenario set includes:

- **Base case:** expected operating conditions
- **Best case:** optimistic assumptions—strong demand, lower costs, favorable financing
- **Worst case:** downside conditions—higher expenses, lower revenue, regulatory friction

Because scenario analysis reflects complex interactions, it is especially powerful for strategic planning, risk assessment, and capital allocation decisions in EST projects where multiple drivers shift together [Phoenix 2025].

8.3.1.3 When to Use Each Method

Sensitivity analysis is most effective when it is necessary to isolate and rank the individual assumptions that exert the greatest influence on model outcomes. In contrast, scenario analysis is the preferred tool for exploring holistic, multivariable environments that require narrative framing and integrated treatment of interacting risks.

In professional-grade modeling, both techniques work together. Sensitivity analysis sharpens pinpoint risk identification. Meanwhile, scenario analysis supports broader strategic resilience planning [CFA 2024b].

EXAMPLE 8.8 SCENARIO VERSUS SENSITIVITY ANALYSIS FOR A NEXT-GEN SOLAR MANUFACTURING PROJECT

A firm evaluating a new photovoltaic (PV) module plant uses both methods:

SENSITIVITY ANALYSIS EXAMPLE

Analysts vary:

- Silicon feedstock price ±10%
- Line throughput (modules/hour) ±5%
- Yield loss rate ±2%

They find that a 10% increase in feedstock cost reduces gross margin by 350 bps, which is the most sensitive driver. This highlights procurement volatility as a key risk.

SCENARIO ANALYSIS EXAMPLE

The team constructs three scenarios:

1. **Best Case:**
    - Accelerated permitting
    - Strong EV-adjacent demand
    - Lower-than-expected energy costs
2. **Base Case:**
    - Market demand grows steadily
    - Input prices stable
    - Moderate automation capex
3. **Worst Case:**
    - Trade restrictions reduce export volume
    - 12-month equipment delivery delays
    - Labor shortages drive wage inflation

The results show NPV remains positive in the base and best cases but drops significantly negatively in the worst case. This confirms a staged investment approach and early supply-chain diversification.

Selecting the correct analysis technique ensures that models not only identify the levers that matter but also remain resilient when whole environments shift, providing a clearer foundation for strategic judgment.

8.3.2 Building a Scenario Manager in a Spreadsheet

A scenario manager transforms a static financial model into a dynamic decision tool. Instead of maintaining multiple files for base, upside, and downside cases, a well-designed scenario manager centralizes assumptions, automates switching between cases, and ensures consistency across all linked financial statements. In EST-finance environments—where uncertainty spans regulatory timelines, cost curves, R&D productivity, commodity prices, and adoption rates—scenario managers provide a disciplined way to evaluate how multi-variable shifts propagate through a model.

8.3.2.1 Structure: Centralizing Assumptions and Defining Scenarios

The foundation of a robust scenario manager is a clean assumptions table containing all scenario-dependent variables. Each scenario, Base, Optimistic, Pessimistic, or project-specific variants, occupies a column, while each assumption occupies a row. This architecture ensures that all inputs are auditable and eliminates the risk of "hidden overrides" scattered throughout the workbook. Financial modeling guidance emphasizes defining scenarios in a structured table that lists the case names and the full set of underlying assumptions [EIR 2025].

Scenarios should be limited to those that add analytical value. Too many variants hinder clarity and introduce maintenance cost; best practice is to maintain between three and five strategically distinct cases.

8.3.2.2 Case Selection: Enabling User-Controlled Switching

The scenario manager becomes interactive once a selector is added. This selector is typically a dropdown menu in the inputs or dashboard sheet. Data validation lists work well here, but form-controls (combo boxes) offer enhanced UI visibility for executive-facing dashboards. The selector acts as the "master switch" that drives the entire model. Online tutorials highlight the importance of a simple scenario selection cell that allows users to instantly toggle results across all modeled cases.

Professional models in Excel use lookup-based functions such as @XLOOKUP, @INDEX, @MATCH, or @CHOOSE to retrieve the selected case's assumptions and map them into the model's input section. This technique ensures backward compatibility and avoids brittle formulas such as @VLOOKUP [EIR 2025]. Once linked, every downstream calculation, including revenue build, cost schedules, CapEx rollout, and cash-flow forecasts, dynamically updates based on the active scenario.

8.3.2.3 Linkage and Integration: Ensuring Full-Model Consistency

To maintain model integrity, all scenario-dependent inputs must flow through a consistent mapping layer. This prevents "half-switched" models where only some assumptions are updated. Financial modeling best-practice guides emphasize centralized, clearly labeled inputs and real-time linkage to KPIs and financial statements.

Key guidance includes:

- Maintain one live scenario output area, not multiple fragmented versions.
- Use named ranges for clarity when mapping assumptions.
- Avoid referencing scenario tables directly in multiple places; use an intermediary "active assumptions" block.
- Test by switching scenarios repeatedly to confirm outputs change coherently [EIR 2025].

By ensuring that all scenario-dependent inputs flow through a single, coherent mapping layer, the model remains consistent, auditable, and fully responsive to changes in assumptions. This is exactly the discipline required for reliable scenario analysis.

8.3.2.4 Auditing and Visualization

Scenario managers are only valuable if stakeholders can interpret the results quickly. Excel's built-in **Scenario Summary** report remains a useful tool for side-by-side comparison of outputs (e.g., revenue, EBITDA, and NPV).

More advanced models integrate:

- **Tornado charts** to highlight sensitivity across variables (Figure 8.2).
- **Dashboard-level indicators** that display active scenario, key drivers, and resulting valuation bands.
- **Formula tracing** (precedents/dependents) to audit scenario flow and confirm clean logic.

EXAMPLE 8.9 SCENARIO MANAGER FOR A HYDROGEN-ELECTROLYZER PROJECT

A company evaluating a 200-MW green-hydrogen electrolyzer facility builds a scenario manager to assess cost and adoption uncertainty.

STEP 1—DEFINE SCENARIOS

Create a table with three cases:

- **Base:** \$450/kW capex, 70% utilization, electricity at \$35/MWh
- **Optimistic:** \$390/kW capex, 80% utilization, electricity at \$28/MWh
- **Pessimistic:** \$520/kW capex, 60% utilization, electricity at \$45/MWh

STEP 2—ADD THE SELECTOR

Insert a dropdown titled *Active Scenario* using data validation. The user selects "Base," "Optimistic," or "Pessimistic."

STEP 3—LINK ACTIVE ASSUMPTIONS

Use **@XLOOKUP** (or **@INDEX/@MATCH**) to pull values from the scenario table into the model's input layer. When the user chooses "Optimistic," the model automatically loads lower capex, higher utilization, and lower electricity prices.

STEP 4—UPDATE FINANCIAL STATEMENTS AND KPIS

The revenue model, OpEx schedules, and hydrogen production curves update automatically. The output sheet immediately shows how the levelized cost of hydrogen (LCOH) and NPV shift under each scenario.

STEP 5—GENERATE SUMMARY

Use Excel's "Scenario Summary" feature to produce a side-by-side table of LCOH, EBITDA, IRR, and peak funding across all three cases for investor presentations.

The scenario manager centralizes change, enabling instant, audit-ready toggling across cases.

8.3.3 Probability Modeling Foundations

Probability modeling provides the mathematical backbone for Monte Carlo simulation, scenario analysis, and all risk-aware financial modeling. In EST-finance contexts, having a disciplined approach to probability is essential for translating raw assumptions into credible stochastic models. Three pillars define this foundation: random variables and distributions, correlation structures, and copulas for multivariate dependence. Let's explore these further.

8.3.3.1 Random Variables and Distributions

Probability models begin with *random variables*, which formally represent uncertain quantities such as project yield, material costs, component failure rates, or commodity prices. Their behavior is described by *probability distributions*, which quantify the likelihood of different outcomes.

Core univariate distributions used in financial and engineering models include:

- **Normal** (Gaussian): appropriate for variables influenced by many small, independent effects.
- **Exponential** and **Poisson**: for modeling lifetimes, failures, and count processes.
- **Uniform**: useful for simple ranges or sampling from bounded intervals.
- **Mixed or empirical distributions**: used when historical data exhibit fat tails or structural discontinuities.

Standard probability theory covers essential tools such as expectations, variance, correlation, and joint distributions. These statistics support constructing simulation-ready models.

8.3.3.2 Correlation and Dependence Structures

Real-world variables rarely move independently. For example, a solar project's output depends on irradiance patterns; battery-degradation rates depend on operating temperature; component pricing depends on commodity cycles. Modeling such linkages requires *correlation structures*, which represent how variables co-vary.

Traditional correlations (Pearson's ρ) assume linear relationships and symmetry in the distribution of co-movements. This can be problematic when dealing with operational or market variables that have nonlinear interactions or asymmetric tail behaviors. Literature in risk and decision analysis emphasizes that joint distributions can be formed using marginal distributions combined with correlation-based dependence models, often using multivariate normal structures when the marginal behavior is tractable.

8.3.3.3 Copulas: Capturing Complex Multivariate Dependency

Modern quantitative finance often relies on copulas, which allow analysts to model dependence structures separately from the marginal distributions. A *copula* is a multivariate cumulative distribution with uniform marginals. Copulas let you model relationships between variables separately from their individual distributions.

Sklar's theorem states that any multivariate joint distribution can be written in terms of its marginal distributions linked together by a copula. Conversely, any set of marginals combined with a copula defines a valid joint distribution. In the continuous case, this copula is unique, allowing the dependence structure to be modeled separately from the marginals. This decoupling allows flexible modeling of:

- Non-linear dependencies
- Asymmetric relationships
- Tail dependence.

Tail dependence (the tendency of variables to become more correlated during extreme events) is especially important during stress periods when multiple risk factors become highly correlated.

Copulas such as the Gaussian, t-copula, and Archimedean families are widely used for portfolio risk, credit modeling, and scenario-dependent stress testing. They also support generating multivariate pseudo-random samples for Monte Carlo simulations. It is beyond this text to discuss these further, but interested readers can consult [Nelsen 2006].

EXAMPLE 8.10 MULTIVARIATE MODELING FOR A CLEAN-ENERGY PORTFOLIO

A clean-energy investment fund needs to model uncertainty in three key drivers:

1. Annual solar irradiance variability
2. Lithium carbonate spot-price volatility
3. Battery-module degradation rates

STEP 1—CHOOSE MARGINAL DISTRIBUTIONS

- Irradiance modeled as a normal distribution (historical data shows near-Gaussian seasonality).
- Lithium pricing modeled as a *fat-tailed t-distribution*.
- Degradation is modeled as a *lognormal distribution* reflecting multiplicative decay processes.

STEP 2—ESTIMATE CORRELATIONS

Historical analysis shows a moderate correlation between lithium prices and degradation (supply-chain tightness affects product quality/usage patterns). A correlation matrix is constructed using rank correlations (Pearson's ρ).

STEP 3—APPLY A COPULA TO CAPTURE NONLINEAR DEPENDENCE

A *Gaussian copula* is used to bind the three marginals into a single joint distribution. This allows the fund to simulate thousands of combined states. This captures both "normal" years and extreme joint events like low irradiance coinciding with supply-chain shocks.

STEP 4—GENERATE MONTE CARLO SAMPLES

Using the copula-driven joint distribution, the fund produces multivariate samples feeding into revenue models, cost models, and expected battery-replacement timing. This approach accurately reflects not only the uncertainty of individual drivers but also their interactions.

Using a random variable as a basis for observation, and with distributions and dependence in place, we can simulate coherent futures rather than isolated variables.

8.3.4 Monte Carlo Simulation Workflow

Monte Carlo simulation is the workhorse of quantitative risk modeling in EST-finance contexts, where uncertainty arises from technical performance, commodity prices, regulatory timing, and interactions among multiple stochastic drivers. The method enables analysts to propagate uncertainty through a financial model by repeatedly sampling from probability distributions and aggregating the resulting outcomes. A rigorous workflow typically involves

1. generating random draws,
2. setting seeds and defining the number of trials,
3. propagating uncertainty through the model, and
4. capturing and interpreting outputs such as NPV/IRR distributions, VaR, and CVaR.

TABLE 8.2
Summary Statistics From 10,000 Monte Carlo Trials

Statistic	Value
Trials	10,000
Mean NPV ($M)	–47.1619
Median NPV ($M)	–47.2528
Std Dev ($M)	12.108694472679002
P10 ($M)	–62.4821
P90 ($M)	–31.7521
Min ($M)	–93.92
Max ($M)	0.3518783862229725
Skewness	0.058036
Kurtosis (Fisher)	0.081404
Pr[NPV < 0]	0.9999
VaR 95% ($M) = 5th pct NPV	–66.6909
CVaR 95% ($M) = mean below VaR	–71.911

These distributions quantify upside and downside risk (Table 8.2).

Table 8.2 summarizes the statistical output of a 10,000-trial Monte Carlo simulation, providing a quantitative picture of the uncertainty surrounding a fictitious project's NPV. It reports central tendency measures such as the mean and median NPV, along with standard deviation, which reflects overall dispersion in simulated outcomes.

The table also lists *P10* and *P90* values, which are percentile cutoffs indicating the NPV that is only exceeded 10% or 90% of the time; these percentiles help frame downside and upside boundaries within the simulated distribution. The minimum and maximum NPVs observed across all trials further illustrate the full range of possible outcomes produced by the simulation.

Beyond central and percentile statistics, the table includes shape-diagnostic metrics such as *skewness,* which measures asymmetry in the distribution, and *kurtosis,* which indicates whether the distribution has heavier or lighter tails than a normal distribution. The entry Pr[NPV < 0] specifies the probability of the project destroying value. This is an essential risk metric for capital-intensive EST projects.

Finally, the table reports Value at Risk (VaR 95%), defined here as the 5th-percentile NPV, and *conditional value at risk* (CVaR 95%), which is the average NPV of all outcomes worse than the VaR threshold. These two tail-risk measures highlight the severity of extreme downside scenarios, offering a more complete view of potential losses than point estimates alone.

8.3.4.1 Generating Random Draws

The first step is defining the uncertain variables and assigning appropriate probability distributions, such as normal, log-normal, *t*-distribution, or empirical distributions, depending on the behavior of the underlying driver. Random sampling from these distributions forms the engine of simulation. Monte Carlo methods rely on large numbers of pseudo-random inputs to produce a representative distribution of outcomes.

In financial contexts, common applications include simulating asset returns, commodity prices, interest-rate paths, or project-level operational drivers such as throughput or failure rates. These simulations can incorporate correlations between variables or more advanced dependence structures (e.g., copulas).

8.3.4.2 Setting Seeds and Trials

To ensure reproducibility and allow auditors or collaborators to replicate simulations, Monte Carlo engines typically allow for seeding the random-number generator. Fixing the seed guarantees that random draws occur in the same sequence, making results fully repeatable. This is an essential requirement for regulatory and engineering-grade modeling.

The number of trials (iterations) influences convergence. As the law of large numbers dictates, increasing the number of iterations improves the stability of the simulated distribution and reduces sampling error. Finance tutorials emphasize choosing iteration counts appropriate to the model's complexity and desired statistical precision.

8.3.4.3 Propagating Uncertainty through the Model

Once random draws are generated, the simulation engine pushes each set of sampled inputs through the model, including cash-flow logic, operational curves, valuation structures, schedule dependencies, or scenario-driven cost frameworks. Each iteration produces a different outcome, forming an ensemble of possible futures.

This approach transforms the model from a point-estimate calculator into a probability-based forecasting tool. Monte Carlo simulation is particularly powerful for modeling multi-variable uncertainty where analytical solutions do not exist or are intractable.

Examples in finance include option-pricing simulations, portfolio return projections, and multi-period project-finance evaluations.

8.3.4.4 Capturing Outputs: NPV/IRR Distributions, VaR, and CVaR

After thousands (or millions) of iterations, analysts compile output distributions such as:

- **NPV/IRR histograms**—show the likelihood of different project valuations.
- **Value at Risk (VaR)**—identifies the loss threshold not exceeded with a chosen confidence level.
- **Conditional VaR (CVaR)**—measures the expected loss *beyond* the VaR cutoff, providing insight into tail risk.

Additional outputs may include fan charts, cumulative probability curves, and stress-case percentiles.

EXAMPLE 8.11 MONTE CARLO SIMULATION FOR AN OFFSHORE WIND PROJECT

A developer evaluates a 500-MW offshore wind installation with uncertainty in turbine availability, wind-speed distributions, and long-term electricity prices.

STEP 1—DEFINE INPUTS AND DISTRIBUTIONS

- Wind speed modeled using a Weibull distribution
- Electricity prices modeled using a correlated geometric Brownian motion
- Turbine downtime modeled as a Bernoulli process (availability vs. forced outage)
  (General Monte Carlo distribution assignment follows standard guidance.)

STEP 2—GENERATE RANDOM DRAWS AND CONTROL REPRODUCIBILITY

Random samples are generated for all variables each period. The random-number generator is seeded (e.g., SEED=1521158126) to ensure reproducibility for investor review.

STEP 3—RUN 50,000 TRIALS THROUGH THE FINANCIAL MODEL

Each simulation produces a unique revenue, OPEX, and net cash-flow sequence, ultimately yielding a distribution of NPVs. Iteration volume is selected to satisfy the convergence criteria emphasized in the stochastic-methods guidance.

STEP 4—ANALYZE OUTPUTS

- P50 NPV: $68M
- P10 downside NPV: −$20M
- Portfolio VaR (95%): −$35M
- CVaR (95%): −$44M

These outputs demonstrate an asymmetric risk profile dominated by low-wind and low-price regimes and are used to structure a risk-sharing PPA and reliability guarantees with the OEM.

By summarizing thousands of simulated futures, the results provide a structured, transparent foundation for evaluating uncertainty in EST-finance models.

This kind of analysis quantifies upside potential and downside risk more realistically than deterministic modeling [Statology 2025].

8.3.5 Communicating Uncertainty

Effective communication of uncertainty is essential in EST-finance modeling, where decision-makers must evaluate not only expected outcomes but also the distribution of possible futures. Because uncertainty touches capital allocation, risk tolerance, regulatory posture, and technical feasibility, clarity in how uncertainty is expressed can make the difference between informed decision-making and misinterpretation.

Three core tools shape best-practice communication: fan charts, prediction/uncertainty intervals, tornado charts and decision thresholds. Let's look at these.

8.3.5.1 Fan Charts and Forecast-Density Visualizations

Fan charts provide a visual summary of uncertainty around future projections by displaying a range of plausible outcomes rather than a single point forecast. As predictions extend further into the future, the dispersion widens, producing the characteristic "fan" shape (Figure 8.2).

Fan charts communicate multiple probability bands simultaneously (e.g., 50%, 70%, and 90% confidence regions). These bands highlight asymmetry, fat tails (extreme outcomes happen more often than a normal distribution would predict), or skewness, which is noteworthy when EST project outcomes hinge on nonlinear technical or commodity-price risks.

Methodology research demonstrates fan charts' ability to quantify risk, display a balance-of-risk metrics, and clarify the likelihood of falling within ranges relevant to policy or investment thresholds. Fan charts, originally popularized by the Bank of

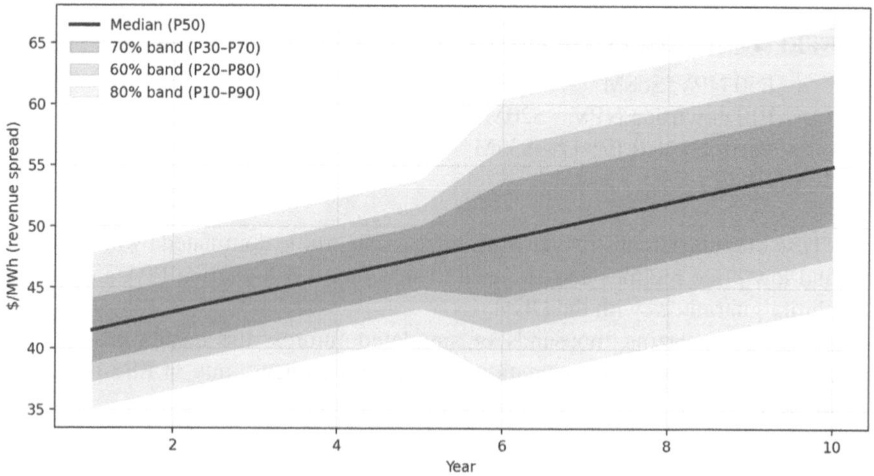

FIGURE 8.2 Fan chart of free cash flow under uncertainty.

England, are now widely used in finance and policy analysis for inflation forecasts, market-risk bands, and macro projections.

8.3.5.2 Prediction Intervals, Confidence Intervals, and Error Bars

In addition to fan charts, forecast uncertainty can be communicated through *confidence intervals, prediction intervals*, and *error bars*. Confidence intervals quantify the plausible range for a model's estimated parameters while prediction intervals reflect uncertainty in future observations and are therefore always wider (Figure 8.3). These tools visually express sampling uncertainty, model variance, and residual noise.

Prediction intervals incorporate both estimation error and irregular shocks, making them indispensable for forecasting operational metrics such as output, degradation, or cost escalation in EST projects.

Error bars provide an intuitive visual cue for variability and help audiences compare uncertainty across scenarios or technologies (Figure 8.4).

Error bars are common in both scientific and financial visualizations.

8.3.5.3 Tornado Charts, Decision Thresholds, and Risk Appetite Alignment

While fan charts and intervals communicate distributional uncertainty over time, *tornado charts* communicate *cross-sectional* uncertainty. Tornado charts rank drivers by their impact on key outputs such as NPV or IRR (Figure 8.5).

These charts help executives and engineers assess which variables meaningfully influence outcomes, and which risks deserve active mitigation.

Decision-makers often struggle with uncertainty messaging, yet research shows that transparency about uncertainty increases trust rather than undermines it. Effective communication requires explicitly linking uncertainty bands or ranges to *decision thresholds*. For example, minimum acceptable IRRs, maximum tolerable downside scenarios, or probability-of-breach metrics.

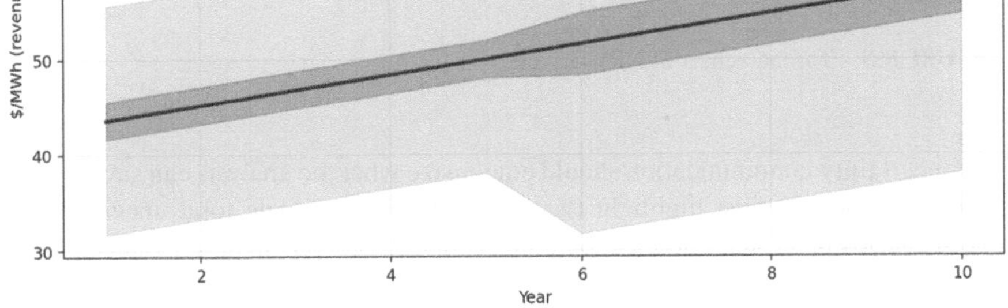

FIGURE 8.3 Prediction and confidence intervals.

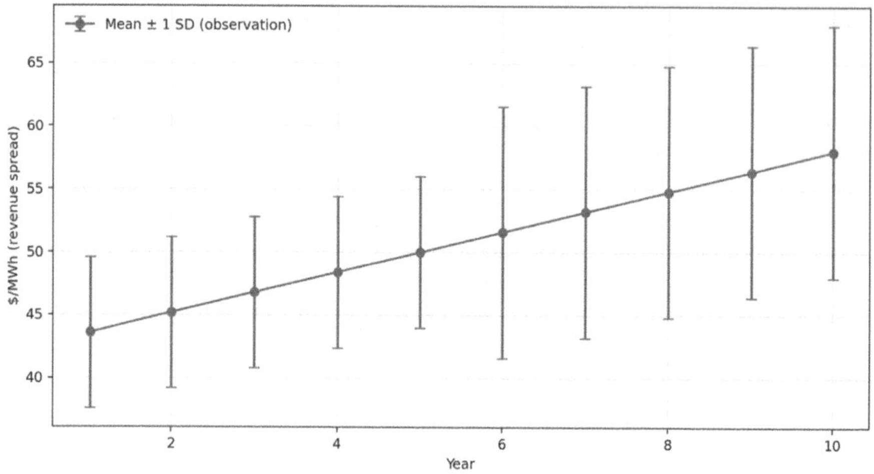

FIGURE 8.4 Error bars.

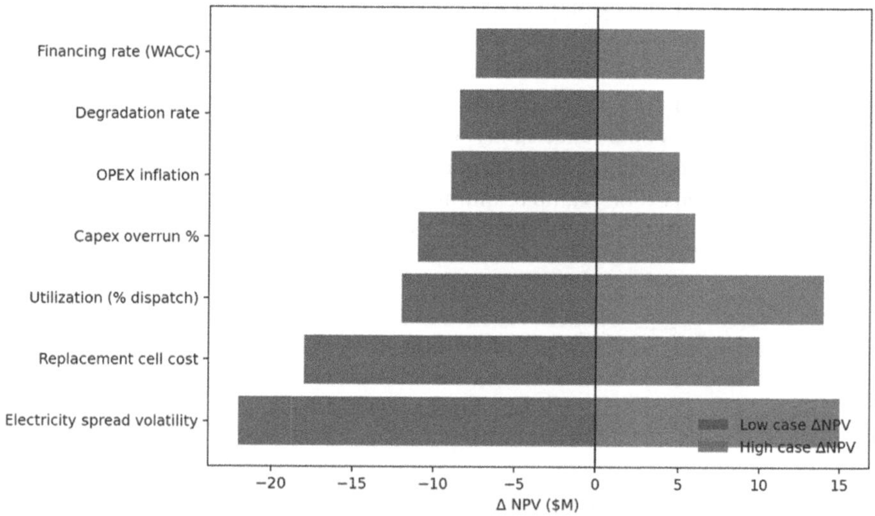

FIGURE 8.5 Tornado chart of NPV drivers.

Uncertainty communication should emphasize what the analysis can say, not only what it cannot. It must highlight the variables that dominate total uncertainty or the scenarios most likely to push outcomes outside acceptable performance bounds [BoE 1997].

EXAMPLE 8.12 COMMUNICATING UNCERTAINTY FOR A GRID-STORAGE INVESTMENT FORECAST

A grid-storage developer presents a 10-year forecast for a 200-MWh battery system under uncertain degradation rates, electricity spreads, and cell replacement costs. Figures 8.2, 8.3, and 8.4 were generated using simulated data for this case study. Let's look at those figures more closely.

FAN CHART FOR REVENUE SPREAD

The fan chart in Figure 8.2 illustrates widening uncertainty in annual revenue spreads as wholesale-market volatility grows over time. Darker central bands show the 50% probability region, while lighter outer bands represent 70% and 90% probability ranges. Decision-makers immediately see that while median revenues remain stable, the downside tail widens significantly after year five. This mirrors findings from central-bank density-forecast applications.

PREDICTION INTERVALS FOR DEGRADATION FORECASTS

Prediction intervals (95%) around degradation curves show substantial variability under different cycling conditions (Figure 8.3). Error bars at discrete inspection points visually highlight risks of crossing the replacement threshold too early (Figure 8.4).

TORNADO CHART FOR NPV DRIVERS

The tornado chart in Figure 8.4 demonstrates that electricity-spread volatility and replacement-cell cost assumptions dominate uncertainty in project NPV. This guides management to focus diligence and hedging strategies on these high-impact levers.

DECISION-THRESHOLD HIGHLIGHT

A probability overlay shows a 22% risk of NPV falling below the firm's investment hurdle rate. This is a transparent communication of downside risk that aligns with research on clear messaging to decision-makers.

These distributional outputs convert point estimates into decision-ready risk profiles.

8.3.6 Implementation Notes

Monte Carlo simulation is only as reliable as the implementation behind it. In EST-finance modeling, careful attention to performance, structure, formula discipline, and validation is essential. Three technical pillars shape robust implementation: performance optimization, model structure and formula hygiene, and validation against analytic or simplified solutions.

8.3.6.1 Performance Tips: Vectorization, Structure, and Avoiding Volatile Formulas

Large simulations frequently push spreadsheet tools such as Excel to their computational limits. Performance best practices emphasize modularity, structured data organization, and selective use of formulas. Leading modeling guidance highlights the importance of organizing workbooks into logical sections, namely, Inputs, Calculations, and Outputs to simplify recalculation paths and prevent redundant or circular dependencies that slow computation.

To deal with these problems, avoiding certain volatile formulas (e.g., @OFFSET, @INDIRECT, @TODAY, and @RAND in Excel) is crucial. Volatile formulas recalculate every time Excel updates, multiplying the computational load across thousands of simulation iterations. Instead replace volatile lookups with stable references (e.g., @INDEX/@MATCH or @XLOOKUP in Excel), use named ranges for clarity and minimize unnecessary dynamic arrays. These recommendations align with advanced Excel modeling best-practice guidelines that warn about fragile or overly complex formulas.

Computational efficiency improves significantly when simulation engines batch calculations or use vectorized operations rather than cell-by-cell logic. Where feasible, pre-computing intermediate values, like discount factors, degradation multipliers, or time-indexed capacity curves, can reduce repeated overhead across trials.

8.3.6.2 Structuring the Model for Auditability and Speed

A well-organized model is faster and easier to debug. Best-practice literature emphasizes modularity, scalability, and transparency. Inputs should be clearly labeled and documented. Calculations should be grouped by functional blocks. And outputs should be centralized for ease of review.

Proper structuring also prevents accidental overwriting or misalignment during simulation runs. Named ranges help maintain formula readability and reduce the risk of referencing the wrong cells, especially in multi-scenario Monte Carlo setups. This approach is recommended in advanced modeling guidance as a key tool for avoiding common structural pitfalls.

Additionally, segregating large tables, historical datasets, and simulation matrices onto separate tabs improves both navigation and performance. Grouping and collapsing sections streamline workflows and reduce cognitive load for teams reviewing or maintaining the model.

8.3.6.3 Validation against Analytic Solutions and Reasonableness Checks

No simulation should be accepted without validation. Financial-modeling tutorials emphasize verifying logic by:

1. comparing Monte Carlo results against closed-form or analytic solutions when available (e.g., discounted cash-flow formulas and binomial approximations),
2. running simplified test cases, and
3. confirming that outputs behave monotonically when inputs are shifted in predictable directions.

Validation also includes traceability and stress testing. *Traceability* means ensuring that all formulas can be traced through precedents and dependents for audit purposes. *Stress testing* means that extreme parameter values are used to help reveal hidden bugs, such as divide-by-zero errors, incorrect sign conventions, or omitted constraints.

Many industry guidelines emphasize the importance of reproducibility, that is, using fixed seeds for testing sequences. Guidelines also advocate checking whether simulation variance stabilizes when trial counts increase. These checks ensure convergence and prevent over-interpretation of noisy or under-sampled results [CFI 2024a].

EXAMPLE 8.13 IMPLEMENTATION PITFALLS AND FIXES IN A TURBINE-LIFETIME RISK SIMULATION

A team building a Monte Carlo model to assess turbine-lifetime variability encounters slow calculation speeds and inconsistent outputs. Applying the principles above leads to measurable improvements:

ISSUE 1—VOLATILE FORMULAS SLOWING COMPUTATION

The original model uses @OFFSET-based range definitions for degradation curves, causing recalculation during every trial.

Fix: Replace @OFFSET with indexed named ranges backed by @INDEX/@MATCH. Performance improves by 40%.

ISSUE 2—POOR STRUCTURE LEADING TO LOGIC ERRORS

Inputs, assumptions, and embedded calculations spread across 12+ worksheets create misalignment when scenario toggles change.

Fix: Rebuild using a three-sheet structure (Inputs → Calculations → Outputs). (Modular structure recommended by advanced Excel modeling frameworks.)

ISSUE 3—LACK OF VALIDATION CREATING OVER-OPTIMISTIC RESULTS

Simulated failure rates diverge from manufacturer-spec curves due to incorrect unit scaling.

Fix: Validate the Monte Carlo output by comparing it against closed-form reliability formulas and running deterministic test points.

Collectively, these improvements produce faster iterations, clearer logic, and results that align with both engineering intuition and analytical expectations.

This example shows that performance, structure, and validation together keep stochastic models fast, transparent, and fit for purpose.

8.4 MODEL GOVERNANCE, DOCUMENTATION, AND ETHICS

Modern financial models influence high-stakes decisions, like capital deployment, regulatory reporting, risk management, and long-range planning. As complexity grows, so does the need for disciplined *model governance, transparent documentation*, and *ethical safeguards*.

In EST-finance contexts, where models integrate technical, operational, and financial assumptions, governance failures can lead not only to economic loss but also to safety, compliance, and societal risks.

A robust framework encompasses

1. documentation standards,
2. controls and independent review,
3. ethical handling of data and modeling techniques, and
4. awareness of key regulatory touchpoints.

8.4.1 Documentation Standards: Model Intent, Scope, and Ownership

Clear documentation is an essential part of responsible modeling. Documentation is not a bureaucratic exercise; it is a core risk-mitigation tool that improves sustainability, interpretability, and institutional memory. Models without documented assumptions, logic, and limitations often generate errors, misaligned expectations, or flawed decisions. Model documentation should include:

- **Model purpose and scope**—what decisions the model supports, and what it does *not* cover.
- **Assumptions and methodologies**—drivers, scenarios, statistical methods, data sources, and parameter choices.
- **Limitations and caveats**—known weaknesses, stability boundaries, and areas of high uncertainty.
- **Versioning and change logs**—recording updates, rationale, and impacted areas.
- **Ownership and stewardship**—model owner, developer, validator, and accountable executive roles.

Regulatory and practitioner guidance highlights that documentation must be detailed enough so that a qualified practitioner unfamiliar with the model could understand its foundations and reproduce key results.

8.4.2 Controls: Peer Review, Periodic Backtesting, and Audit Trails

Strong controls reduce model risk by ensuring models are trustworthy before and after deployment. SR 117 (the US Federal Reserve's Model Risk Management guidance) and equivalent global standards emphasize comprehensive controls across the entire model lifecycle, including development, validation, ongoing monitoring, and retirement [Fed 2011].

Typical control elements include:

- **Independent validation**—testing conceptual soundness, performance, data integrity, and implementation accuracy.
- **Peer review**—ensuring formulas, logic, and structure align with modeling principles.
- ***Backtesting* and benchmarking**—comparing predicted versus realized outcomes to detect drift or misspecification.
- **Ongoing monitoring**—alert systems, error logs, and performance dashboards.
- **Audit trails**—traceable histories of assumptions, changes, and user interactions.

Model governance frameworks increasingly incorporate automated model inventories, monitoring dashboards, and workflow systems to maintain consistency and reduce manual effort.

8.4.3 Ethical Use of Data and Models: Privacy, Fairness, and Transparency

As models become more complex, ethical risks grow. The increased use of machine-learning and AI components only increases these risks. Therefore, practitioners must ensure models adhere to standards of fairness, transparency, and privacy, particularly when models touch sensitive data or inform consequential decisions.

Key ethical considerations include:

- **Data minimization and privacy**—using only the data necessary for the model's purpose and ensuring secure handling of personal or proprietary information.
- **Bias detection and mitigation**—testing for algorithmic drift or unintended discrimination embedded in assumptions, training data, or structural choices.
- **Explainability**—defining how results are generated, especially for probabilistic or ML-based models, and ensuring decisions can be justified to stakeholders or regulators.
- **Avoiding misuse**—ensuring models are operated within their intended domain; documented limitations prevent inappropriate deployment.

Regulatory expectations (e.g., model governance statements from central banks) increasingly highlight the need to manage risks arising from AI-driven modeling, including transparency and accountable oversight.

8.4.4 Regulatory Touchpoints (High-Level): SOX, Model-Risk Principles, and Assurance

Financial models increasingly fall under regulatory oversight, especially when used in reporting, forecasting, or risk assessment. Key frameworks include:

- SOX (Sarbanes-Oxley)—requiring internal controls and documentation for financial reporting processes, including spreadsheet models used for disclosures [SOX 2002].
- SR 117 (US Federal Reserve)—foundational model-risk management guidance covering governance, validation, and oversight (widely adopted beyond banking) [Fed 2011].
- PRA SS1/23 (UK)—formalizes principles for model identification, governance, independent validation, and oversight of AI-enabled models [PRA 2023].
- Basel & international model risk management standards—emphasize documentation, inventory management, and control environments for model-driven institutions [Basel 2019].

Across frameworks, common regulatory expectations include model inventories, assigned accountability, validation independence, regular performance assessments, and clear audit trails [BoE 1997].

EXAMPLE 8.14 GOVERNANCE AND ETHICS REVIEW FOR AN ENERGY-STORAGE DISPATCH MODEL

A utility deploys a battery-dispatch optimization model to schedule charging and discharging across a fleet of grid-storage assets. During a quarterly governance review, several issues emerge:

DOCUMENTATION GAPS

The model's degradation-cost function is undocumented, making it unclear whether the cost curve matches the latest vendor data. Reviewers flag this as a material risk requiring immediate update. The importance of documentation is emphasized in model-risk literature.

VALIDATION FINDINGS

Backtesting reveals systematic deviations between predicted and realized peak-shaving savings. Validators identify a flawed price-forecast input that drifts during high-volatility periods. This is consistent with recommended lifecycle validation practices.

ETHICAL CONSIDERATION

The optimization model begins favoring dispatch strategies that shift emissions to off-peak fossil-generation windows. The ethics review notes unintended environmental consequences and recommends adding emissions-weighted cost factors. This aligns with broader governance emphasis on fair, transparent, purpose-aligned modeling.

REGULATORY ALIGNMENT

Since the model informs revenue forecasts used in public filings, the SOX compliance team mandates improved audit trails and sign-off procedures. This is reflective of typical reporting-control requirements.

Following these interventions, the model is updated, re-validated, re-documented, and re-approved according to the organization's model-risk framework. This improves reliability and stakeholder trust.

By integrating these visual and statistical techniques, analysts can communicate uncertainty with the same rigor used to model it, ensuring results remain transparent, interpretable, and actionable for stakeholders. Taken together, these methods elevate financial modeling from a computational exercise to a structured decision-support discipline that can guide long-horizon investments in complex EST environments.

8.5 EXERCISES

8.1 Design a three-sheet financial model (Inputs → Calculations → Outputs) for a small engineering project. Your model must:
   a. clearly separate assumptions, inputs, and outputs,
   b. use color-coding conventions,
   c. avoid hard-coding values in formulas,
   d. and include at least three error-check cells (e.g., circularity check, balance check, and sign convention check).

8.2 Create a table with three columns—Assumptions, Inputs, Outputs—and fill it with at least five examples per column for a hypothetical clean-energy manufacturing plant. Then explain why each item belongs in its specific category.

8.3 Download or construct a dataset that contains synthetic schema drift—for example, a column that switches units mid-series (kWh → MWh), category labels that change spelling, or timestamp formats that shift.
   1. Identify and document at least three drift events.
   2. Correct the dataset using data validation, transformation logic, and a data-assumptions table.
   3. Produce a before/after summary showing how schema drift would have distorted downstream analytics.
   4. Briefly explain why data lineage and drift detection are essential in EST-finance ingestion pipelines.

8.4 Convert a block of historical revenue data into an Excel Table, then build a formula using structured references that automatically updates as you add new months of revenue.
   Demonstrate the difference between using a normal range versus a Table reference.

8.5 Construct a modular Price × Volume revenue model for a robotics hardware line. Include:
   a. a base demand forecast,
   b. a churn/retention component for service contracts,
   c. step-cost logic (e.g., add a technician every 200 units sold).
   Then perform a short paragraph analysis of what cost driver is likely to dominate margins.
8.6 Build a simplified 3-statement model (P&L, Balance Sheet, Cash Flow Statement) for a five-year forecast.
   Your model must correctly:
   a. link net income → retained earnings,
   b. flow depreciation into PP&E roll-forward,
   c. connect working-capital changes to cash flow,
   d. tie ending cash between statements.
   Check for consistency using a "Cash Balance Error" cell.
8.7 Develop three scenarios—Base, Optimistic, Pessimistic—for an industrial automation project.
   For each scenario, define at least:
   a. revenue growth,
   b. unit costs,
   c. utilization,
   d. capex intensity,
   e. discount rate.
   Then write a 4–6 sentence narrative describing how the macro, regulatory, and technical conditions differ across scenarios.
8.8 Create a WACC module where capital structure, cost of equity, and cost of debt each vary by scenario (Optimistic/Base/Downside).
   a. Link the WACC calculation to your scenario manager.
   b. Demonstrate how the updated WACC affects valuation metrics (e.g., NPV, terminal value, and IRR).
   c. Write three to five sentences explaining why scenario-linked discount rates are critical when modeling technologies exposed to regulatory, commodity, or adoption-curve uncertainty.
8.9 Build a data table that evaluates NPV sensitivity to discount rate ranging from 7% to 14%.
   Then add a second data table that evaluates sensitivity to unit price ranging from −10% to +10% of the base case.
   Summarize which variable is more influential and why.
8.10 Using Excel's Scenario Manager, create named scenarios and map them to your Inputs block.
   Demonstrate that switching scenarios dynamically updates:
   a. revenue,
   b. EBITDA,
   c. free cash flow,
   d. NPV.

Export a Scenario Summary and highlight one insight that would matter to an investor.

8.11 Given three uncertain variables—demand growth, commodity prices, and failure rates—estimate pairwise correlations using synthetic or historical data.
Then discuss (5–8 sentences) why linear correlations fail in tail events, and when a firm should consider moving to copulas instead.

8.12 Select two marginal distributions (e.g., lognormal demand × *t*-distributed cost inflation) and combine them using a Gaussian copula.
Simulate at least 1,000 joint samples and plot the scatter of outcomes.
Explain in 3–5 sentences how the dependence structure differs from using a basic correlation matrix.

8.13 Define five uncertain inputs with associated probability distributions (e.g., capex overrun %, utilization %, pricing, downtime, and OpEx inflation).
Run a 5,000-iteration Monte Carlo simulation and produce:
a. a histogram of NPV,
b. mean, median, P10, P90,
c. and a simple Value-at-Risk calculation.
Interpret the distribution's shape and tail behavior.

8.14 Create a fan chart for a 10-year free-cash-flow forecast using your Monte Carlo output.
Then create a tornado chart showing which inputs contribute most to NPV variance.
Write a paragraph comparing which visualization is more compelling for executives and why.

8.15 Select three uncertain variables relevant to an EST context (e.g., commodity price inflation, degradation rate, and dispatch hours).
a. Assign non-normal marginal distributions (e.g., lognormal, *t*-distribution, and empirical).
b. Use a Gaussian or t-copula to model dependence among the variables.
c. Run a 5,000-trial Monte Carlo simulation and plot the resulting NPV distribution.
d. Write a short analysis comparing the copula-based results to what you would have obtained using simple linear correlations.

8.16 Perform a mock audit of your model using a governance checklist.
Your audit must comment on:
a. documentation completeness,
b. version control,
c. formula transparency,
d. assumption provenance,
e. validation tests,
f. and whether model risk controls are appropriate for the model's importance.
Write a 1-page memo summarizing your findings.

8.17 Design a 20-item audit checklist suitable for a capital-intensive EST financial model.
Your checklist must cover:
a. documentation and assumptions transparency,
b. version control and change logs,
c. formula structure and separation of concerns,
d. error-check coverage,
e. scenario/sensitivity consistency,
f. validation and reasonableness tests, and
g. governance or ethical considerations.
Apply your checklist to a model you have built and write a one- to two-paragraph summary of the audit outcomes, focusing on the most material risks you identified.

8.18 Write a short (one to two pages) analysis of an ethical dilemma involving a financial model—for example:
a. using biased historical data,
b. omitting uncertainty,
c. knowingly deploying a model beyond its domain,
d. or embedding assumptions that systematically disadvantage a stakeholder.
Clearly articulate:
1. the ethical breach,
2. who is harmed?
3. how governance mechanisms should prevent the issue.

8.19 Using a three-statement model of your choice, select four interacting assumptions (e.g., utilization, pricing, OPEX inflation, and discount rate).
1. Run a four-variable sensitivity grid (each variable at Low/Base/High).
2. Decompose the total change in NPV into:
    - main effects (single-variable contribution), and
    - interaction effects (joint contribution from variable pairs).
3. Write a short paragraph explaining why interaction terms matter in EST models where technical and market drivers reinforce (or offset) each other.

NOTE

1 A *Sharpe Ratio* = (Return—Risk-Free Rate)/Standard Deviation of Returns, which measures how efficiently a strategy converts risk into return.

REFERENCES

Bailey, D. H., & López de Prado, M. (2021). How backtest overfitting in finance leads to false discoveries. *Significance*, *18*(6), 22–25.

Bank of England. (1997). *Inflation report: Use of fan charts in uncertainty communication.* https://www.bankofengland.co.uk/inflation-report/1997/november-1997

Basel Committee on Banking Supervision. (2019). *SRP30—Risk management* (Supervisory Review Process). Bank for International Settlements. https://www.bis.org/basel_framework/chapter/SRP/30.htm

Berkowitz, J. (2019). *Financial modeling: A practical guide for applied research and teaching* (2nd ed.). Wiley.

Bloomberg, L. P. (2025). *Portfolio & risk analytics: A Bloomberg professional service offering.* https://data.bloomberglp.com

Board of Governors of the Federal Reserve System & Office of the Comptroller of the Currency (Fed). (2011). *Supervisory guidance on model risk management* (SR 117). https://www.federalreserve.gov/supervisionreg/srletters/sr1107a1.pdf

Corporate Finance Institute (CFA). (2024a). *Documenting Excel models: Best practices.* https://corporatefinanceinstitute.com

Corporate Finance Institute (CFA). (2024b). *Scenario analysis vs. sensitivity analysis.* https://corporatefinanceinstitute.com

Corporate Finance Institute. (2025). *Data sources in financial modeling.* https://corporatefinanceinstitute.com/resources/financial-modeling/data-sources-in-financial-modeling/

DashboardsExcel. (2025). *How to choose between DCF and relative valuation.* https://dashboardsexcel.com

Data Science Council of America. (2024). *A comprehensive guide to mastering exploratory data analysis.* https://dasca.org

EIR Experts. (2025). *Financial modeling: Building a scenario manager, step by step.* https://eir-experts.com

General Electric Company. (2024). *Annual report: Industrial technology, energy systems, and financial performance.* GE.

Lockheed Martin Corporation. (2024). *Annual report: Advanced engineering, aerospace systems, and long-cycle program management.* Lockheed Martin.

Nelsen, R. B. (2006). *An introduction to copulas* (2nd ed.). Springer.

NVIDIA Corporation. (2024). *Annual report: Accelerated computing and data-driven engineering.* NVIDIA.

Phoenix Strategy Group. (2025). *Scenario analysis vs. sensitivity analysis: Key differences.* https://phoenixstrategy.group

Prudential Regulation Authority (PRA). (2023). *SS1/23: Model risk management principles for banks.* Bank of England. https://www.bankofengland.co.uk/prudential-regulation/publication/2023/may/model-risk-management-principles-for-banks-ss

Siemens, A. G. (2024). *Annual report: Engineering, digital industries, and financial performance.* Siemens.

Statology. (2025). *The concise guide to Monte Carlo simulation.* https://www.statology.org/the-concise-guide-to-monte-carlo-simulation/

Texas Instruments. (2024). *Annual report: Semiconductor engineering, manufacturing systems, and financial governance.* Texas Instruments.

United States Congress. (2002). (SOX). *Sarbanes–Oxley act of 2002* (Public Law 107-204). https://www.congress.gov/107/plaws/publ204/PLAW-107publ204.pdf

9 Next Generation Financial Concepts and Strategies

This chapter explores emerging technologies shaping the financial landscape, with a focus on their implications for engineers, scientists, and technologists. As digital transformation accelerates, these "next-gen" tools, such as blockchain, AI, and quantum computing, offer unprecedented opportunities for efficiency, innovation, and risk management in finance. However, they also introduce new challenges such as cybersecurity threats, regulatory hurdles, and ethical dilemmas.

In this chapter, we will examine each concept's core principles, financial applications, integration with EST fields, and strategic considerations, equipping you to leverage them in your career or ventures while mitigating downsides.

9.1 BLOCKCHAINS AND DISTRIBUTED LEDGERS

Blockchains and distributed ledgers are already reshaping identity, tokenized assets, market infrastructure, and post-trade operations. These are areas where provenance and policy controls dominate the cost curve. For engineers, scientists, and technologists, blockchain provides a shared basis for secure data sharing across IoT, AI workflows, and supply chains, with governance rules expressed as code. Properly engineered, blockchains and distributed ledgers reduce reconciliation work, increase auditability, and make "who did what, when" objectively verifiable across firms and devices.

First, let's look at the basic principles, issues, and applications of blockchains and distributed ledgers for EST.

9.1.1 Core Principles and Mechanics

At its heart, a blockchain is a *distributed ledger*, that is, a digital record that many computers share and update together, without relying on a central owner or database. In a distributed ledger system, every participating node keeps its own copy of the data, and the group agrees on updates through a consensus mechanism, rather than trusting a single authority. This structure makes the ledger tamper-evident and tamper-resistant because no single party can quietly rewrite history without everyone noticing.

Blockchain takes this idea a step further by arranging transactions into "blocks" and linking those blocks in chronological order using cryptographic hashes. Figure 9.1 illustrates the core steps of blockchain technology in a simplified flow.

DOI: 10.1201/9781003733850-9

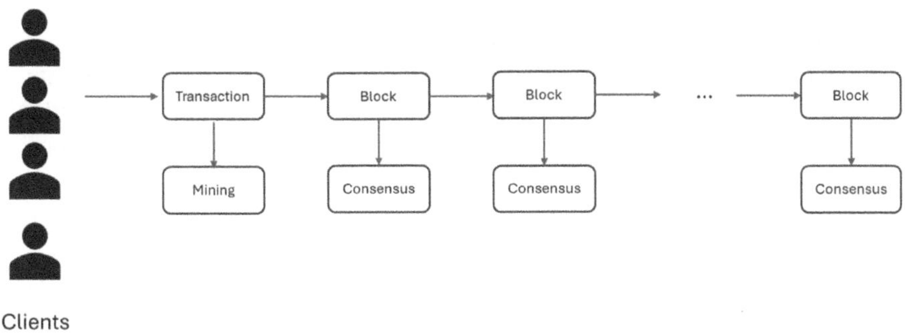

FIGURE 9.1 Conceptual view of a blockchain.

The "chain" begins with a "transaction" block, representing a user or client-initiated action, which leads to "mining," where transactions are validated and grouped. This is followed by "hashing," where cryptographic hashes secure the block's contents. The validated block is then added to the chain, and "Consensus" ensures agreement across the network before finalizing the addition. Each block is linked sequentially, forming a secure, tamper-resistant ledger.

Think of each block as a sealed container that includes both new transactions and the fingerprint of the previous block. Because each fingerprint depends on all earlier ones, even a small change to past data would break the chain and be rejected by the network. This design is why blockchains are considered *immutable* under normal operation. That is, because rewriting history becomes computationally or organizationally infeasible.

To keep the ledger synchronized, the network needs a way to decide who gets to add the next block. This is where *consensus models* come in. Public networks such as Bitcoin and Ethereum originally used *proof-of-work,* which requires expensive computing power. Newer systems, like Ethereum, use *proof-of-stake,* which selects validators based on the cryptocurrency they secure. Other systems, especially those built for businesses, use permissioned models where only vetted organizations operate validator nodes. These choices matter because they influence energy use, throughput, governance, and even regulatory friendliness.

Modern blockchains do more than store transactions; they can also run small programs, called *smart contracts,* which automatically enforce business rules. These contracts are deterministic, meaning every node must get the same result when executing them, which helps participants trust the outcome. And because blockchains cannot natively fetch outside information, they rely on oracles to bring in real-world data such as prices, sensor readings, or identity attestations.

EXAMPLE 9.1 PREDICTIVE MAINTENANCE ON A PERMISSIONED BLOCKCHAIN IN WIND FARM OPERATIONS

An energy cooperative uses permissioned blockchain to support predictive maintenance in wind farms. In this setup, each turbine's IoT sensors continuously produce high-frequency telemetry data such as temperature, vibration, load cycles, and other operating variables. These sensor events are logged onto a consortium blockchain, where participating organizations (the wind-farm operator, component suppliers, maintenance firms, and auditors) share a synchronized, tamper-resistant history of equipment data.

Machine-learning models then analyze this on-chain telemetry to predict failures and automatically trigger the ordering of spare parts. Every component involved in the repair has its own blockchain-recorded "chain of custody," tracing it from the supplier to installation. This makes it easy for engineers to quickly reconstruct a part's complete lifecycle, from materials, manufacturing parameters, transport conditions, and installation events, when diagnosing issues like a blade delamination.

The result is a maintenance process where *data provenance* is guaranteed, repairs are faster, and root-cause analysis becomes more accurate because no party can dispute or alter historical records. This integration of IoT + AI + blockchain creates a trustworthy, multiparty engineering environment. This is something traditional databases struggle to provide when multiple firms must share critical operational data.

Paired with emerging identity tools like *decentralized identifiers* and *verifiable credentials*, these features let systems verify people, devices, or organizations without exposing unnecessary personal data. Thus, they create a secure foundation for engineering, finance, and cross-organizational collaboration [Yaga et al 2018].

9.1.2 Financial Applications

Blockchains are increasingly used to streamline how financial records are created, transferred, and audited. One major trend is the *tokenization of financial instruments*, where assets such as securities, fund shares, or even real-world items are represented on-chain so that lifecycle events, for example, issuance, transfers, and corporate actions, are coordinated by shared code. In January 2026, the US SEC clarified that tokenized securities remain fully regulated securities and must comply with existing registration and disclosure requirements, which gives banks and brokers clearer rules for adopting on-chain recordkeeping.

EXAMPLE 9.2 TOKENIZED NOTES FOR AUTOMATED INTEREST PAYMENTS

An engineering firm issuing *tokenized notes* (*corporate paper*[1] or bonds) on a permissioned blockchain can restrict ownership to approved wallets using on-chain whitelists backed by verifiable credentials. Smart contracts can automate monthly interest payments using a stablecoin price oracle, while auditors later confirm all transfers against the issuer's on-chain master securityholder file. This is a pattern that the SEC recognizes for issuer-sponsored tokenization.

Blockchains can also be used to maintain cap-tables. A *capitalization table (or cap-table)* is a structured record that shows who owns what percentage of a company, including founders, employees, investors, option holders, and any other equity stakeholders. It is usually stored in a spreadsheet or database. Blockchain-based cap-tables, however, work by tokenizing equity so that each ownership unit (share, option, warrant) is represented as a digital token on a distributed ledger. This enables real-time updates, automated accuracy, transparent auditability, and lower administrative overhead because every transaction is recorded immutably and can be verified without reconciling multiple systems.

Beyond equity, securities, and identity, blockchains improve *post-trade operations*, reporting, and *auditability* by providing a shared, append-only ledger that reduces reconciliation errors and accelerates regulatory attestations. Some banks are also exploring how ledgers can support new openbanking frameworks (see Section 9.1.4) by tracking data lineage, consent, and revocation as consumers gain expanded rights to authorize data sharing under rules like the CFPB's final Personal Financial Data Rights rule [Grand View Research 2025].

9.1.3 Engineering, Science, and Technology Integrations

Blockchains are increasingly appearing in EST settings as part of a larger trust framework that connects devices, data, and decision systems. When paired with IoT sensors and AI models, a distributed ledger provides a shared, tamper-resistant record of events that multiple organizations can rely on. This is particularly important in complex environments where data flows across company boundaries and where integrity, timing, and provenance matter just as much as the data itself.

One major area of adoption is *supply-chain provenance*, especially for high-value or safety-critical components. Aerospace maintenance, repair, and overhaul stakeholders, for example, are piloting blockchain-based systems that trace serialized parts from manufacturing through installation to end-of-life. Because every handoff is recorded on a shared ledger, engineers and auditors can instantly review a complete "back-to-birth" history of each component, reducing counterfeits and speeding maintenance. Similar approaches are emerging in clinical research, where permissioned ledgers help track consent, protocol adherence, and data lineage while preserving privacy through selective-disclosure techniques.

Another important integration point is the use of blockchain as a *trust fabric* for cyber-physical systems. Here, IoT devices generate authenticated telemetry, AI systems analyze that data, and the ledger stores an immutable record of what happened and when. This pattern reduces single points of failure and supports strong audit trails in environments such as energy systems, autonomous manufacturing lines, or distributed research facilities.

EXAMPLE 9.3 BLOCKCHAIN-ENABLED PREDICTIVE MAINTENANCE IN WIND TURBINE OPERATIONS

A wind-turbine operator logs high-frequency sensor data, for example, temperature, vibration, and load cycles, to a consortium blockchain so manufacturers, maintenance contractors, and auditors all share the same trusted history. AI models use this data to predict failures and trigger part orders automatically, while each component's on-chain chain-of-custody record allows engineers to trace defects back to specific material batches or transport conditions. This creates a far more reliable and transparent engineering workflow than traditional siloed databases.

By anchoring identity and data integrity at the infrastructure level, teams can coordinate across organizations with significantly lower reconciliation overhead.

9.1.4 Risks and Strategies

Blockchains offer powerful trust and automation benefits, but they also introduce a set of technical, privacy, and regulatory risks that must be understood and managed thoughtfully. For example, proof-of-work systems have drawn criticism for high energy use, but design choices matter. For example, Ethereum's Merge (September 15, 2022) moved the network to *proof-of-stake* and cut estimated energy consumption by ~99.95%. This event shows that consensus is a controllable variable when selecting or designing a platform.

Another problem is that even without storing personal identifying information on-chain, metadata (e.g., timestamps, transaction links, and addresses) can leak patterns. To address this problem the system should employ pairwise identifiers, off-chain storage of sensitive payloads, and selective-disclosure or zero-knowledge techniques for attribute checks and revocation [Ethereum 2025].

Another risk is that smart-contract defects, insecure wallets, or compromised continuous integration and development pipelines can cascade across an ecosystem. To address this problem, the system should treat signers, oracles, and build systems as Tier-1 (critical) assets. It should also enforce key management with hardware security modules, maintain a software bills of materials,[2] and require audits and upgrade discipline (with transparent governance and time delays).

A third risk is in regulatory fit and security economics. When using blockchain, the design should be mapped to the applicable policy regime (e.g., US, EU, or other) regime early.

Finally, organizations must budget for the risk of breach. For example, the IBM Cost of a Data Breach 2024 report estimates the global average breach cost at USD 4.88M, higher in financial services [IBM 2024].

EXAMPLE 9.4 PRIVACY-PRESERVING PARTS TRACEABILITY WITH SALTED HASHES

An OEM deploying a permissioned parts-traceability ledger stores only salted hashes and revocation states on-chain. A *salted hash* is a security technique used to protect stored data by adding a random value (called a *salt*) to the input before hashing it. Salted hashes are frequently used when storing passwords.

The OEM also keeps rich records in encrypted data vaults and uses verifiable credentials plus zero-knowledge proofs so buyers can check "this serialized part is genuine and unrevoked" without learning supplier pricing or other confidential attributes. This meets privacy constraints while preserving auditability.

The conclusion is clear. Shared ledgers do not remove cyber risk, but they can reduce reconciliation overhead and central data custody in multiparty settings. Simultaneously, they can improve time-to-detect/contain when instrumented well.

9.1.5 Implementation Guidance for EST Organizations

With the major risks and mitigation strategies in view, the next step is translating these principles into a practical playbook. This playbook should provide clear, actionable guidance EST teams can follow when deciding how to design, deploy, and govern blockchain-based systems.

Designing a blockchain system for engineering, science, or technology work starts with choosing the right problem. The most successful projects focus on specific, measurable pain points rather than adopting blockchain for its own sake. Typical areas ripe for blockchain include reconciliation delays, counterfeit risks, or slow audit cycles. Once the problem is clearly framed, teams can decide whether a shared, tamper-resistant ledger is the best tool for coordinating data, identity, and workflow across organizations.

The next step is selecting the appropriate ledger model. Public proof-of-stake chains offer openness and strong economic security, while permissioned platforms provide tighter control, higher throughput, and clearer compliance paths. This is often a better fit for regulated or sensitive EST environments.

After choosing the ledger, teams should model identity and data flows with care with verifiable credentials for actors and assets, and keep sensitive information off-chain. This approach preserves privacy and performance while still delivering auditability.

Finally, good engineering discipline is essential. Wallets and signing keys must be treated like production infrastructure, with formal processes for key rotation, recovery, and separation of duties. Smart-contract development should follow secure-coding practices, include independent audits, and allow for controlled upgrades with transparent governance. Compliance checks to government regulations and sector-specific rules should be incorporated early. Once the system is live, teams should track outcomes such as cycle time, break rates, audit hours, defect escape, and MTTR to decide whether to iterate or scale [Lesavre et al 2020].

EXAMPLE 9.5 PERMISSIONED LEDGER FOR SECURE PARTS TRACKING

An EST team building a parts-tracking network chooses a permissioned ledger, assigns digital identifiers to manufacturers and components, stores detailed records in encrypted off-chain vaults, and anchors only hashed proofs on-chain. Smart contracts automatically enforce chain-of-custody rules, while verifiable credentials let auditors confirm authenticity and revocation status without exposing competitive or proprietary data.

Using best practices, blockchain technology gives EST teams a practical, repeatable roadmap for building application systems that are not only secure and compliant but also aligned with real engineering and operational needs.

9.2 CRYPTOCURRENCIES, NFTS, AND OTHER ASSETS

Blockchains enable more than shared databases. They can also be used to support *native digital assets* (e.g., cryptocurrencies), *unique digital items* (e.g., NFTs), and *tokenized representations* of real-world or financial instruments. These assets reside on distributed ledgers where ownership is proven with asymmetric cryptography, transfers are recorded by consensus, and rules can be encoded in smart contracts [EU 2023].

9.2.1 Fundamentals

Cryptocurrencies are digital, ledger-native assets secured by cryptographic keys and maintained on decentralized networks that achieve agreement through consensus protocols rather than relying on any central issuer. They function as digitally native forms of money or value, with transactions recorded on a tamper-evident chain of blocks, ensuring transparency and resistance to unauthorized modification.

Common example cryptocurrencies include Bitcoin, Ethereum, Solana, Ripple, and Cardano. Each of these differs in consensus design and network permissioning, which in turn shape openness, throughput, and governance. EST companies may leverage cryptocurrencies for cross-border micropayments, machine-to-machine settlements, or automated escrow mechanisms in multiparty engineering workflows.

A different kind of digital asset, a non-fungible token *(NFT),* is a unique digital identifier that represents ownership, authenticity, or rights to a digital or physical item such as real estate, art, luxury good, or collectible. EST companies can leverage NFTs in many ways, including:

- tokenizing patents, research findings, code, or designs;
- tying equity-like stakes to projects or discoveries to attract investors or revenue share with other engineers, inventors and other stakeholders;
- securing datasets, publications, or experimental results;
- associating them with parts or systems for tracking provenance from manufacturing to end-use;
- to represent virtual replicas of physical assets (digital twins) for simulation, maintenance, or virtual collaboration in industrial metaverses;
- using badges, access passes, or rewards for contributions, enhancing engagement and loyalty.

Functionally NFTs are smart-contract-managed entries that assert singular ownership, often combined with metadata or off-chain storage pointers; the "uniqueness" comes from the contract plus consensus, not from the file itself [EU 2023].

EXAMPLE 9.6 NFT-BASED DEVICE LICENSING ON A PERMISSIONED CHAIN

An engineering team issues per-device licenses as NFTs on a permissioned chain. Each token's metadata stores a hardware fingerprint and expiration; firmware checks the token state through a gateway and enables features only if the NFT is valid. Therefore, no internet-wide license server is required.

There is also growing interest in *tokenized securities* and other real-world assets that use ledgers for transfer restrictions and corporate actions.

EXAMPLE 9.7 SMART-CONTRACT ESCROW FOR SUPPLIER PAYMENTS

A manufacturer escrows a supplier payment on a permissioned chain. When test data (hashed on-chain) meets tolerances, and an inspector's verifiable credential cosigns, the contract releases funds automatically. No manual reconciliation is needed.

Companies also use crypto assets both as assets (treasury, collateral, settlement medium) and as *rails* (faster, programmable settlement and conditional payouts). Smart contracts can also automate escrow, milestone releases, and royalty split calculation and distribution.

In practice, "crypto assets" differ by fungibility, rights attached *permissioning, consensus mechanism,* and where the "truth" resides. *Fungibility* means an asset's units are interchangeable, that is, each unit is identical in value and function, a like dollars or Bitcoin.

Rights refer to the specific permissions or entitlements a token or asset grants, such as payment ability, access, royalties, or governance authority. *Permissioning* defines who is allowed to participate in a blockchain network or validate transactions. *Where the truth resides* describes whether the authoritative record of ownership or state is stored fully on-chain or depends partly on off-chain legal or data sources. The *consensus mechanism* determines how a blockchain network reaches agreement on the validity and ordering of transactions. Understanding those distinctions helps EST teams choose the right pattern for payments, licensing, provenance, or automated workflows.

Table 9.1 summarizes the three major categories of blockchain-based assets. These three classes are distinguished by comparing their fungibility, typical uses, where authoritative records ("truth") reside, and key engineering-related risks.

TABLE 9.1
Digital Asset Types

Asset type	Fungible?	Typical rights/uses	Where truth lives	Primary risks (engineering view)
Cryptocurrency	Yes	Payments, settlement asset, staking/gas	On-chain ledger	Key loss, consensus and fee volatility, regulatory treatment for treasury use [eur-lex.europa.eu]
NFT	No (unique)	Provenance, licensing, access, digital twins	On-chain token + metadata (often off-chain pointer)	Off-chain link rot, IP ambiguity, contract bugs [eur-lex.europa.eu]
Tokenized security/RWA	Usually yes (but may be unique)	Ownership/entitlement to off chain instrument or asset	On-chain record + off chain legal record	Compliance mapping, custody, accurate linkage to legal rights

Together, these distinctions give EST professionals a clear framework for selecting the appropriate digital-asset model when building secure, interoperable, and audit-ready engineering and financial systems or for use in one of the many different applications mentioned [EU 2023].

9.2.2 Engineering, Science, and Technology Applications

NFTs and tokens are increasingly applied to *provenance and IP management* in labs and factories. A token can mark the first item from a process run, bind it to specific machine parameters (as hashed metadata), and carry forward royalties or usage rights. Combined with decentralized identity, devices, staff, and artifacts can each present proof of who/what/when without over-sharing underlying data.

EXAMPLE 9.8 TOKENIZED DATA ACCESS AND DYNAMIC WARRANTY MANAGEMENT

1) A lab publishes a dataset's chain and issues NFTs granting de-identified access tiers. Reviewers redeem a token to view data through an enclave; revocation is handled via on-chain status, while proofs of access are audit-ready.
2) A compressor ships with a warranty token whose coverage adjusts by verified run-hours (IoT hash anchors). If hours exceed the free threshold, the token prompts a micro-payment stream; an auditor later verifies the entire history from the shared ledger.

In cyber-physical systems, programmable assets can make machine-to-machine transactions possible. For example, a robot "purchases" spare capacity, an instrument "subscribes" to a calibration service, or a microgrid settles energy among peers. Smart contracts can also encode the policy, consensus secures the record, and identity standards gate access [EU 2023].

9.2.3 Risks and Opportunities

Managing risk for crypto assets involves strong architectural design and governance practices such as key custody, smart-contract quality, oracle trust, and data privacy. Even when payloads stay off-chain, metadata can leak patterns, so it is wise to use pairwise identifiers, off-chain storage with on-chain hashes, and selective disclosure or zero-key techniques for attributes. It is also imperative to choose consensus and permissioning that match security and sustainability goals.

On the opportunity side, tokens enable fine-grained royalties, conditional warranties, usage-based pricing, and transparent audit trails, which are especially attractive for multi-party engineering environments where reconciliation is costly. When assets represent regulated instruments, they need to align early with applicable governance structures [EU 2023].

9.3 FINTECHS

A *fintech* is a software-driven financial technology firms. Fintechs apply engineering discipline, data analytics, and AI to re-architect traditional financial services to deliver faster, cheaper, and more user-centric products and services. These include mobile banking platforms and automated investment advisors. Over the past four decades, fintechs have pushed markets away from manual, human-mediated processes, and toward highly automated, algorithmic ecosystems. This evolution mirrors broader advances in computing, communications infrastructure, data-exchange standards, and real-time decision automation that now define modern financial operations.

9.3.1 From Human Trading Floors to Electronic Markets

For most of the 20th century, financial markets ran on human judgment and paper-based workflows. Trades were shouted across crowded floors, recorded on slips, and reconciled at the end of the day. This environment produced frequent errors and required armies of staff to reconcile books manually. The shift toward automation began in the late 1980s and early 1990s with systems like NASDAQ's Small Order Execution System (SOES) and the Chicago Board Options Exchange's (CBOE's) Retail Automatic Execution System (RAES). These systems guaranteed automated execution for small orders and introduced the first wave of engineering discipline into market operations.

The real transformation accelerated when the FIX protocol emerged in 1992. FIX standardized electronic message formats and made real-time connectivity between brokers, exchanges, and trading systems possible. With common interfaces and machine-readable workflows, markets could finally move away from bespoke integrations and toward resilient, fault-tolerant digital infrastructure. Through the 1990s and 2000s, trading pits gave way to fully electronic limit-order books and industrial-grade matching engines, laying the groundwork for algorithmic and ultra-low-latency trading.

By the early 2000s, financial markets had evolved into some of the world's most complex autonomous-agent ecosystems. Billions of quotes and orders flowed through tightly engineered systems each day, with algorithms (not humans) providing most liquidity. Regulators responded with rules emphasizing software quality, version control, and operational integrity, recognizing that markets now operated more like high-reliability computing environments than traditional financial venues.

Today's electronic markets stand as the most mature example of large-scale, real-time automation. The resulting ecosystem was built on decades of engineering discipline and continuous technological refinement [Kumiega 2026].

9.3.2 Regulatory Evolution Shaping Fintech

Blockchains offer meaningful trust, automation, and multi-party coordination benefits for FinTech, but they also introduce a distinct risk surface that EST teams must manage deliberately. Consensus models, for example, shape not only performance and governance but also energy consumption. For example, when—Ethereum's 2022—transitioned to proof-of-stake, it cut its estimated energy use by 99.95%. Even with careful design, metadata such as timestamps and transaction links can still reveal patterns, making privacy a systems-level concern rather than something solved by simply keeping PII off-chain [Ethereum 2025].

Security engineering is equally critical. Smart-contract defects, compromised wallets, and insecure CI/CD pipelines can all propagate rapidly across an ecosystem because blockchains remove many of the "friction points" that normally slow down failure. This means signers, oracles, build systems, and key-management workflows must be treated as Tier-1 assets. Strong audit practices, transparent upgrade paths, and disciplined operational controls are not optional; they are foundational to safe deployment in any engineering or financial environment.

Finally, every blockchain design must align with the applicable regulatory and economic context. Jurisdictions now differ sharply on tokenization rules, identity requirements, and treatment of digital records, making early regulatory mapping essential to avoid re-architecting later. And while shared ledgers can reduce reconciliation overhead and improve auditability, organizations must still budget for breach response and ongoing risk monitoring. The goal is not blind adoption, but deliberate, engineering-grade integration that preserves privacy, ensures security, and supports long-term operational resilience [Kumiega 2026].

9.3.3 Evolution and Types

Fintechs evolved from point solutions that digitized narrow tasks (online bill-pay and card gateways) into end-to-end platforms that recompose financial services as software (Figure 9.2).

The early wave focused on payments and personal finance; subsequent waves layered identity, risk, and data-science pipelines on standardized APIs, enabling real-time underwriting, automated compliance, and embedded finance inside non-bank products.

Today, the landscape is best understood by function rather than charter. These functions include payments and money movement, lending and credit, wealth/

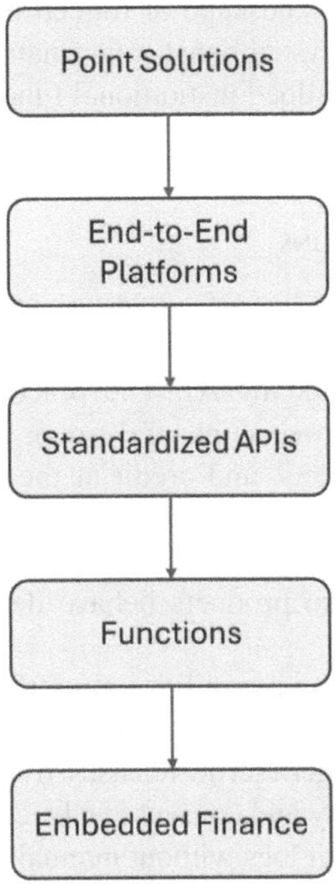

FIGURE 9.2 Fintech evolution.

markets, *insurtech,* and infrastructure providers. What unifies these types is a software supply chain that treats finance as an API. Identity is verified at the edge, risk scored in milliseconds, decisions enforced by code, and records synchronized across counterparties.

Neobanks, productize this stack into consumer and subject matter expert accounts. P*ayment processors* expose this stack as software development tools. Wealth platforms wrap it in compliant order-routing and custody. Insurtech binds sensors and actuarial models to dynamic premiums. Finally, in *business as a service,* vendors furnish the ledgers, controls, and audits needed to ship regulated features at software speed.

EXAMPLE 9.9 EMBEDDED FINANCE AUTOMATING SUPPLIER PAYMENTS

A mid-market manufacturer uses embedded finance to automate supplier payments and early-pay discounts. As IoT delivery events close out, the payment API initiates settlement, and a lending API offers just-in-time working capital against approved invoices. This approach reduces reconciliation effort and funding friction while keeping engineers out of "banking" operations.

Taken together, the evolving landscape of fintech illustrates a steady shift toward highly modular, API-driven systems that treat financial capabilities as composable software services rather than siloed institutional functions, enabling real-time, automated financial workflows across EST environments.

9.3.4 Financial Innovations

Fintech innovation has progressed from digitizing point tasks to recomposing finance as software primitives, namely, payments, identity, risk, treasury, exposed through APIs and orchestrated by AI. The practical results are programmable money movement (account-to-account, real-time), instant underwriting based on verified data, embedded wallets, and credit at the point of need, and automated compliance that rides the same event streams as transactions. Providers now integrate these primitives into end-to-end flows, for example, onboarding → KYC → fraud screening → payout, so products behave deterministically and scale with code rather than headcount.

For EST organizations, these capabilities close the loop between physical events and financial actions. Sensor-verified milestones (goods received, quality passed, and uptime achieved) can trigger escrow releases, dynamic discounts, or just-in-time working-capital offers. Identity and consent can be checked at the edge. Settlements can post back to ERP and audit logs without manual reconciliation.

The operating pattern is consistent, namely, instrument the event, authorize with standards, decide in milliseconds, execute via API, and maintain an auditable record.

9.3.4.1 Engineering, Science, and Technology Synergies

As payment rails, data-sharing standards, and AI mature, providers now compose lending, payments, identity, and treasury as services. This shifts execution from single-function apps to embedded capabilities inside operational software. This API-first stack, often delivered via BaaS, supports real-time credit, automated compliance, and context-aware payments across sectors, with AI raising the bar on fraud detection, personalization, and operating leverage in payments and lending.

EXAMPLE 9.10 SENSOR-TRIGGERED EMBEDDED FINANCE FOR MANUFACTURERS

A mid-market manufacturer connects shop-floor telemetry and supplier portals to an embedded-finance stack. When sensors confirm goods-received and quality control tolerances, a payment-initiation API releases partial settlement, and an AI-driven risk service adjusts dynamic credit terms for the supplier.

This pattern reduces days-payable/receivable cycles and shrinks reconciliation effort, while preserving auditability through event logs and standardized interfaces.

For EST firms, the practical impact of fintech is a tighter loop between events in the physical world and financial actions. The result is faster reconciliation, lower capital friction, and fewer handoffs. Thus, finance becomes an automated subsystem of engineering operations rather than a separate, manual back office [Grand View Research 2025].

9.3.4.2 Challenges and Strategies

As fintech capabilities broaden across payments, lending, and embedded finance, the operational risks grow accordingly. Cybersecurity remains a top constraint as firms expose more surface area through APIs, event streams, and third-party integrations. Regulatory scrutiny has increased as well. Global payment systems now operate under tighter expectations around data governance, authentication, auditability, and fraud controls. These pressures mirror trends identified in recent global payments analyses, which note that providers must balance innovation with escalating compliance and resilience requirements in a more fragmented regulatory environment.

EXAMPLE 9.11 HARDENED PAYMENT APIS FOR ROBOTICS WORKFLOWS

A robotics manufacturer connects its supplier-financing API to shop-floor scheduling software. If the integration is not hardened, for example, missing endpoint authentication or insufficient rate-limits, a compromised device or misconfigured controller could trigger erroneous payment requests. To mitigate this, the firm enforces signed requests, event-level traceability, and fail-closed payment initiation rules, ensuring that financial automation inherits the same safety discipline applied to physical automation.

For engineering-centric organizations adopting fintech services, the challenge is integrating financial automation into systems that already depend on reliable telemetry, deterministic workflows, and well-governed identity frameworks. Poorly instrumented APIs, misaligned consent flows, or insufficient audit logs can create blind spots that undermine both operational integrity and regulatory posture. Strategies increasingly emphasize a zero-trust approach to API interactions, strong key-management practices, and continuous monitoring of payment and data-sharing events These treat financial workflows with the same rigor as production control systems [Grand View Research 2025].

9.4 OPEN BANKING

Open banking (OB) refers to a financial ecosystem that allows third-party providers secure, permission-based access to consumer banking, transaction, and related financial data through standardized *application program interfaces (API).* Rather than relying on traditional, predefined relationships among financial institutions, OB enables customers to direct how various banking entities interact on their behalf.

This flexibility supports "finance as a service," where a cloud-based collection of banking participants offers both microservices and macroservices. Microservices include payments, deposits, and transfers, while macroservices include loan and mortgage origination (Figure 9.3).

Within the OB ecosystem, banks and nonbank financial institutions operate as interconnected service clouds, all adhering to common security profiles, interfaces, and customer-experience guidelines. By opening access to proprietary data, OB promotes a cashless, digital-first economy while empowering consumers to mix and match best-of-breed financial products. A customer might hold a checking account with one provider, a savings account with another, and credit or loan services with yet others. Open banking allows the customer to move funds seamlessly among these entities through dashboards that support transactions, analytics, alarms, and optimization tools.

This increased openness fosters competition, enables smaller players to enter the financial services market, and expands customer choice. Aggregated accounts provide users with clearer financial insight and greater convenience, while the broader ecosystem encourages innovation and reduces anticompetitive barriers. EST companies can leverage open banking to streamline financial workflows, integrate real-time financial data into digital systems, and enable more automated, data-driven product and service innovations [Laplante and Kshetri 2021].

9.4.1 Principles and Regulations

There are different approaches to open banking across the world, based on the regulatory and political environment where the system operates. For example, the EU and the United Kingdom are fully committed to a regulatory approach, requiring banks to open customer-authorized access to payments data. The United Kingdom created a dedicated implementation body—the Open Banking Implementation Entity (OBIE). This body published detailed API specifications, security profiles, and customer experience guidelines, which helped ensure every player built standardized systems.

Australia took a slightly different route with its Consumer Data Right (CDR) legislation. Instead of limiting data sharing to banking, it designed one framework intended to reach multiple industries—banking first, then energy and telecoms. This means consumers can authorize a trusted third party to pull their data using standardized APIs, all under an accreditation model built into the law [ACCC 2026].

The United States has been more market-driven, letting fintechs and banks figure out interoperability themselves, which led to years of screen-scraping and patchwork data-sharing agreements. That situation is changing—in 2024—the CFPB's Personal Financial Data Rights rule set out clearer expectations for data access, security, and third-party obligations. While phased in over several years, it establishes a national baseline that moves the United States closer to standards-based

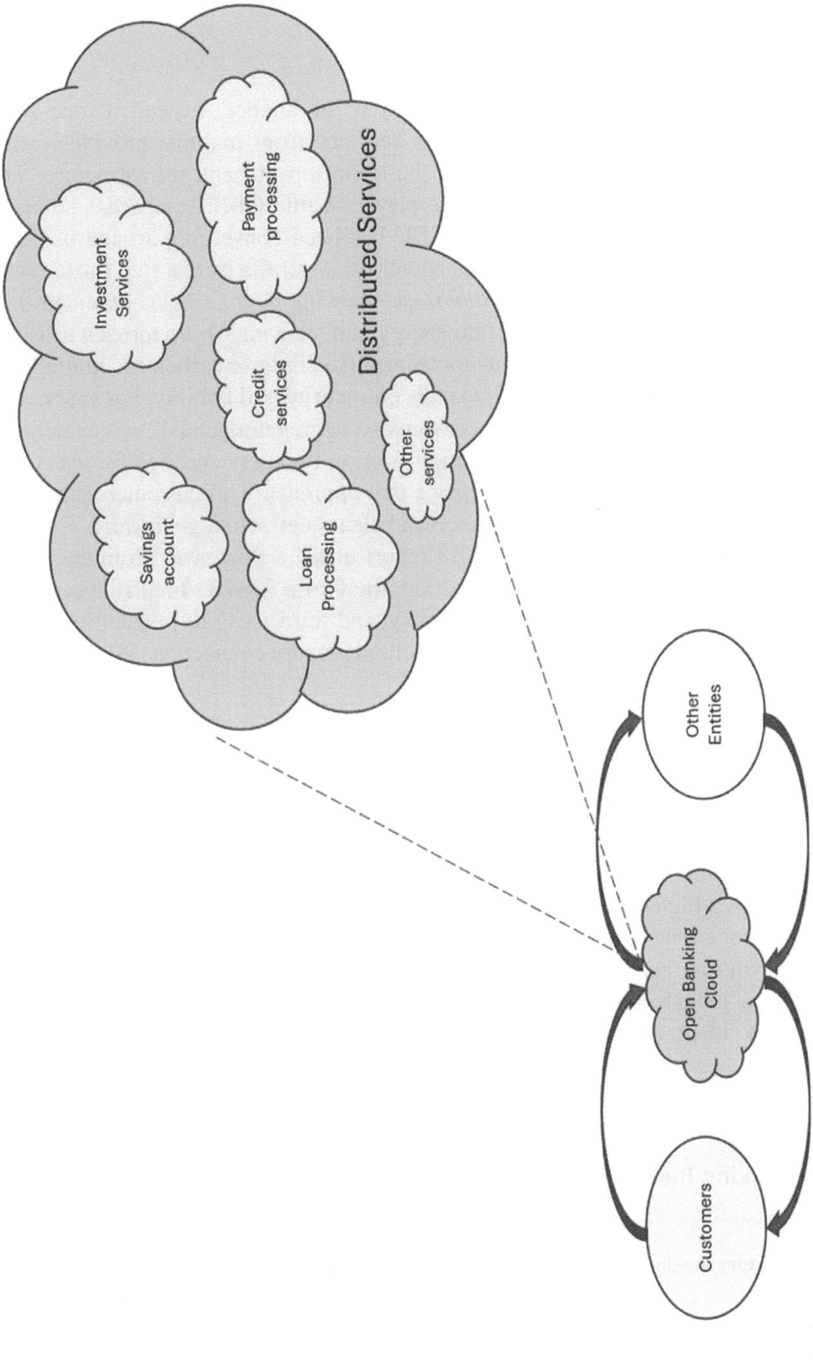

FIGURE 9.3 Open banking ecosystem.

open banking—without fully copying the EU model. Across all these regions, security rules are the anchor [FERC 2021].

9.4.2 Financial Ecosystem Impact

Open banking changes the "distribution layer" of finance. Instead of one bank providing everything, customers can mix services from multiple providers—and fintechs can plug into standard APIs to build on top of bank infrastructure. The UK's program helped by publishing one playbook (the OBIE standards), so products look and feel consistent, while the EU has been converging around use case profiles to reduce fragmentation across countries. Australia went a step further with the *Consumer Data Right (CDR) framework,* showing how a single consented-data model can extend beyond banking into energy and telecom. This approach is useful when seeking to consolidate insights for things like credit or carbon accounting.

A large part of the ecosystem's impact is on security and liability. For years, data sharing often meant handing over passwords so aggregators could "screen-scrape" bank sites. This approach introduces high-friction for users and risk for everyone. Policymakers have pushed hard to replace that approach with customer-authorized APIs protected by modern authentication and clear contractual guardrails.

In the United States, Treasury's 2018 report urged a shift away from credential sharing toward standardized APIs; in 2024, the CFPB's §1033 final rule set baseline national expectations for access, privacy, and third-party responsibilities. This moves the market decisively toward interoperable, secure connections [Skadden et al 2024].

9.4.3 Engineering, Science, and Technology Applications

Table 9.2 summarizes the core building blocks of open banking. The table covers the legal baseline, API classes, security profiles, governance structures, and risk frameworks that define how data and payments are shared securely across institutions. It also provides a high-level view of the components required to implement standards-based, interoperable open-banking ecosystems.

Open banking becomes useful when treated as infrastructure for data and payments. This includes clean APIs, clear security profiles, and a shared vocabulary for consent. In practice, most programs rely on a few core API classes. Table 9.3

TABLE 9.2
Open Banking Building Blocks

Building block	Function
Legal/regulatory baseline	Consumer's right to access/share data; secure third-party access; authentication
API classes	Read/write, open data, directory/registration, management/MI
Security profiles	Authorization & token flows; transport security
Program governance	Accreditation, directories, conformance, and incident reporting

TABLE 9.3
Open Banking API Classes and Typical Operations

API class	Example operations
Read-only	Balances, transactions, product catalogs
Write/Initiation	Payment initiation, standing orders
Directory/Registration	Onboarding third-party providers, keys, certificates
Management/Management Information	Uptime, error codes, consent logs

summarizes the main open-banking API classes and lists representative operations each class supports [Yaga et al 2018].

Open banking is still maturing at various rates internationally. Still, it holds promise for EST in a variety of use cases, for example.

- A clinical lab's information system requests read-only account data to post automated refunds; authorization scopes are limited to a specific account set, access is time-boxed, and events are logged for audit.
- A micro-grid platform uses payment initiation APIs to settle peer-to-peer energy trades, with the directory API verifying that counterparties are accredited participants.
- A mid-sized manufacturer connects multiple banks via open-banking APIs to aggregate receivables and payables, reconcile invoices automatically, and trigger variable recurring payments to suppliers when IoT delivery events close out. This approach cuts reconciliation time and improves cash-flow predictability.
- A micro-grid platform uses payment initiation APIs to settle trades between participants. The directory/registration service checks that each counterparty is accredited, and SCA is handled by the bank, thus keeping the operator out of the credential business.

In EST open banking can connect identity, data, and payments, enabling products to read balances, initiate payments, or verify attributes with customer consent. Properly designed, OB can replace many clunky tactics, such as screen-scraping with secure, auditable API calls and common data schemas, enabling automation, analytics, and embedded finance in labs, factories, field operations, and digital services [Voas et al 2022].

9.5 ARTIFICIAL INTELLIGENCE

Artificial Intelligence (AI) encompasses a family of computational techniques that enable software systems to learn from data, detect patterns, and make predictions or decisions at scales and speeds unattainable by humans. AI techniques include classical machine learning, neural networks, and modern generative model

Figure 9.4 shows the relationship between types of AI. Artificial intelligence describes machines and software programmed to think, work, and react like humans.

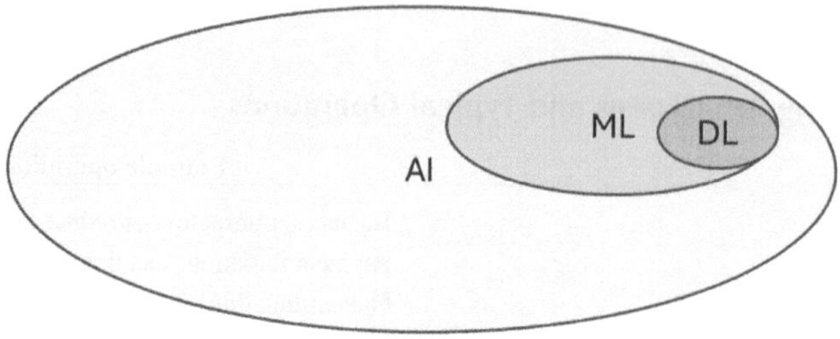

FIGURE 9.4 Relationship between Artificial Intelligence (AI), Machine Learning (ML), and Deep Learning (DL).

Machine learning (ML), is a form of AI in which a system is trained by giving it data so it can "learn" over time. *Deep learning (DL)* is a kind of machine learning that is designed to function like the human brain using artificial neural networks.

AI is reshaping financial systems, engineering workflows, scientific research, and digital operations. Just as importantly, responsible deployment requires risk-aware governance aligned with frameworks such as the NIST AI Risk Management Framework (AI RMF), which provides structured guidance for trustworthy and safe AI design, development, and use [Tabassi 2023].

9.5.1 Foundations

AI systems generate outputs, such as predictions, recommendations, or decisions, by learning from large volumes of structured and unstructured data. According to NIST's AI RMF, these systems operate with varying degrees of autonomy and can influence both physical and digital environments. The framework highlights that AI risks emerge because models may drift, be trained on biased or rapidly changing data, or operate within socio-technical contexts that make failures difficult to detect.

EXAMPLE 9.12 DEFECT-CLASSIFICATION MODEL FOR COMPOSITES QA

A manufacturing quality-assurance team deploys a supervised learning model to classify defects in composite turbine-blade materials. Over time, shifts in lighting conditions and supplier material batches cause a drift in the model's accuracy. This problem requires continuous monitoring, periodic retraining, and rigorous traceability in line with AI RMF guidance.

Generative AI represents a class of models that can synthesize new content by learning patterns from massive training corpora. This approach contrasts with classical supervised learning, which predicts labels or numerical outputs, generative systems produce text, images, designs, code, and simulations that mimic the structure of their training data. Generative AI includes *large language models,* diffusion models, and multimodal architectures. Their power stems from scale: billions of parameters, trained on diverse data, enabling them to generalize across domains and assist with complex reasoning, planning, and content creation.

As emphasized by frameworks such as the NIST AI Risk Management Framework, generative models introduce both significant opportunities and new risk surfaces, including hallucination, data-provenance uncertainty and the need for careful governance when outputs influence operational or financial decisions.

In EST environments, generative AI is becoming a practical co-processor for high-complexity tasks. For example, an aerospace composites lab may use generative models to propose candidate layups that satisfy weight, stiffness, and fatigue requirements, while a structural solver validates each candidate before entering the design cycle. In semiconductor manufacturing, generative agents can synthesize test recipes or diagnostic prompts based on historical failure signatures, reducing engineering cycle time while improving coverage.

In financial–EST workflows, a generative model might prepare draft supplier-risk narratives or convert raw telemetry into standardized audit explanations that are then validated through established controls before entering compliance systems. These are not fully autonomous systems; they are productivity and reasoning accelerators that help domain experts move faster while maintaining engineering discipline.

As these tools mature, their real value lies not in replacing technical expertise but in augmenting EST professionals with rapid, context-aware synthesis and ideation. When combined with strong guardrails, domain-specific validation, and responsible governance, generative AI becomes a leveraging technology. It can act as an intelligent layer that transforms how EST organizations design, diagnose, and explain their systems, while keeping human judgment firmly in the loop [Tabassi 2023].

9.5.2 Financial Transformations

AI is now a structural component of modern finance, powering fraud-detection engines, credit-scoring pipelines, *robo-advisors*, and algorithmic trading systems. For example, AI-enabled fraud detection systems use behavioral pattern analysis to identify anomalies within milliseconds, reducing fraudulent transactions and false positives. Robo-advisors apply ML algorithms to produce personalized, low-cost automated investment advice, helping democratize access to financial planning.

EXAMPLE 9.13 AI-DRIVEN REAL-TIME SUPPLIER CREDIT RISK SCORING

A mid-market engineering firm uses an AI-driven credit-risk engine to evaluate supplier financing requests. The model analyzes operational telemetry (delivery success rates, machine-usage consistency, historic defect rates) combined with traditional financial metrics to generate real-time, risk-adjusted credit terms delivered through the firm's open-banking APIs.

Beyond traditional fraud-detection and robo-advisory, AI is increasingly becoming used to link operational telemetry, financial decision systems, and risk-aware automation across engineering-centric industries.

EXAMPLE 9.14 AI-DRIVEN FINANCIAL–OPERATIONAL SYNCHRONIZATION

Many engineering organizations now deploy multimodal machine-learning systems that ingest sensor traces, ERP activity, workflow states, and maintenance histories to produce risk-adjusted financial signals. Models evaluate not only the financial posture of a counterparty but also how well that counterparty performs operationally in terms of delivery timeliness, process capability, equipment uptime, defect rates, and other reliability indicators.

These signals flow directly into embedded-finance APIs, enabling systems that make dynamic, conditional financial decisions in milliseconds, rather than in weekly or monthly cycles.

In modern EST environments, financial actions are most valuable when they occur in the same loop as physical or digital operations. For example, in credit decisions, settlement events, warranty adjustments, or risk scoring, AI synchronizes these pieces [Dushyant 2025].

9.5.3 Prognostics and Health Management

AI enables deeper integration across EST domains by converting sensor data, simulation outputs, and laboratory results into predictive insights. *Prognostics and Health Management (PHM)* is an AI-enabled approach that uses sensing, diagnostics, and physics-/data-driven prognostics to estimate the state of health and *remaining useful life (RUL)* on-based maintenance and risk-aware decisions.

EXAMPLE 9.15 AI-BASED EARLY DETECTION OF GEARBOX FAILURES

Wind energy is a canonical case for PHM. A wind-turbine operator deploys an AI model to detect early-stage gearbox vibration signatures. Coupled with secure IoT telemetry, the model predicts impending failures days in advance, allowing technicians to schedule maintenance windows during low-wind periods. This improves uptime and reduces operational risk.

Here is another example of PHM being used in EST.

EXAMPLE 9.16 AUTOMATED TURBINE-BLADE INSPECTION VIA COMPUTER VISION

A commercial aircraft engine manufacturer deploys an AI-enabled blade-inspection system to enhance both the speed and reliability of its quality-assurance workflows. Traditionally, turbine-blade inspections require meticulous manual review under varying lighting conditions, making results dependent on inspector experience and prone to inconsistency. By integrating computer-vision models trained on thousands of annotated defect images, the manufacturer automates the identification of cracks, delamination patterns, and surface anomalies.

The AI system analyzes high-resolution inspection images in real time, flagging potential defects and prioritizing them for human validation. This reduces total inspection time by approximately 50% while also improving the consistency of defect detection across shifts and facilities.

Because the model is continuously retrained on new inspection data, it adapts to changes in materials, tooling, or lighting. This mitigates drift and maintains high accuracy over time. The result is a more reliable, repeatable, and efficient inspection pipeline that strengthens overall engine safety and performance assurance.

PHM can also support grid transformers, industrial generators, batteries, pipelines, and other critical infrastructure, where diagnostics and RUL prediction improve reliability and optimize maintenance planning in complex engineering systems.

9.5.4 Risks and Ethical Strategies

AI introduces unique risks. NIST's AI RMF emphasizes governance functions such as *Map, Measure, and Manage* to identify system risks, assess trustworthiness, and implement controls that reduce harmful outcomes. It also highlights the growing need for safeguards against evolving threats, such as memory-poisoning attacks on agentic AI systems and unreliable generative outputs [Tabassi 2023].

Figure 9.5 highlights the major categories of AI risk by placing the model at the center and surrounding it with five classes of vulnerabilities that can degrade trustworthiness.

Bias refers to systematic distortions in model outcomes arising from unbalanced or non-representative data. For example, a predictive-maintenance model trained mostly on data from newer equipment may systematically underestimate failure risk in older machines, leading to biased maintenance decisions that reduce reliability.

Hallucination denotes syntactically plausible but factually incorrect outputs produced by generative models. For instance, a generative-AI assistant for factory

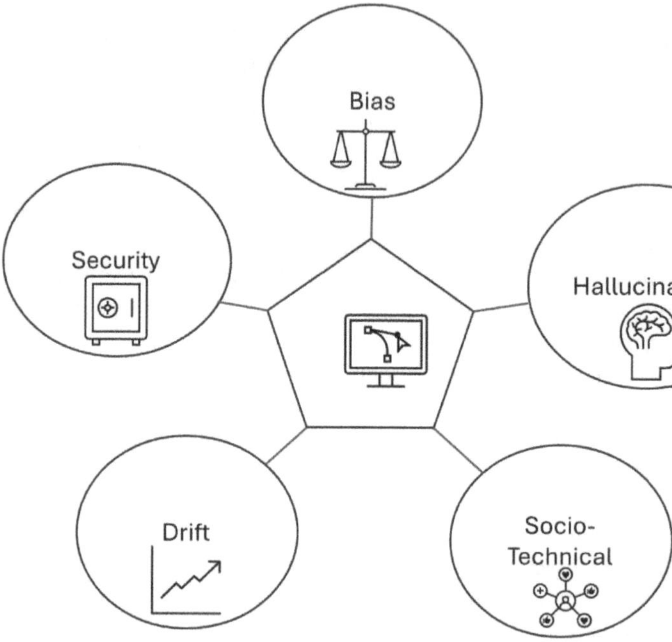

FIGURE 9.5 AI risk landscape.

technicians hallucinates a nonexistent calibration protocol, producing an incorrect, but confident-sounding, instruction that could mislead operators if not caught.

Drift describes performance degradation over time as real-world data distributions shift. For example, a predictive-maintenance model for turbine-blade inspections gradually becomes less accurate as lighting conditions and material batches shift over time, causing performance drift that requires retraining to stay reliable.

Finally, *socio-technical risk* encompasses failures that arise from human–AI interaction, organizational context, or deployment environments rather than the model alone. For instance, a factory's AI-driven scheduling tool causes production delays because workers misinterpret its recommendations and override them inconsistently.

EXAMPLE 9.17 GOVERNED GENERATIVE MODELS IN BIOMEDICAL DISCOVERY

A biomedical research laboratory uses a generative model to propose new compound structures. Without proper governance, the model produces chemically invalid candidates and occasionally unsafe molecular suggestions. The lab mitigates these risks using a domain-specific validation pipeline and an AI-RMF-aligned "Manage" function to restrict unverified outputs.

Together, these risks emphasize that trustworthy AI requires continuous monitoring, validation, and governance across both technical and human systems.

9.5.5 AI Governance

AI governance refers to the policies, accountability structures, and lifecycle oversight mechanisms that ensure AI systems are designed, deployed, and monitored responsibly. NIST's AI RMF frames governance as a *cross-cutting function*. Cross-cutting means establishing transparent policies, roles, and risk-management culture that apply across mapping, measuring, and managing AI risks, including controls for supply-chain data and third-party tools.

EXAMPLE 9.18 AI GOVERNANCE BOARD FOR SAFETY-CRITICAL SYSTEMS

A national laboratory deploying AI-enhanced predictive-maintenance models for its particle-accelerator cooling systems establishes an *AI Governance Board* that defines approved use cases, risk tiers, data-quality requirements, and mandatory validation steps before any model can influence operational decisions. The board requires every model to undergo documented performance testing, bias assessment, and change-control review, and it mandates human-in-the-loop verification for any recommendations affecting safety-critical equipment.

Through these policies, accountability structures, and lifecycle oversight, the organization ensures its AI systems remain traceable, compliant with engineering standards, and aligned with mission-assurance expectations.

For EST organizations operating safety-critical systems, strong governance provides the strategic backbone, ensuring AI aligns with engineering requirements, regulatory expectations, and mission assurance [Tabassi 2023].

9.5.6 AI Guardrails

AI guardrails are the technical enforcement mechanisms that apply governance rules in real time by constraining behavior, preventing unsafe actions, and ensuring outputs comply with organizational policies. Guardrails are implemented in models, APIs, or agent workflows. Unlike *governance*, guardrails are operational controls embedded in the system architecture and designed to mitigate risks such as misuse, hallucinations, privacy exposure, or unsafe automation, especially relevant in agentic or autonomous EST environments.

In EST applications, robust guardrails help ensure that AI-driven diagnostics, simulations, or control recommendations cannot silently drift into unsafe or non-compliant states, preserving both safety and engineering integrity. Guardrails enforce policies by embedding the organization's rules directly into the AI system's technical architecture, ensuring those rules are applied automatically and in real time. They work by intercepting inputs, outputs, or tool calls, applying safety constraints (e.g., content filters, PII detection, and permission checks), and blocking, modifying, or routing requests whenever they violate the governance policies those guardrails embody [NIST CSF].

EXAMPLE 9.19 AI GUARDRAILS FOR SAFETY-CRITICAL SYSTEMS

A large aerospace manufacturer deploys an AI assistant that helps technicians troubleshoot avionics issues. To enforce safety and compliance, the system includes guardrails that block any action that could suggest unsafe maintenance steps, such as bypassing required lock-out/tag-out procedures or recommending changes to certified flight-control parameters.

When a technician asks, "Can I override the torque-limit setting to speed up the repair?", the guardrail layer intercepts the request, detects that it violates operational-safety policy, and returns a safe alternative, such as directing the technician to approved procedures and linking to the relevant maintenance manual instead of passing the question to the model for free-form generation.

Guardrails are an important part of any financial or other AI strategy as the "enforcement layer," that is, the code-level constraints, filters, validation checks, and access controls that prevent an AI model or agent from taking actions outside approved boundaries, even if the model attempts to do so.

9.6 QUANTUM COMPUTING

Quantum computing is a new way to compute using qubits. These are tiny physical systems that can be in a "mix" of 0 and 1 states at the same time. This mix is called *superposition*. Based on this principle, a small quantum register can explore many possibilities in parallel before measurement (establishing a single result). This approach is valuable for certain problems in chemistry, finance, and logistics that overwhelm classical computing machines.

While practical quantum computing systems may be a few years or even decades from reality, they are worth discussing for their profound potential in EST financial applications.

9.6.1 Basic Principles

A *qubit,* ψ, is a two-state quantum system represented by Equation (9.1):

$$|\psi\rangle = \alpha|0\rangle + \beta|1\rangle \quad (9.1).$$

Unlike a classical 0/1 bit, the qubit exists in a *superposition* or blend of two basis states, $|0\rangle$ and $|1\rangle$ (Figure 9.6).

Until the qubit is measured, it exists in that blend; upon measurement, its "collapses" to 0 or 1 with complex probability amplitudes α and β, respectively. α and β must satisfy equation (9.2).

$$|\alpha|^2 + |\beta|^2 = 1 \quad (9.2)$$

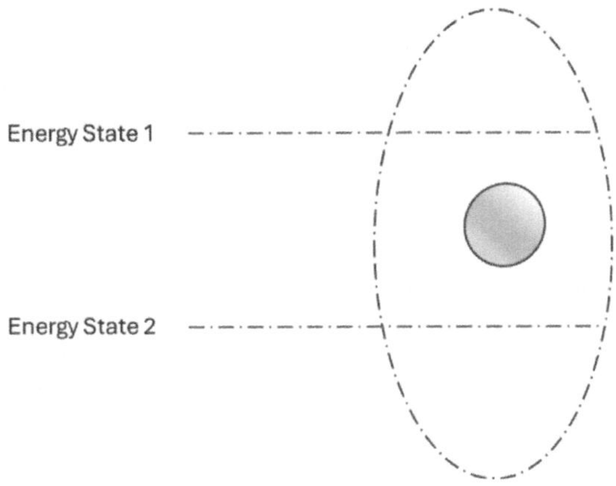

FIGURE 9.6 Depiction of a qubit.

A special kind of gating function establishes the qubit in an even 50/50 superposition. This is usually the first step needed to perform calculations on multiple qubits.

Two landmark algorithms establish the potential for practical quantum computing. *Shor's algorithm* states that, on an ideal fault-tolerant quantum computer, integer factoring can be done in *polynomial* time. *Grover's algorithm,* shows that quantum computing could reliably speed up unstructured search from $O(N)$ to $O(\sqrt{N})$. For example, if a password space has N candidates, Grover cuts the number of oracle calls to about $\sqrt{N}$.

Taken together, it is widely understood that quantum computing threatens modern cryptographic algorithms if large machines arrive and therefore needs to be taken seriously even though practical applications are not imminent [NIST 2024]. This fact has important implications for the cryptographic approaches that are used to secure financial computing applications.

9.6.2 Applications

Though years or even decades from reality, there are many potential applications for quantum computing in finance. For example, researchers are mapping problems like portfolio optimization, risk modeling, and derivative pricing into quantum computing-friendly forms of classical optimization algorithms to reduce computational effort by orders of magnitude.

As noted, quantum computing is expected to be used widely in both defensive and adversarial security of financial and other systems. NIST's recently released post-quantum cryptography standards for computing security describe these in detail [NIST 2024].

There are also potential direct applications of quantum computing in EST. For example, in first-principles molecular simulation, quantum computers can pair with traditional high-performance computers to better estimate molecular ground states and interactions. Such an approach would be useful in a pharma/clinic collaboration.

Beyond chemistry, teams can use quantum (and quantum-inspired) optimization on traffic-flow and routing, for example, framing a city network as a quantum process for solution by hybrid simulated annealing. For example, quantum computing could be used to solve traffic routing problems, deriving reports near-optimal routes with measurable congestion reductions. This is a promising approach for smart-city digital twins.

9.6.3 Challenges and Opportunities

Quantum computers remain constrained by physical error rates, limited coherence times, and the large overhead of *quantum error correction (QEC)*. Recent experimental progress is meaningful but incremental. Given these limits, the most potential value in quantum computing is from *hybrid quantum–classical* workflows. In these "variational" or "co-processing" patterns, narrowly scoped quantum subroutines are executed on current hardware, while classical software orchestrates the optimization loop, error-mitigation, batching, and data movement. Modern toolchains explicitly support this division of labor.

Even before large, fault-tolerant machines arrive, the cryptographic stakes are immediate. Standards bodies have begun publishing *post-quantum cryptography (PQC)* selections and finalizations so organizations can migrate away from quantum-vulnerable schemes on measured timelines. For engineering and financial systems, the actionable takeaway is to begin inventorying cryptographic dependencies and planning for rollouts that align with NIST's PQC standardization and guidance [NIST 2024].

9.7 CYBERSECURITY

Starting in late 2019 and discovered in December 2020, the SolarWinds campaign was a highly sophisticated cyber espionage operation, attributed to Russia's Foreign Intelligence Service. This supply-chain attack was executed by injecting malicious code into software updates. The attack affected 18,000 customers worldwide, including numerous US government agencies and private companies. The attack allowed the intruders to gain persistent access to these systems, steal sensitive data, and conduct long-term espionage on a subset of high-value targets before being detected. The SolarWinds campaign is a reminder that adversaries target build environments and federated identity to pivot laterally [CISA 2021].

Cybersecurity comprises the principles and practices of protecting systems, networks, and data across the full lifecycle of modern digital operations. For example, securing chips and sensors to clouds and consortium-ledgers. Common attack vectors include phishing/credential abuse, vulnerable services, software supply-chain compromise, and ransomware. For EST organizations, the stakes are not only confidentiality and privacy but also safety, availability, and integrity of cyber-physical processes.

Therefore, we need to establish the foundational threats, frameworks, and architectural principles in cybersecurity that EST professionals must understand before applying modern zero-trust security practices. The following sections cover these topics.

9.7.1 Threats, Frameworks, and Zero Trust

Financial systems face several common cyber-attack vectors. *Phishing* and *credential abuse* occur when attackers trick users into revealing passwords, enabling unauthorized access to sensitive financial accounts; for example, a spoofed email may steal an employee's login credentials. *Vulnerable services* refer to outdated or misconfigured software exposed to the internet, allowing attackers to exploit flaws and enter financial infrastructure. *Software supply-chain compromise* involves attackers inserting malicious code into trusted software updates, thus enabling wide, stealthy infiltration across financial organizations.

Ransomware is another major threat in which attackers encrypt systems and demand payment; for instance, a financial firm may lose access to transaction systems until a ransom is paid. These vectors collectively highlight how human error, technical weaknesses, and trusted-software dependencies create systemic risk across financial environments [CISA 2021].

Modern cybersecurity practices use layered protection mechanism (*defense-in-depth),* telemetry for monitoring, and rehearsed incident response. A disciplined program organizes controls and outcome. Typical approaches use the NIST Cybersecurity Framework (CSF) 2.0, which includes a governance function and strengthens supply-chain risk management [NIST CSF]. A *zero-trust architecture* is also highly recommended (Figure 9.7), which eliminates implicit trust, enforces per-session authentication/authorization, and continuously monitors posture, and a comprehensive security program is in place [NIST CSF].

In the access model shown in Figure 9.7, a client needs access to certain computing resource or application. Access is granted through a *policy decision point (PDP)*—an authentication algorithm—and corresponding *policy enforcement point (PEP)*—a rules-based computation.

EXAMPLE 9.20 ZERO-TRUST ACCESS FOR NANOFABRICATION FACILITIES

A university nanofabrication facility adopts the NIST CSF 2.0 for governance and risk registers and deploys a zero-trust architecture (policy engine + enforcement points) to gate instrument access by device posture and user role, rather than virtual local area network boundaries. The result is that any compromised workstation cannot reach proprietary tools or recipes.

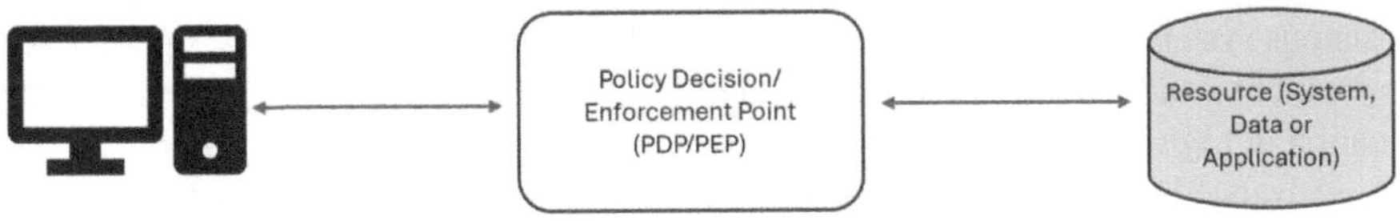

FIGURE 9.7 Zero-trust access high-level model.

Collectively, these practices reinforce that security must function as a core engineering discipline. That is using zero-trust principles, continuous telemetry, and rigorous operational controls to ensure resilient and trustworthy EST environments.

9.7.2 Financial Ramifications: Breach Economics and Compliance

Security breaches impose direct and indirect costs related to forensic analysis tools and services, downtime, decreased response, lost business, and additional legal and compliance costs. IBM's 2024 report estimates the global average breach cost at $4.88 M, with AI-driven security/automation associated with significant savings when present; costs tend to be higher in financial services. Under the GDPR, severe violations can incur fines up to €20 M or 4% of global turnover, whichever is higher [Bonderud 2024].

EXAMPLE 9.21 GDPR-DRIVEN RESPONSE TO MEDICAL DATA EXFILTRATION

A medical device lab faces an *exfiltration* (extraction of proprietary information) event. Because customer data from EU trials is in scope, counsel triggers GDPR Article obligations: the firm's lack of data classification prolonged detection, increasing both reporting complexity and risk of enforcement.

Collectively, the practices described in this section underscore that effective cybersecurity requires treating breach economics and regulatory exposure as core engineering inputs, ensuring that detection, response, and compliance disciplines are embedded directly into operational planning.

9.7.3 Threats to EST Systems

The threat landscape most relevant to EST environments centers on intellectual-property theft, supply-chain compromise, and ransomware operations that increasingly target interconnected engineering and scientific systems.

Intellectual-property theft poses a significant risk to EST organizations because proprietary designs, materials, data, and manufacturing processes can be exfiltrated long before a breach is detected, eroding competitive advantage and compromising safety-critical systems. For example, an aerospace supplier's compromised continuous systems integration pipeline could allow an attacker to quietly siphon CAD assemblies and simulation models, enabling counterfeit production or foreign replication of high-value components.

Advanced intrusions exploit federated identity and supply chains and can bypass multifactor authentication. Meanwhile, ransomware is always an ongoing threat [CISA 2021].

EXAMPLE 9.22 CONSORTIUM RESPONSE TO AEROSPACE IP THEFT

An aerospace supplier consortium shares certain digital assets, such as design documentation. An attacker compromises one of the partner systems and steals design assemblies. In response, the consortium changes system access keys and rebuilds the continuous integration pipeline from trusted baselines.

EXAMPLE 9.23 RANSOMWARE RESPONSE AT A REGIONAL WATER UTILITY

A regional water utility's historian server is encrypted; the actor threatens a data leak and a denial-of-service attack. The utility coordinates with FBI/CISA guidance and accelerates segmentation and immutable backups.

As a practical matter EST organizations must treat identity infrastructure as a critical asset and regularly rehearse supply-chain and ransomware scenarios to maintain resilient, well-prepared security operations.

9.7.4 Strategies for Professionals: Controls, Engineering Discipline, and Supply-Chain Security

Effective cybersecurity in EST environments depends not only on strong controls but on disciplined engineering practices that make those controls operational across complex systems. These practices must be anchored in in recognized frameworks such as NIST's CSF 2.0 outcomes and Zero Trust tenets. Where applicable, organizations also need to observe secure software development practices to reduce vulnerabilities and improve supply-chain transparency [NIST CSF].

EXAMPLE 9.24 SECURE SDLC AND SBOM DISCIPLINE FOR ROBOTICS

A robotics manufacturer formalizes secure software development practices, threat modeling every firmware release. It also uses secure architectural and software bill of materials rigor for third-party libraries, reproducible builds, and signed releases. Procurement requires vendor attestations consistent with applicable government regulations.

Following these practices establishes the engineering rigor necessary to harden EST systems against evolving threats while maintaining predictable, auditable operational performance.

9.7.5 Security AI and Automation: Promise and Caveats

Security AI and automation are becoming essential force multipliers in EST environments, improving detection accuracy across complex operational networks. By correlating signals from hosts, identity systems, and industrial telemetry, modern security detection platforms can identify weak indicators that would be invisible to manual analysis. Security AI/automation can reduce breach cost and dwell time when integrated into detection and response. Industry data highlights significant cost reductions when security AI is extensively used [CrowdStrike n.d.].

EXAMPLE 9.25 AI-ASSISTED DETECTION: PREVENTING RECIPE THEFT

A semiconductor test floor streams electronic design records and network telemetry into an AI-assisted detection and response that baselines "normal" job dispatch and data egress. Early-stage anomalies in a tool vendor account indicate that recipe exfiltration (theft) is occurring, triggering containment before recipe theft completes.

Other best practices include combining AI detection with strict identity controls and playbook automation, measurement of MTTD/MTTR deltas, and validation of vendor AI claims with controlled *red-team* (simulated cyberattack) exercises.

Used appropriately, security AI tightens the detection–response loop and strengthens overall operational resilience. That is, provided organizations pair automation with disciplined identity controls, continuous validation, and well-governed playbooks.

9.7.6 Privacy, Data Governance, and Open Banking Interfaces (for EST FinTech Integrations)

As EST organizations integrate open-banking–style APIs and financial data flows into their products, privacy and data-governance obligations become central to secure system design. Draft NISTIR 8389[3] frames OB's security and privacy considerations, including API abuse risks and governance models; industry responses emphasize comprehensive cybersecurity and accreditation [Voas et al 2022].

EXAMPLE 9.26 OPEN-BANKING SECURITY FOR MICROGRID PAYOUTS

A microgrid platform initiates account-to-account payouts through bank APIs. The operator uses open banking security profiles, consent logs, and directory/accreditation checks; scopes and expirations are enforced per device/user.

Other best practices include mandating consent granularity, token scopes, and setting API access rate limits. Strong privacy and data-governance practices ensure that EST systems can exchange financial information securely while maintaining regulatory compliance and preserving user trust.

9.7.7 Quantum-Driven Cryptographic Readiness and Transition

As quantum-capable adversaries become more plausible, EST organizations must begin preparing for cryptographic transition long before large fault-tolerant quantum machines arrive. Even before large fault-tolerant quantum computers arrive, the potential risks justify planning.

EXAMPLE 9.27 PREPARING DEVICES FOR POST-QUANTUM CRYPTOGRAPHY

An instrument vendor inventories crypto dependencies in firmware and front-end software, tests a hybrid quantum/conventional computing architecture in the lab, and schedules field upgrades for devices with update channels; contract language with labs requires post quantum computing-capable components after 2027.

It is also advisable to build a *crypto bill of materials (CBOM)*, that is, a list of artifacts that are dependent on or can be affected by quantum computing technology. Then, prioritize high-value data and long-lived secrets for early migration; pilot post quantum computing in test environments aligned to applicable standards. These preparations ensure EST systems can withstand future cryptanalytic breakthroughs while maintaining the integrity and continuity of critical operations [NIST 2024].

9.7.8 Incident Response and Recovery: From Tabletop to Eviction

Incident response in EST environments must integrate preparation, containment, and recovery practices that account for the unique coupling of digital systems with physical processes. Effective incident response blends preparation (roles, communications, legal) with *playbooks* for identity compromise, ransomware, and supply-chain intrusions. CISA's SolarWinds eviction guidance underscores the need for *out-of-band communications* (i.e., uncompromised channel), identity reinstitution, and staged eviction that may require multi-day isolation.

EXAMPLE 9.28 INCIDENT RESPONSE WITH OUT-OF-BAND COORDINATION

A biotech determines that a sensitive cluster is suspected of lateral movement. The incident response team uses out-of-band (off the compromised channel) communications, snapshots clusters for forensics, rotates federation keys/certificates, rebuilds affected domains, and restores from trusted backups after validation.

In practice, organizations strengthen their resilience by maintaining clean rooms, preserving golden images, and keeping trusted (immutable) backups ready for use. These are supported by regular restore drills to ensure everything works when it matters.

Companies can also benefit from lining up incident-response retainers in advance by addressing forensics, legal counsel, and public relations issues. They should also establish clear communication pathways with law enforcement so that, if an incident occurs, the response can move immediately rather than scrambling to assemble support. These practices ensure EST organizations can respond decisively to complex incidents while restoring operations safely and with full forensic accountability [CISA 2021].

9.8 EXERCISES

9.1 Explain, in your own words, why a distributed ledger eliminates the need for a central authority and how consensus mechanisms ensure tamper-resistance in multi-party environments. Reference the role of cryptographic hashes as described in your explanation of block formation.

9.2 Compare proof-of-work and proof-of stake as described in Section 9.1.1. Discuss how these different consensus mechanisms influence energy consumption, governance, and suitability for EST-focused use cases.

9.3 Using the wind-farm predictive maintenance scenario provided in the chapter, identify three engineering pain points (e.g., sensor integrity, reconciliation delays, and provenance guarantees) that blockchain helps mitigate.

9.4 Describe how smart contracts work and why determinism is a required property in multiparty engineering workflows. Explain the role of oracles in bringing real-world data into blockchain systems.

9.5 Illustrate how tokenization of financial instruments changes post trade operations. Reference regulatory guidance mentioned in Section 9.1.2 and discuss implications for engineering firms issuing tokenized notes.

9.6 Define a capitalization table (cap-table) and evaluate how a blockchain-based cap-table improves accuracy and transparency in comparison to traditional spreadsheet-based approaches.

9.7 Describe how blockchain supports supply-chain provenance for serialized components. Using the aerospace example, identify at least two benefits for auditors or engineers.

9.8 Summarize the three major risk categories described in Section 9.1.4 (energy use, metadata leakage, and smart-contract/wallet vulnerabilities). Provide one mitigation for each.

9.9 Explain why a software bill of materials becomes essential when deploying blockchain-based systems in EST environments. Discuss the connection to historical supply-chain compromises.

9.10 Based on the implementation guidance provided, outline the steps an EST organization should take when deciding whether blockchain is the correct technology for a given problem.

9.11 Differentiate "fungible," "non fungible," and "tokenized securities/RWAs" using the definitions and Table 9.1. Provide one engineering-relevant application for each asset type.

9.12 Describe how cryptocurrencies function as ledger-native assets and identify two ways EST firms might use them in automated workflows (e.g., machine-to-machine payments and escrow).

9.13 Explain how NFTs can encode device-level licensing or warranty metadata. Use the examples provided to show how this pattern eliminates the need for centralized licensing servers.

9.14 Evaluate how tokenized securities enforce transfer restrictions and corporate actions via smart contracts. Tie these features to the "where the truth resides" framework described in the chapter.

9.15 Discuss how NFT-based provenance or IP management can support laboratory or factory workflows. Analyze one benefit and one potential risk (e.g., off-chain link rot).

9.16 Summarize the role of programmable assets in enabling machine-to-machine financial transactions in cyber-physical systems. Provide one example inspired by the text.

9.17 List the primary risks associated with crypto-asset use (metadata leakage, key custody, consensus design, oracle trust). For each, state one engineering mitigation strategy.

9.18 Explain how fine-grained royalties, conditional warranties, or usage based pricing become feasible through tokenization. Connect your explanation to the automation themes in Section 9.2.3.

9.19 Reflect on the intersection of digital-asset models and regulatory regimes (e.g., MiCA and SEC tokenization guidance). Discuss why early regulatory mapping is especially important for EST organizations.

9.20 Using insights from Sections 9.1 and 9.2 collectively, justify when an EST organization should use:

a. a cryptocurrency,
b. an NFT, or
c. a tokenized security.

to solve an engineering, licensing, or financial-workflow challenge

NOTES

1 A kind of long term financial obligation issued by a company.

2 A *software bill of materials (SBOM)* is a repository of software components, their origin, version, and other important identifying information.

3 This draft has been officially withdrawn but is available in perpetuity and contains valuable information.

REFERENCES

Australian Competition and Consumer Commission (ACCC). (n.d.). *The consumer data right.* https://www.accc.gov.au/by-industry/banking-and-finance/the-consumer-data-right, accessed 2/2/2026.

Bonderud, D. (2024). *Cost of a data breach 2024: Financial industry.* IBM. https://www.ibm.com/think/insights/cost-of-a-data-breach-2024-financial-industry

CrowdStrike. (n.d.). *CrowdStrike signal transforms AI-powered threat detection.* https://www.crowdstrike.com/en-us/blog/crowdstrike-signal-transforms-ai-powered-threat-detection/, accessed 2/2/2026.

Cybersecurity and Infrastructure Security Agency (CISA). (2021, May 14). *Remediating networks affected by the SolarWinds and Active Directory/M365 compromise.* https://www.cisa.gov/news-events/news/remediating-networks-affected-solarwinds-and-active-directorym365-compromise

Ethereum Foundation. (2025, November 18). *The merge.* https://ethereum.org/roadmap/merge/

European Parliament, & Council of the European Union (EU). (2023). Regulation (EU) 2023/1114 of the European Parliament and of the Council of 31 May 2023 on markets in crypto-assets and amending Regulations (EU) No 1093/2010 and (EU) No 1095/2010 and Directives 2013/36/EU and (EU) 2019/1937. *Official Journal of the European Union, L 150*, 40–205. https://eur-lex.europa.eu/eli/reg/2023/1114/oj/eng

Dushyant, K. (2025, August 5). *How AI is powering fintech: Fraud detection to robo-advisors.* Accounting Byte. https://accountingbyte.com/ai-powering-fintech/

Federal Energy Regulatory Commission, & Electricity Information Sharing and Analysis Center (FERC). (2021). *SolarWinds and related supply chain compromise: Lessons for the North American electricity industry* (White paper). https://www.ferc.gov/sites/default/files/2021-07/SolarWinds%20and%20RelatedSupply%20Chain%20Compromise%20White%20Paper_1.pdf

Grand View Research. (2025). *Blockchain technology market size, share & trends analysis report, 2025–2030.* https://www.grandviewresearch.com/industry-analysis/blockchain-technology-market

IBM Security, & Ponemon Institute. (2024). *Cost of a data breach report 2024.* IBM. https://cdn.table.media/assets/wp-content/uploads/2024/07/30132828/Cost-of-a-Data-Breach-Report-2024.pdf

Kumiega, A. (2026). From risk control to platform stability: A reliability engineering approach to fintech. *Reliability, 3*(1), 44–49.

Laplante, P., & Kshetri, N. (2021). Open banking: Definition and description. *Computer, 54*(10), 122–128.

Lesavre, L. D., Varin, P. C., Mell, P. M., Davidson, M. S., & Shook, J. (2020). *A taxonomic approach to understanding emerging blockchain identity management systems* (NIST Cybersecurity White Paper No. NIST.CSWP.01142020). National Institute of Standards and Technology. https://doi.org/10.6028/NIST.CSWP.01142020

National Institute of Standards and Technology (NIST). (2024, August 13). *NIST releases first 3 finalized post-quantum encryption standards.* https://www.nist.gov/news-events/news/2024/08/nist-releases-first-3-finalized-post-quantum-encryption-standards

Skadden, A., Slate, M., & Flom, L. L. P. (2024, December 6). *CFPB releases final open banking rule on personal financial data rights.* https://www.skadden.com/insights/publications/2024/12/cfpb-releases-final-open-banking-rule

Tabassi, E. (2023). *Artificial intelligence risk management framework (AI RMF 1.0)* (NIST AI 100-1). National Institute of Standards and Technology. https://doi.org/10.6028/NIST.AI.100-1

Voas, J., Laplante, P., Kassab, M., Lu, S., Ostrovsky, R., & Kshetri, N. (2022). *Cybersecurity considerations for open banking technology and emerging standards* (NIST Internal Report 8389, Initial Public Draft). National Institute of Standards and Technology. https://doi.org/10.6028/NIST.IR.8389-draft

Yaga, D., Mell, P., Roby, N., & Scarfone, K. (2018). *Blockchain technology overview* (NIST Interagency/Internal Report No. 8202). National Institute of Standards and Technology. https://doi.org/10.6028/NIST.IR.8202

Glossary

A/B offer—a pricing or marketing experiment where two different versions of an offer—labeled "A" and "B"—are presented to separate groups of customers to compare perceived value.

AC—see *actual cost.*

acceleration—in stock options, a mechanism for triggering the vesting of the option. See also *single-trigger acceleration* and *double-trigger acceleration.*

accounting equation—the fundamental relationship between assets, equity, and liabilities. Also known as the *universal law.*

accounting rate of return—see *average rate of return.*

accrual accounting—an approach to accounting where revenues and expenses are recorded when they are earned or incurred, not when cash changes hands. Contrast with *cash accounting.*

acid test—see quick ratio.

activity-based budgeting—a budgeting technique that assigns costs based on the activities required to produce outputs, linking expenses to engineering or operational drivers.

actual cost (AC)—in earned value management, the actual expenditures incurred, tracked via invoices, payroll, or accounting systems.

AI—see *artificial intelligence.*

AI governance board—a formal oversight group that defines approved use cases, risk tiers, data-quality requirements and validation processes for AI systems.

AI guardrails—technical and operational controls that restrict or guide AI system behavior to ensure safety, compliance, and reduce unintended outcomes.

AI RMF—see *Artificial Intelligence Risk Management Framework.*

amortization—an accounting way of expensing the cost of intangible assets (like patents, copyrights, trademarks, software, or licenses) over the period they provide economic benefit. Contrast with *depreciation.*

analogous estimation—estimation based on the actual cost of similar projects.

angel (investor)—an individual who invests personal funds in very early-stage companies.

API—see *application programming interface.*

arbitrage—the practice of simultaneously (or near-simultaneously) buying and selling related assets in different markets to exploit temporary price discrepancies.

ARR—see *average rate of return.*

artificial intelligence (AI)—computational systems that learn, reason, and perform tasks that normally require human intelligence.

asset—anything of value.

asset-backed lending—loans secured by specific company assets, such as receivables or inventory, often used when revenue is more stable or predictable.

average rate of return **(ARR)**—a financial metric that estimates profitability using average accounting income over the investment's life, often as a percentage of initial or average investment. Sometimes called *accounting rate of return.*

BaaS—see *business as a service.*

BAC—see *budget at completion.*

backlog—represents work (products or services) that has been contracted but not yet delivered (nor billed). Also known as *book-to-bill.*

backtesting—a validation method that compares a model's historical predictions with realized outcomes to detect drift, misspecification, or instability.

balance check—an automated test that verifies that core statements reconcile (e.g., the balance sheet balances and cash ties out), preventing mechanical errors from propagating through the model.

balance sheet—*a financial statement that reports the assets, liabilities, and owner's equity for an entity.*

Balanced Scorecard (BSC)—a performance measurement framework.

bank credit line—a committed borrowing facility from financial institutions that provide flexible, revolving access to cash.

base case—in *scenario analysis* a case tied to expected system performance. Contrast with *upside case* and *downside case.*

bias—Systematic distortion in AI model outputs resulting from unbalanced, incomplete, or non-representative training data.

black swan—a rare and unpredictable outlier event with extreme impact.

bonding company—a company or agency that issues a bond, promising to complete the work or pay damages if the contractor fails. Also known as a *surety company.*

book-to-bill—see *backlog.*

bottom-up estimation—an estimation determined summing the detailed task costs from the bottom up of a WBS.

break-even analysis—the point at which project or investment costs equal associated revenues.

Brinson attribution model—a performance-attribution framework that decomposes active return into allocation, selection, and interaction effects relative to a benchmark.

BSC—see *Balanced Scorecard.*

bubble—an overhyped market leading to a crash.

budget at completion (BAC)—the total approved budget for completing all project-related work.

budgeting—the process of translating estimates into a financial plan to guide project execution.

business as a service (BaaS)—a model in which regulated financial infrastructure is delivered through modular APIs.

business valuation—an assessment of the value of a company by experts based on numerous factors.

CapEx—see *capital expenditure.*

capital budget—funding allocated for long-lived assets such as facilities, specialized equipment, test infrastructure, and other major investments that support multi-year technical development.

capital expenditure/expense **(CapEx)**—funds used to purchase or maintain physical assets.

capital gains—profits realized when an entity or person sells assets for more than their purchase price.

capitalization table (cap-table)—a ledger detailing ownership of shares, tokens, options, and warrants.

capture—an externally funded opportunity to be pursued such as a contracts, grant, or bid to bring in near-term revenue. Contrast with *investment campaigns.*

carryforward—a tax attribute (such as a net operating loss) that can be used in future periods to reduce taxable income, influencing projected earnings and cash flow.

cash accounting—an accounting approach where transactions are recorded when cash changes hands, not when they are earned or expensed. Contrast with *accrual accounting.*

cash flow—refers to the movement of money into and out of a business (or individual finances) over a period.

cash ratio—a strict liquidity test that measures only cash and cash equivalents divided by current liabilities,

CBOM—see *crypto bill of materials.*

CCB—*see change control board.*

C-Corporation or C-Corp—a business structure that is the most sophisticated and most suitable for medium-to-large and complex businesses. Also known as a "public/private limited company" or "corporation."

CDR—see *consumer data right.*

central bank—*the* banking entity of a country that manages banking policy.

CFPB—see US *Open Banking* and *Personal Financial Data Rights* regulation.

chain of custody—a documented record of a component or asset's provenance across manufacturing, transport, usage, and installation.

change control board (CCB)—an organized group of stakeholders responsible for reviewing, evaluating, and approving proposed changes to a project.

chart of accounts—an organized list of all financial accounts used by an organization to record transactions, categorized into assets, liabilities, equity, revenues, and expenses.

circularity management—the set of techniques used to handle a circular reference (e.g., interest depending on ending debt, which depends on cash flow) using controlled iteration or toggles to maintain model stability.

co-development—see *partnering.*

COGS—see *cost of goods sold.*

commercial paper—short-term, unsecured corporate debt used by established companies to meet near-term liquidity needs.

commodity—a product with little or no differentiation.

conditional value at risk (CvaR)—the expected loss given that the outcome is already beyond the VaR threshold, quantifying tail severity.

confidence interval—a statistical range showing uncertainty in an estimated model parameter, narrower than a prediction interval because it excludes randomness in future outcomes.

consensus mechanism—the algorithm by which blockchain participants agree on the ordering and validity of transactions.

consensus model—in a distributed ledger, the method by which all participating nodes agree on the validity and ordering of transactions without relying on any central authority.

Consumer Data Right (CDR)—Australia's data-sharing framework enabling consumer-authorized access across banking, energy, and telecom.

consumer price index (CPI)—an economic indicator of the average change in consumer prices, over time, for a selection of goods and services. It is used to measure inflation.

contingency—a reserve fund or time buffer added to cost estimates to account for uncertainties, risks, and unforeseen events that could increase project costs. Contrast with *management reserve.*

contra-asset—an account that is paired with a regular asset account to reduce its recorded value on the balance sheet.

copula—a multivariate function that links marginal distributions into a joint distribution, enabling modeling of nonlinear dependence and tail behavior.

corporate paper—a kind of bond issued by a company used for long-term financing.

correlation structure—the representation of how uncertain variables co-move, typically through Pearson or rank correlations, forming the basis for joint distribution modeling.

cost estimation—the process of forecasting the expenses associated with labor, materials, equipment, overhead, and contingencies for some project.

cost leadership—a competitive strategy that centers on becoming the lowest-cost producer in an industry while maintaining acceptable quality, often through scale, efficiency, or process innovation.

cost of goods sold (COGS)—direct costs related to the creation of the products and services that a company sells.

cost performance index[1] **(CPI)**—a measure of a project's cost efficiency. *CPI = EV/AC.*

cost–benefit trade-off—an analysis used to evaluate whether benefits justify costs for some financial decision.

cost-plus pricing—a pricing model where the client agrees to reimburse the vendor for all allowable project costs. Also known as *cost-reimbursable.*

cost-plus-award-fee—in cost-plus pricing, an award tied to specific criteria.

cost-plus-fixed-fee—in *cost-plus pricing*, a performance-based incentive. Also known as a *cost-plus-incentive-fee.*

cost-reimbursable—see *cost-plus pricing.*

CvaR—*see conditional value at risk.*

covenant—a contractual requirement or constraints tied to loans or credit agreements that borrowers must comply with.

CPI—see *cost performance index* and *consumer price index.*

credit—an entry recorded on the right side of a ledger.

credential abuse—a type of cyberattack in which attackers use information (often through *phishing* to gain unauthorized access to systems, accounts, or sensitive financial resources).

crowdfunding—fundraising from a broad public audience through equity or preorder campaigns, commonly used to validate demand and raise early technical development funds.

crypto bill of materials (CBOM)—a catalog of cryptographic components and dependencies used to evaluate quantum-readiness and cryptographic risk.

cryptocurrency—*a* digital, ledger-native assets secured by cryptographic keys and maintained on decentralized networks that achieve agreement through consensus protocols.

current ratio—a metric that measures overall short-term liquidity.

data provenance—a verifiable history of where data originated and how it has changed over time.

data-snooping bias—in financial modeling when re-testing strategies on the same historical dataset inflates the likelihood of finding misleadingly "good" results that reflect chance rather than skill.

DCF—discounted cash flow.

debit—an entry recorded on the left side of a ledger that usually increases assets or expenses and decreases liabilities or equity.

debt-to-equity ratio **(D/E)**—a metric calculated as total debt divided by shareholders' equity.

decentralized identifier—a cryptographically verifiable identity artifact used to authenticate individuals, devices, or organizations. Also known as a *verifiable credential.*

decision threshold—a defined cutoff (e.g., minimum IRR or maximum downside probability) used to judge whether an uncertain outcome remains within acceptable bounds.

deep learning (DL)—a subset of machine learning using multi-layer neural networks to recognize and extract complex patterns.

defense in depth—a security strategy using multiple overlapping layers of controls to reduce the probability of a successful attack.

deflation—an unusual situation where there is a general decrease in prices across an economy, It is the opposite of *inflation.*

Depreciation—a systematic way to account for the reduced usefulness of a tangible fixed asset over its useful lifetime.

derivative—*a* contract whose value depends on an underlying asset, index, or benchmark, for example, a stock, bond, commodity, currency, or interest rate.

DID—see *decentralized identifier.*

differentiation—a competitive strategy that involves creating unique products or services that command a premium price, emphasizing innovation, branding, or superior performance.

discount rate—the interest rate charged by a central bank when lending to other banks. It is also a general term for any interest or inflation rate used in financial calculations.

discounted payback period—a *payback period* calculation that uses discounted cash flows to account for the time value of money.

disruption risk—a *risk* due to a paradigm-shifting change, for example, technical or logistical innovations rendering current technology obsolete.

DL—see *deep learning.*

double entry accounting—an accounting system where every financial transaction is recorded in at least two accounts: one as a debit and the other as a credit.

double taxation typically—when both the profits of the company and the earnings that are passed through to the shareholders are subject to applicable taxes.

double-trigger acceleration—a kind of stock option *acceleration* that requires both a change of control and a qualifying termination (e.g., without cause) to accelerate.

downside case—in *scenario analysis,* a case that reflects delays, failures, or external shocks to the system. Contrast with *base case* and *upside case.*

drift—degradation of a model's predictive quality caused by changes in real-world data distributions over time.

due diligence—a rigorous evaluation of technical, financial, operational, and IP aspects to verify claims and identify risks before an acquisition or partnership.

earned value (EV)—in earned value management, the budgeted cost of work completed, measured by task completion percentages.

earned value management (EVM)—a technique that meshes cost, schedule, and scope to provide real-time insights into project health.

EBIT—Earnings Before Interest and Taxes. See *operating profit.*

EBITDA—Earnings Before Interest, Taxes, Depreciation, and Amortization.

EBITDA ratio—EV/EBITDA (enterprise-value-to-EBITDA) multiple. It answers, "how many times EBITDA investors are paying for the company."

EDA—see *exploratory data analysis.*

efficient market hypothesis—a financial theory which suggests that stock prices reflect all available information and expectations about future performance.

elastic—see price *elasticity.*

embedded finance—financial services such as payments, credit, or risk scoring integrated directly into non-financial software through APIs.

equity—the stored value of the company after all liabilities are paid.

equity crowdfunding—a kind of crowdfunding that sells securities to a broad base of investors under regulated exemption. Contrast with *pre-order crowdfunding.*

equity stake—see *partnering.*

error bar—a graphical indicator of variability or uncertainty around a point estimate, often used to compare the uncertainty of different scenarios or assumptions.

estimate validation—the process of ensuring that cost estimates for engineering projects are accurate, reliable, and aligned with project objectives.

ethical safeguard—a control ensuring models avoid misuse, bias, privacy violations, or unintended harm, particularly when using ML or sensitive data.

EVM—see *earned value management.*

***exercise* price**—the price at which a *stock option* permits the owner to purchase. Also known as *strike price.*

exfiltration—the unauthorized removal or theft of data from a computer system or network.

expense—a cash outflow needed to run the business.

***exploratory data analysis* (EDA)**—a critical first pass over any dataset that reveals structure, irregularities, and directional signals before more formal modeling begins.

exploration phase—the earliest stage of M&A or partnership analysis in which potential opportunities are screened for strategic and technical fit.

factor-based analysis—a portfolio method that attributes risk and return to exposures such as value, momentum, size, or sector factors rather than individual securities.

false discoveries—high-performing simulations that collapse in real-world securities trading.

fan chart—a visual that displays increasing uncertainty over time using multiple probability bands, making the distribution of possible future outcomes easy to interpret.

fat-tailed t-distribution—a probability distribution with heavier tails than the normal distribution, suitable for modeling extreme price or cost shocks.

fintech—a software-driven firm that modernize financial services through automation, digital workflows, and data analytics.

fixed expense (cost)—an expense that is independent of output. Contrast with *variable expense.*

fixed price—a pricing approach where the client agrees to pay a predetermined amount for the entire project or a specific deliverable, regardless of the actual time or resources consumed.

focus—a competitive strategy that narrows scope to a specific segment, applying either cost leadership or differentiation within that niche

free cash flow—the cash in hand left after spending on operations and capital expenditures.

FTE—full-time equivalent.

fundamental data—company-reported information such as financial statements, guidance, and disclosures that anchor valuation and forecasting models.

fungible—the property of interchangeability, where one unit is equivalent to any other identical unit.

gainsharing—see *shared savings.*

Gartner Hype Cycle—a graphical model mapping technology maturity through five phases Technology Trigger, Peak of Inflated Expectations, Trough of Disillusionment, Slope of Enlightenment, and Plateau of Productivity.

gatekeeping process—a structured review points–often tied to engineering milestones–used to approve or defer additional spending based on technical evidence.

general ledger—see *ledger.*

generative AI—AI systems that produce new content—text, images, code, or data—based on learned patterns.

geopolitical risk—any *risk* due to the complex relationships between competing and cooperating nations.

goodwill—an accounting term that refers to the monetary value of a company's reputation in the market.

gross margin—(*revenue*—cost of goods sold) ÷ *revenue.*

gross profit—*revenue*—*cost of goods sold.*

gross revenue—see *revenue.*

Grover's algorithm—a quantum algorithm that speeds up unstructured search tasks, reducing search complexity from O(N) to O(√N).

hallucination—When an AI system generates output that appears coherent but is factually incorrect.

hedge fund—an investment mechanism that uses a broad range of strategies, particularly derivatives. Hedge funds are typically open only to high-net-worth or institutional investors and are less regulated, giving them more flexibility.

historical data analysis—using data from similar projects (analogous estimation), estimate a measured outcome such as cost or duration.

HSM—hardware security module.

hyperinflation—when the rate of inflation increases by at least 50% per month.

immutable—in blockchain, the property where each block contains a cryptographic hash that depends on all previous blocks, making any attempt to modify past data impossible.

incentive stock options—see *stock option.*

incentive-based pricing—see performance-based pricing. or *value-based pricing.*

income statement—shows if the business entity made or lost money during the accounting period. Sometimes referred to as *profit and loss statement or P&L.*

incremental budgeting—a method that starts with last year's budget and adjusts it up or down by defined increments such as inflation or expected growth. Contrast with *zero-based budgeting.*

inflation is the general increase in prices across an economy over time. It is the opposite of *deflation.*

insurtech—a *fintech* segment applying digital technologies to improve insurance products and operations.

integration—the post-transaction process of combining teams, systems, processes, and technologies following an acquisition, partnership, or merger.

internal rate of return (IRR)—the discount rate in the NPV equation that causes the calculated NPV to be zero.

interest coverage—*a metric that tests* whether normal business operations can easily handle interest payments on debt.

investment campaigns—an internal, longer-horizon bet to develop differentiated capabilities. Contrast with *captures.*

IP—intellectual property.

IPO—initial public offering.

IRR—see *internal rate of return.*

joint venture—see *partnering.*

kurtosis—a statistical measure describing the tail heaviness of a distribution relative to normality.

KYC—*know your customer.*

large language model (LLM)—a generative AI model trained on a massive text corpus capable of reasoning, synthesis, and natural language generation.

ledger—the central book or digital record where all financial transactions are organized into accounts, showing debits, credits, and balances. The ledger is often called the *second book of entry* because transactions are first recorded in journals and then posted to the ledger for classification and summarization. Also known as *general ledger.*

letter of intent (LOI)—a type of contract of prospective action.

liability—money owed to other entities.

limited liability company (LLC)—a hybrid business structure that provides the legal protections of a corporation with the flexibility of a partnership for ownership and management structure.

line of credit—a flexible short-term arrangement that lets a business borrow money as needed up to a preset limit.

liquid—a term that is applied to an asset that can be easily and quickly spent or converted to some other asset.

liquidity—the tendency of an asset to be *liquid.*

LLC—see *limited liability company.*

LLM—see *large language model.*

lognormal distribution—a distribution appropriate for multiplicative processes, such as degradation behavior, where the logarithm of the variable is normally distributed.

LOI—see *letter of intent.*

look-ahead bias—an error in model construction where future information inadvertently contaminates historical inputs, artificially inflating performance.

machine learning (ML)—A family of algorithms that learn patterns from data to make predictions or decisions without explicit programming.

management reserve—*a* separate fund for major, unanticipated scope changes controlled by senior management. Contrast with *contingency.*

market capitalization—the total of the outstanding shares (at current price) of a company.

market regimes—a generic term for macro cycles, volatility regimes, policy shocks, and technological adoption curves.

market risk—*risk* that is *a* function of volatile supply and demand in a particular market.

MFA—see *strong customer authentication.*

milestone-linked forecasting—a forecasting approach where financial projections (cost, revenue, cash flow) adjust automatically based on engineering milestone outcomes rather than calendar timing.

ML—see *Machine Learning.*

moat—a competitive advantage that competitors cannot easily copy.

model governance—the framework of documentation, controls, independent validation, and oversight that ensures financial models remain reliable, transparent, and compliant.

monitoring—project management activity that compares actual expenditures to the cost baseline to detect deviations, assess performance, and prevent overruns.

Monte Carlo simulation—uses statistical modeling to simulate thousands of scenarios based on some probability distribution of a control variable to estimate a measured outcome such as cost or duration.

multiple—market-based valuation ratios used in relative valuation comparative analysis.

native digital asset—an asset that originates on a blockchain and whose ownership is governed by on-chain rules.

neobank—a digital-only banking platform delivering modern financial services without physical branches.

net income—see *net profit.*

net present value (NPV)—the value of an asset today based on the principle that a currency today is worth more than the same currency in the future due to inflation.

net profit—profit after all expenses, including operating expenses, interest, taxes, and other non-operating items. Also known or *net income.*

net worth—*see equity.*

NFT—see *non-fungible token.*

nondilutive funding/grant—a cash infusion without equity loss.

non-fungible token (NFT)—A unique digital token representing ownership of a specific item or right, typically on a blockchain.

nonqualified stock option—see *stock option.*

non-statutory stock option—see *stock option.*

normal distribution—a symmetric, bell-shaped distribution used for variables driven by many small, independent effects, such as irradiance variability.

NPV—see *net present value.*

NRE—*non-recurring engineering.*

OB—see *open banking.*

OBIE—see *open banking implementation entity* (UK).

OEM—original equipment manufacturer.

open banking (OB)—a system where banks provide customer-authorized financial data via standardized APIs to enable interoperability.

Open Banking Implementation Entity (OBIE)—UK organization responsible for OB API standards.

operating budget—the recurring, day-to-day expenses required to run engineering or technical programs—typically including labor, materials, prototype builds, testing, and cloud or compute resources.

operating cash flow ratio—a metric that evaluates cash generation against obligations.

operating leverage—a measure of how a company's fixed and variable costs affect its profits when sales change.

operating profit (EBIT)—*gross profit—operating expenses.*

operational expenditure/expense (OpEx)—an expense related to the normal running of a business such as payroll, rent, insurance, utilities, marketing, and administrative expenses.

OpEx—*see operational expenditure.*

opportunity cost—the value of the best alternative surrendered in choosing one option over another.

out-of-band communications—*a security technique in which communications are shifted to an* uncompromised channel.

overfitting—a modeling error in which a model captures noise instead of true structure, often due to excessive parameter tuning or repeated strategy testing.

overhead rate—indirect cost multiplier applied to direct labor or project costs to recover expenses that are not directly billable to a specific project.

ownership dilution—when the total number of stock shares increases, reducing the percentage of ownership for all existing owners. Contrast with *value dilution.*

P&L statement—*statement of profit and loss.* See *income statement.*

P/B—see *price-to-book.*

P/E—see *price-to-earnings ratio.*

P/S—see *price-to-sales.*

P10 value—the 10th-percentile outcome in a simulated distribution, representing a downside boundary exceeded 90% of the time.

P90 value—the 90th-percentile outcome in a simulated distribution, representing an upside boundary exceeded only 10% of the time.

parametric estimation—an estimation methodology that uses some statistical models to predict the cost of some project.

partnering—a business collaboration model that creates deeper, ongoing relationships between two companies. Also known as a *joint venture, equity stake,* or *co-development.*

partnership—a business structure where a collection of individuals who choose to pool resources and expertise for a competitive advantage.

pass-through taxation—a tax protocol where profits and losses "pass through" to the owners' personal tax returns instead.

payback period—the time to recover an initial investment in an asset or an initiative.

payment bond—see *performance bond.*

payment processor—a technology provider offering API-based payment initiation, settlement, and fraud risk scoring.

PDP—see *policy decision point* (zero-trust architecture).

PEP—see *policy enforcement point* (zero-trust architecture).

percentage-based contingency determination—adds a fixed percentage to the base estimate, based on project complexity and historical data.

performance appraisal—the process of determining whether observed performance reflects true skill or random variation using statistical significance tests.

performance bond—an insurance-like guarantee required on many construction, infrastructure, and engineering projects to protect owners from contractor default. Also known as a *payment bond.*

performance-based pricing—a model that ties payment to achieving specific performance metrics, outcomes, or value delivered. Also known as *incentive-based pricing* or *value-based pricing.*

permissioning—rules governing who can access, participate in, or validate transactions on a blockchain network.

personally identifiable information (PII)—data that can identify an individual; referenced in privacy and open-banking security.

PHM—see *prognostics and health management.*

phishing—a type of cyberattack in which an attacker pretends to be a trustworthy entity to trick someone into revealing sensitive information.

PI—see *profitability index.*

PII—see *personally identifiable information.*

planned value (PV)—in earned value management, the budgeted cost of work scheduled up to a given time, based on the WBS baseline.

playbook—a documented, step-by-step set of actions, procedures, or best practices that an organization follows to handle a specific task, process, or scenario.

PPP—see *public–private partnership.*

PQC—see *post-quantum cryptography.*

prediction interval—a forecasting range that incorporates both parameter uncertainty and unpredictable shocks, making it wider than a confidence interval.

pre-order (rewards/commerce) crowdfunding—a form of crowdfunding which exchanges early product commitments for working capital and market signal. Contrast with *equity crowdfunding.*

price elasticity—when a small change in price changes demand significantly.

price-to-book (P/B)—a ratio that compares a company's market value to its book value of equity, calculated as market capitalization divided by shareholders' equity.

price-to-earnings (P/E) ratio—a metric that shows how many years of current earnings investors are willing to pay for the stock.

price-to-sales—a ratio that measures investor willingness to pay per dollar of revenue.

probability distribution—a mathematical function describing the likelihood of possible values for an uncertain variable, underpinning sampling in *Monte Carlo simulation.*

profit and loss statement—see *income statement.*

***profitability index* (PI)**—the ratio of a project's net present value (NPV) to its investment cost.

prognostics and health management (PHM)—AI-powered analytics estimating equipment health and predicting remaining useful life (RUL).

program trading—the use of massive computing power to conduct high-volume trading such as that needed for arbitrage.

proof-of-stake (PoS)—a consensus mechanism where validators are selected based on the amount of value they stake.

proof-of-work (PoW)—a consensus mechanism using computational effort to validate and secure blockchain transactions.

public–private partnership (PPP)—a hybrid partnering model where government and private entities co-finance, build, and operate infrastructure.

PV—see *planned value.*

QEC—see *quantum error correction.*

quantitative easing—when a *central bank* creates new money and uses it to buy back its bonds to lower long-term interest rates, increase liquidity, stimulate lending and investment, and prevent deflation. It is the opposite of *quantitative tightening.*

quantitative tightening—when *a central bank* reduces the money supply. It is the opposite of *quantitative easing.*

quantum computing—computing that leverages quantum mechanical principles such as superposition and entanglement to perform certain tasks more efficiently.

quantum error correction (QEC)—techniques used to protect qubits from noise and maintain coherence in quantum computing.

quick ratio—a liquidity metric that provides a stricter view focused on the most liquid assets. *(Cash + Marketable Securities + Accounts Receivable)/Current Liabilities.* Also known as the *acid test.*

random variable—a formally defined uncertain quantity (e.g., yield, cost, and failure rate) whose outcomes follow a probability distribution.

ransomware—a cyberattack in which attackers lock systems through encryption and demand payment to release the lock.

real option—the right, but not the obligation, to make strategic business decisions regarding a tangible asset, allowing for flexibility in investment and project management.

reasonableness test—a logic check ensuring that a model's outputs remain plausible relative to economic intuition, comparable benchmarks, or operational constraints.

red-team—a group of authorized cybersecurity professionals who simulate real-world adversary attacks.

refresh grants—new stock awards given to existing employees after a financing round.

remaining useful life (RUL)—an estimate of how long a component or system will operate before requiring replacement or servicing.

reserve—in a contract, a portion of the payment amount withheld until a certain performance, deliverable, or time threshold is met.

restricted stock—*a* type of stock incentive that *r*equires no purchase price; value is delivered when units *vest.* Also known as *restricted stock units (RSU).*

retainer—*a* pricing model that involves the client paying a fixed recurring fee, such as a monthly payment, to secure ongoing access to the vendor's services or resources.

retention risk—refers to the possibility that employees will leave the company before their equity incentives (such as stock options or RSUs) fully vest.

revenue—the total earnings from selling products or services before deducting any expenses.

revolving credit—a banking mechanism that allows a borrower to repeatedly draw, repay, and re-borrow funds up to a limit.

RFR—see *risk-free rate.*

right—permission or entitlement associated with a digital asset, potentially including access, governance, or revenue shares.

risk—any event or condition that could interfere with objectives.

risk register—the set of risks and their probabilities of occurrence for some project.

risk-based contingency determination—uses an expected monetary value (EMV) of each risk by assigning probabilities and cost impacts to those risks.

risk-free rates (RFR)—a financial loan rate benchmark based on a historical analysis of actual transactions.

robo-advisor—an automated investment platform that generates portfolio recommendations using machine learning and quantitative strategies.

rolling budgeting—a continuously updated budgeting method in which forecasts and allocations are revised periodically to align with new information.

RSU—see *restricted stock.*

RUL—see *remaining useful life.*

rule of 40—the growth percentage plus free-cashflow margin must sum to 40 or higher.

runway—the amount of time a company can operate before running out of cash, based on current or forecasted burn rates.

RWA—real-world asset.

salted—a security technique in which a random value is added to some data item before hashing it into a *blockchain.*

salted hash—a security technique used to protect stored data by adding a random value (called a *salt*) to the input before hashing it.

SBOM—see *software bill of materials.*

SCA—see *strong customer authentication.*

scenario analysis—an evaluation of how a model behaves under multiple simultaneous changes across key variables.

scenario toggle—a switch or dropdown that loads a specific assumption set (base, best, worst) into the model for consistent, dynamic scenario switching.

scope creep—an incremental change in project scope.

S-Corporation—a business structure with pass-through tax treatment and ownership limits that is suitable for small businesses. Also known as *S-Corp.*

SEC—see *US Securities and Exchange Commission.*

second book of entry—see *ledger.*

sensitivity panel—a structured table that shows how a key metric changes when a single variable is adjusted, helping identify the most influential input driver.

separation of concerns—a modeling principle that keeps inputs, calculations, and outputs in distinct, clearly labeled areas to improve transparency, auditability, and maintainability.

shared savings—a pricing model designed to align the interests of clients and vendors by sharing the financial benefits of cost reductions achieved through efficiencies or innovative practices. Also known as *gainsharing.*

Sharpe Ratio—a measure of how efficiently a strategy converts risk into return.

single-trigger acceleration—a kind of stock option *acceleration* that vests equity upon a change of control of the company.

skewness—a measure of asymmetry in a distribution, indicating whether outcomes lean toward the upside or downside.

Sklar's Theorem—a theorem stating that any multivariate joint distribution can be decomposed into its marginal distributions and a copula.

smart contract—a self-executing program on a blockchain that enforce rules and automate transactions.

sociotechnical risk—risk that arises from interactions between humans, AI systems, and organizational processes rather than from technical models alone.

software bill of materials (SBOM)—a manifest listing all software components and dependencies used within a system, primarily for supply-chain security.

sprint—a brief period (e.g., 1–4 weeks) during which an engineering team focuses on completing a clearly defined set of technical work.

software supply-chain compromise—a type of cyberattack where attackers insert malicious code into trusted software updates–enabling wide, stealthy infiltration across financial organizations.

sole proprietorship—the simplest form of business structure.

stage-gate process—a phased development and funding model in which technical progress at each "gate" must be demonstrated before moving to the next stage.

statement of cash flows—a financial report that shows management where money came from and where it went for the accounting period.

stock dilution—the reduction in existing shareholders' ownership percentage, which occurs when a company issues new shares.

stock option—a contract that gives the holder the right to buy or sell a certain stock at a predetermined price (the *strike price*) over a period.

stress testing—the evaluation of model behavior under extreme or adverse assumptions to reveal hidden errors or instability.

strike price—the predetermined price at which a stock option can be exercised. Also known as *exercise price.*

subsidiary ledger—a decomposed part of the *ledger* showing more detail for a particular account.

sunk cost—an unrecoverable expenditure in an uncompleted project.

supply-chain provenance—traceability of serialized components across the entire supply chain, often coordinated across many organizations.

surety company—see *bonding.*

sweep account—a bank account feature that automatically transfers ("sweeps") excess funds from a checking or operating account into another interest-bearing account.

SWOT—*Strengths, Weaknesses, Opportunities, and Threats.*

systematic risk—the broad economic forces that affect all firms and markets.

T&M—see *time and materials.*

tail dependence—the tendency of variables to become more correlated during extreme events.

tariff—a form of taxation on designated goods and services.

tax—a mandatory financial charge imposed by government.

TCO—*see total cost of ownership.*

teaming—an arrangement with another company that forms a temporary alliance for specific opportunities.

technology readiness level (TRL)—a standardized scale used to assess the maturity of a technology–from early-stage concept (low TRL) to commercially deployable system (high TRL).

time and materials (T&M)—a pricing model is based on billing the client for the actual time spent and materials used during the project.

tokenized note—a debt instrument issued and managed on a blockchain, often with automated compliance and interest-payment functions.

tokenized security—a blockchain-based representations of financial instruments or physical assets.

tornado chart—a ranking visualization that shows which variables have the largest impact on a key output, highlighting the most influential risk drivers.

total cost of ownership (TCO)—a metric used to compute all costs over an asset's lifecycle such as purchase price, maintenance, repairs, supplies, insurance, and disposal.

total debt to EBITDA—a metric showing how many years of cash profit it would take to pay off all debt.

traceability—the ability to track model formulas, assumptions, and data sources through precedents and dependents, supporting audit and validation.

traditional budgeting—an annual budgeting approach with fixed allocations that change little during the fiscal year.

tranche—a portion of an investment or budget allocation released only when predefined technical or financial conditions are met.

transparent documentation—clear, complete model documentation covering purpose, assumptions, methodology, limitations, and version history so that a qualified practitioner can reproduce results.

TRL—see *technology readiness level.*

trust fabric—a shared system, often supported by blockchain, that provides identity, data provenance, authorization, and policy enforcement.

unit price—a pricing approach where the client pays a fixed rate for each unit of work or deliverable.

universal law—see *accounting equation.*

unlimited liability—if a business is sued for any performance issue or malfeasance, all the assets of the owners are at risk.

unsystematic risk—any *risk* tied to the fortunes of a single firm, technology, or sector.

upside case—in *scenario analysis,* a case that reflects accelerated technical or commercial validation. Contrast with *base case* and *downside case.*

valuation—the financial assessment of a company or asset's economic value, often informed by readiness, projected cash flows, and risk. Also, a measure of the *fair market value (FMV)* of common stock that sets the minimum strike price for a *stock options.*

value at risk (VaR)—identifies the loss threshold not exceeded with a chosen confidence level.

value dilution—refers to whether the economic value of the stock shares decreases. Contrast with *ownership dilution.*

value-added tax (VAT)—a type of consumption *tax* applied at each stage of production.

value-based budgeting—a budgeting approach that evaluates spending decisions according to the value they deliver relative to their cost.

value-based pricing—see *performance-based pricing.*

variable expense—an expense that scales with activity. Contrast with *fixed expense.*

VaR—see *value at risk.*

VAT—see value added tax.

VC—see *verifiable credential* or *venture capitalist/capital.*

venture debt—a form of financing used by startups to extend runway between equity rounds.

verifiable credential (VC)—a cryptographically signed attestation used for identity and authorization. Also known as *digital identifier.*

vest—the moment when an employee earns the legal right to their stock or stock units.

vulnerable service—an outdated or misconfigured software exposed to the internet, allowing attackers to exploit flaws and enter financial infrastructure.

WACC—see weighted average cost of capital

WBS—see work breakdown structure.

weighted average cost of capital (WACC)—the blended return required by equity and debt holders, used as the discount rate in DCF valuation and highly sensitive to assumption changes.

where the truth resides—a classification describing whether the authoritative source of record is entirely on-chain or distributed between blockchain and legal systems.

work breakdown structure (WBS)—a hierarchical decomposition of project functionality.

work package—documents the lowest-level tasks in the WBS.

wrap rate—*overhead rate* plus general and administrative expenses.

zero-based budgeting—a budgeting method in which every line item must be justified from zero each cycle, rather than being based on prior-year spending. Contrast with *incremental budgeting.*

zero-knowledge (ZK)—cryptographic methods enabling verification without revealing underlying data.

zero-trust architecture—a security model requiring continuous authentication and authorization for every access request, eliminating implicit trust.

ZK—see *zero-knowledge.*

NOTE

1 Not to be confused with Consumer Price Index (CPI) which is a key economic indicator that measures the average change over time in the prices paid by urban consumers for a market basket of goods and services.

Index

Note: Page numbers in *italic* indicate a figure and page numbers in **bold** indicate a table on the corresponding page.

J

L

M

N

O

P

For Product Safety Concerns and Information please contact our EU representative GPSR@taylorandfrancis.com Taylor & Francis Verlag GmbH, Kaufingerstraße 24, 80331 München, Germany